Cognitive Processes and Spatial Orientation in Animal and Man

NATO ASI Series

Advanced Science Institutes Series

A Series presenting the results of activities sponsored by the NATO Science Committee, which aims at the dissemination of advanced scientific and technological knowledge, with a view to strengthening links between scientific communities.

The Series is published by an international board of publishers in conjunction with the NATO Scientific Affairs Division

A	**Life Sciences**	Plenum Publishing Corporation
B	**Physics**	London and New York
C	**Mathematical and Physical Sciences**	D. Reidel Publishing Company Dordrecht/Boston/Lancaster/Tokyo
D	**Behavioural and Social Sciences**	Martinus Nijhoff Publishers Boston/Dordrecht/Lancaster
E	**Applied Sciences**	
F	**Computer and Systems Sciences**	Springer-Verlag Berlin/Heidelberg/New York
G	**Ecological Sciences**	London/Paris/Tokyo
H	**Cell Biology**	

Series D: Behavioural and Social Sciences – No. 36

Cognitive Processes and Spatial Orientation in Animal and Man

Volume I Experimental Animal Psychology and Ethology

edited by

Paul Ellen

Georgia State University
Atlanta
USA

Catherine Thinus-Blanc

C.N.R.S.
Marseille
France

1987 **Martinus Nijhoff Publishers**
Dordrecht / Boston / Lancaster
Published in cooperation with NATO Scientific Affairs Division

Proceedings of the NATO Advanced Study Institute on "Cognitive Processes and Spatial Orientation in Animal and Man", La-Baume-Lès-Aix (Aix-en-Provence), France, June 27-July 7, 1985

Library of Congress Cataloging in Publication Data

NATO Advanced Study Institute on "Cognitive Processes
and Spatial Orientation in Animal and Man" (1985 :
La Baume-lès-Aix, France)
Cognitive processes and spatial orientation in
animal and man.

(NATO ASI series. Series D. Behavioural and social
sciences ; no. 36-37)
"Proceedings of the NATO Advanced Study Institute on
"Cognitive Processes and Spatial Orientation in Amimal
and Man," La-Baume-lès-Aix (Aix-en-Provence), France,
June 27-July 7, 1985"--T.p. verso.
"Published in cooperation with NATO Scientific
Affairs Division."
Contents: v. 1. Experimental animal psychology and
ethology -- v. 2. Neurophysiology and developmental
aspects.
Includes indexes.
1. Animal cognition--Congresses. 2. Animal
orientation--Congresses. 3. Cognition--Congresses.
4. Orientation (Psychology)--Congresses. I. Ellen,
Paul. II. Thinus-Blanc, Catherine. III. North Atlantic
Treaty Organization. IV. Title. V. Series. [DNLM:
1. Cognition--congresses. 2. Spatial Behavior--
congresses. BF 469 N279c 1985]
QL785.N38 1985 156'.34 86-28478

ISBN-13: 978-94-010-8078-1 e-ISBN-13: 978-94-009-3531-0
DOI: 10.1007/978-94-009-3531-0

Distributors for the United States and Canada: Kluwer Academic Publishers, P.O. Box 358, Accord-Station, Hingham, MA 02018-0358, USA

Distributors for the UK and Ireland: Kluwer Academic Publishers, MTP Press Ltd, Falcon House, Queen Square, Lancaster LA1 1RN, UK

Distributors for all other countries: Kluwer Academic Publishers Group, Distribution Center, P.O. Box 322, 3300 AH Dordrecht, The Netherlands

Softcover reprint of the hardcover 1st edition 1987

ACKNOWLEDGMENTS

This Symposium was sponsored by various organizations, national and international. The Organizing Committee would like to thank them and their representatives :

- North Atlantic Treaty Organization, N.A.T.O.
- Institut de Neurophysiologie et Psychophysiologie, C.N.R.S., Marseilles.
- Màirie d'Aix-en-Provence.
- Roussel-UCLAF Laboratories.
- Rhône-Poulence Laboratories.
- Martinus Nijhoff Publishers.

H O M M A G E

We were in the midst of finishing the edition of this book when, on the 17th of January 1986, Paul Ellen suddenly died. He was a professor of the Department of Psychology of Georgia State University in Atlanta. He was also one of the most outstanding representatives of Cognitive Animal Psychology in the United States. He had received one of the best possible formations, since from 1949 to 1952 he was a research assistant to N.R.F. Maier at the University of Michigan. With N.R.F. Maier, he developed studies on learning, stress-induced and problem solving behavior. In 1953, he was drafted and served during the Korean war in a Military Research Unit, studying the effects of cold weather on human behavior and other assorted problems. Then, he worked for two years as a graduate research assistant at the Vision Research laboratories of the University of Michigan, studying human oculomotor behavior. More specifically, he investigated problems related to the changes in corneal-retinal potential associated with movements of the eye in far-to-near accomodation. Furthermore, he conducted studies of changes in corneal-retinal potential as a function of the adaptation level of the retina. In 1957, he did research for the U.S. Veterans Administration, successively in Massachusetts, Illinois and Mississipi, most of his work related to electrophysiology. Professor of Psychology in Atlanta since 1965, he began at that time to study the septum and limbic system in general in rats, testing first the effects of lesions on timing behavior, conditioning spatial learning, spontaneous alternation and DRL. Then, in the late seventies, he devoted most of his work to the role of the septum in spatial problem solving using Maier's three-table test. In several papers, he thoroughly analyzed the cognitive and spatial mechanisms involved in this task such as the role of exploratory behavior in cognitive mapping. The new and rigorously scientific way he developed this topic provides a remarkable example of what physiological psychology should represent in contemporary Neurosciences. Let's hope that this field will continue to develop along his lines.

Paul Ellen was a hard-working perfectionist, taking great pain in everything he did. He was also extremely open-minded when it came to new projects and collaboration. The NATO ASI meeting we both organized in Aix-en-Provence which was an opportunity for him to promote interdisciplinary exchanges. The last time we worked together was when I spent two weeks in Atlanta, last November, to put the finishing touches on these volumes. We had very heated discussions because he had specific ideas on everything, but he was also tolerant and accepted worthy arguments. This book was very important for him. We are all deeply saddened that he could not see the fruit of his efforts.

Paul was also a very enthusiastic, cheerful and outgoing man, enjoying life and friends. Everybody who met him in Aix-en-Provence will continue to remember him this way. The many saddened letters I have received from the NATO participants are eloquent proof.

I want to express here my deep gratitude for everything he did in the name of our common field of research and also in the name of friendship.

I also want to thank his closest collaborators, Drs. Thom Herrmann from Guelph University (Canada) and Charlene Wages from Georgia State University, as well as Prof. Duane Rumbaugh, the Chairman of the Department of Psychology of G.S.U. : they all helped me to give this last hommage to Prof. Paul Ellen.

Professor Rumbaugh has just established a scholarship fund in his name. May this grant help young students to follow in his footsteps.

C. Thinus-Blanc
Marseille, France : February 1986

PREFACE

These volumes represent the proceedings of NATO Advanced Study Institute on the topic of "Cognitive Processes and Spatial Orientation in Animal and Man" held at La-Baume-lès-Aix, Aix-en-Provence, France, in June-July 1985.

The motivation underlying this Institute stemmed from the recent advances and interest in the problems of spatial behavior. In Psychology, traditional S-R concepts were found to be unsatisfactory for fully accounting for the complexity of spatial behavior. Coupled with the decline in such an approach, has been a resurgence of interest in cognitive types of concepts. In Ethology, investigators have begun to use more sophisticated methods for the study of homing and navigational behaviors. In the general area of Neuroscience, marked advances have been achieved in the understanding of the neural mechanisms underlying spatial behaviors. And finally, there has been a burgeoning interest and body of knowledge concerning the development of spatial behavior in humans.

All of these factors combined to suggest the necessity of bringing together scientists working in these areas with the intent that such a meeting might lead to a cross-fertilization of the various areas. Possibly by providing a context in which members of the various disciplines could interact, it was felt that we might increase the likelihood of identifying those similarities and differences in the concepts and methods common to all groups. Such an identification could provide the basis for a subsequent interdisciplinary research effort.

The concept of "cognitive map", although originated by a psychologist, E.C. Tolman, nearly 40 years ago, has become one of the most current theoretical constructs in each of the disciplines represented at the ASI. As consequence, we are immediately confronted with the question of whether the "cognitive map" that an ethologist invokes in accounting for the homing behavior of pigeons or squirrels is the same kind of "map" that a geographer attributes to an individual learning the layout of a city or a psychologist suggests underlies the spatial problem solving of a rat.

This ASI was divided into four topical areas. In the first volume, spatial behavior is examined from the perspective of animal psychology and ethology ; the second volume considers the neurophysiological mechanisms underlying spatial organization and the development of such spatial function. These traditional approaches to spatial behavior were complemented by contributions dealing with the organization of near, proximal space. Could our view of the development and organization of distal space find its roots in sensori-motor mechanisms underlying eye-hand spatial coordination ? Similarly, the contribution of Neuroethological approaches to species-specific spatial behavior was brought to bear on the general topic of the ASI. Finally, we attempted to gain insight into the problem of spatial behavior by viewing it from the context of the historic controversy between the behaviorist and cognitive psychologies. To the extent that similar controversy still exists today, we have evidence that truth is not to be found on one side or the other ; rather we are dealing with basic differences in philosophy which are not likely to be solved by the methods of science.

We thank all of our participants. They came from the various NATO countries and from Switzerland. Their presence and contribution to the discussions at La Baume provided ten days of intense intellectual stimulation. Special mention should be made of Dr. D. MacDonald of Oxford. His films provided material for excellent discussions. We also wish to make special note of the great and kind assistance rendered to each of the participants by the staff of the INP of CNRS, Marseilles. Their efficient and pleasant attention ensured that our time was devoted to the concerns of spatial behavior. We also wish to thank the participants and also P. Bovet, R. Campan, M. Jeannerod, J. Requin, S. Thorpe and M. Piatelli-Palmarini who kindly consented to review some of the chapters. Their assistance greatly facilitated our efforts in bringing the proceedings of the ASI into book form.

We need to recognize also the members of the International Organizing Committee, Prof. Linda Acredolo of the University of California, Davis, and Prof. Jonathan Winson of the Rockefeller University. Their contribution to the selection of participants played a large role in the success of the ASI.

Finally, mention should be made of Dr. Craig Sinclair of the NATO and Henny Hoogervorst of Martinus Nijhoff, Publishers. They assisted us in a variety of ways in differents stages of this project.

At this time it is too early to evaluate the results of this ASI. We hope that we have begun an interactive process among the various disciplines. At our next meeting we hope to see the results of current efforts.

Paul Ellen
Catherine Thinus-Blanc

Atlanta, Georgia
November, 1985

Volume 1

CONTENTS

SECTION II. Neuroethology and Ethology.

LIST OF THE PARTICIPANTS OF THE NATO ASI

ACREDOLO, L.P. Department of Psychology, University of California, Davis, California 95616, U.S.A.

ALLEN, G.L. Old Dominion University, Norfolk, Virginia 25508, U.S.A.

AMMASSARI-TEULE, M. Istituto di Psicobiologia e Psicofarmacologia, C.N.R., 1, Via Reno, Roma 00198, Italy.

BAKER, R.R. Department of Zoology, University of Manchester, Williamson Building, Manchester M13 OPL, G.B.

BEUGNON, G. Laboratoire de Neuroéthologie, Université Paul Sabatier (Toulouse III), 118, route de Narbonne, 31062 Toulouse cedex, France.

BINGMAN, V. Department of Psychology, University of Maryland, College Park, Maryland 20742, U.S.A.

BOVET, J. Département de Biologie, Université Laval, Québec, Québec G1K 7PA, Canada.

BUHOT, M.-C. Département de Psychologie Animale, Institut de Neurophysiology et Psychophysiologie, CNRS-INP 9, 31 chemin Joseph Aiguier, 13402 Marseille cedex 9, France.

BULLINGER, A. Département de Psychobiologie expérimentale et appliquée, Faculté de Psychologie et des Sciences de l'Education, Université de Genève, 24 rue du Général-Dufour, 1211 Genève 4, Suisse.

CELEBI, G. Department of Physiology, Medical Faculty, University of Ege, Bornova, Izmir, Turkey.

CHAPUIS, N. Département de Psychologie Animale, Institut de Neurophysiologie et Psychophysiologie, CNRS-INP9, 31 chemin Joseph Aiguier, 13402 Marseille cedex 9, France.

CHENG, K. Biology Building, The University of Sussex, Falmer Brighton, Sussex BN1 9QG, G.B.

DOWNS. R.M. The Pennsylvania State University, 302 Walker Building, University Park, Pennsylvania 16802, U.S.A.

DURUP, H. Département de Psychologie Animale, Institut de Neurophysiologie et Psychophysiologie, CNRS-INP9, 31 chemin Joseph Aiguier, 13402 Marseille cedex 9, France.

DURUP, M. Département de Psychologie Animale, Institut de Neurophysiologie et Psychophysiologie, CNRS-INP9, 31 chemin Joseph Aiguier, 13402 Marseille cedex 9, France.

EINON, D. Department of Psychology, University College London, Gower Street, London WC1E 6BT, G.B.

ELLEN, P. Department of Psychology, Georgia State University, University Plaza, Atlanta, Georgia 30303, U.S.A.

ETIENNE, A. Université de Genève, Faculté de Psychologie et des Sciences de l'Education, 24 rue Général-Dufour, 1211 Genève 4, Suisse.

EWERT, J.-P. Gesamthochschule Kassel, Universitat des Landes Hessen, Fachbereich 19, Biologie/Chemie, Heinrich-Plett-Strasse 40, D-3500 Kassel-Oberzwehren, F.R.G.

FABRE-THORPE, M. Laboratoire de Neurophysiologie comparative. 9, Quai Saint-Bernard, 75005 Paris, France.

FABRIGOULE, C. Laboratoire de Psychologie Animale, Institut de Neurophysiologie et Psychophysiologie, CNRS-INP 9, 31 chemin Joseph-Aiguier, 13402 Marseille cedex 9, France.

FOCARDI, S. Istituto di Biologia della Selvaggina, Via Stradelli Guelfi 23A, 40064 Ozzano Emilia, Italy.

FOREMAN, N. Department of Psychology, The University, Leicester LE1 7RH, G.B.

GERBRANDT, L.K. Research and Development, Wang Laboratories Inc., Lowell, Massachusetts 01851, U.S.A.

GLICKSMAN, M. University of Minnesota, Institute of Child Development, 51 East River Road, Minneapolis, Minnesota, U.S.A.

HERRMANN, T. University of Guelph, College of Social Sciences, Department of Psychology, Guelph, Ontario N1G 2W1, Canada.

HONIG, W.K. Department of Psychology, Dalhousie University, Halifax, Nova Scotia B3H 4J1, Canada.

JAMON, M. Département de Psychologie Animale, Institut de Neurophysiologie et Psychophysiologie, CNRS-INP 9, 31 chemin Joseph Aiguier, 13402 Marseille cedex 9, France.

LEPECQ, J.-C. Laboratoire de Psychobiologie de l'Enfant, Ecole Pratique des Hautes Etudes, 41 rue Gay Lussac, 75005 Paris, France.

LEVESQUE, F. Laboratoire de Neurophysiologie comparative, 9 Quai Saint-Bernard, 75005 Paris, France.

LIBEN, L.S. The Pennsylvania University, 302 Walker Building, University Park, Pennsylvania 16802, U.S.A.

LOCKMAN, J.J. Department of Psychology, 2007 Percival Stern Hall, Tulane University, New Orleans, Louisiana 70118, U.S.A.

MAKTAV-YILDIRIM, S. Ayranci, Guvenlik Cad., Fuar apt 89/3, Ankara, Turkey.

MENZEL, E. State University of New York at Stony Brook, Department of

Psychology, Stony Brook, N.Y. 11794, U.S.A.

MITCHELL, J. University of Manchester, Department of Psychology, Manchester 1I3 IPL, G.B.

MORROW, L. University of Pittsburgh, Western Psychiatric Institute and Clinic, Neuropsychology, 3811 O'Hara Street, Pittsburgh, Pennsylvania 15261, U.S.A.

NEISSER, U. Emory University, Department of Psychology, Atlanta, Georgia 30322, U.S.A.

OLIVIER, E. Laboratoire de Neurophysiologie, Université de Louvain, UCL 5449, 1200 Bruxelles, Belgique.

OLTON, D. The John Hopkins University, Department of Psychology, Baltimore, Maryland 21218, U.S.A.

PACTEAU, C. Université Louis-Pasteur, Laboratoire de Neurobiologie comportementale, 7 rue de l'Université, 67000 Strasbourg, France.

PAILHOUS, J. Laboratoire de Psychologie de l'Apprentissage, IBHOP, Avenue des Géraniums, 13014 Marseille, France.

PAILLARD, J. Département de Psychophysiologie Générale, Institut de Neurophysiologie et Psychophysiologie, CNRS-INP 4, 31 chemin Joseph Aiguier, 13402 Marseille cedex 9, France.

PARKO, E. 16 Kingstone Road, Avondales Estates, Georgia 30002, U.S.A.

PERRUCH, P. Laboratoire de Psychologie de l'Apprentissage, IBHOP, Rue des Géraniums, 13402 Marseille cedex 9, France.

POTEGAL, M. New York State Psychiatric Institute, 722 West 168th Street, New York, N.Y. 10032, U.S.A.

POUCET, B. Département de Psychologie Animale, Institut de Neurophysiologie et Psychophysiologie, CNRS-INP 9, 31 chemin Joseph Aiguier, 13402 Marseille cedex 9, France.

POUCET, H. Département de Psychologie Animale, Institut de Neurophysiologie et Psychophysiologie, CNRS-INP 9, 31 chemin Joseph Aiguer, 13402 Marseille cedex 9, France.

RASHOTTE, M.E. The Florida State University, Department of Psychology, Tallahassee, Florida 32306, U.S.A.

RATCLIFF, G. Harmarville Rehabilitation Center, Inc., P.O. Box 11460, Guys Run Road, Pittsburgh, Pennsylvania 15238, U.S.A.

REQUIN, J. Département de Psychobiologie expérimentale, Institut de Neurophysiologie et Psychophysiologie, CNRS-INP 3, 31 chemin Joseph Aiguier, 13402 Marseille cedex 9, France.

RIEHLE, A. Département de Psychobiologie expérimentale, Institut de Neurophysiologie et Psychophysiologie, CNRS-INP 3, 31 chemin Joseph Aiguier, 13402 Marseille cedex 9, France.

SCHENK, F. Institut de Physiologie, Faculté de Médecine, Université de Lausanne, 7 rue de Bugnon, CH-1011 Lausanne, Suisse.

SHERRY, D.F. Erindale Campus, University of Toronto in Mississauga, Mississauga, Ontario L5L 1C6, Canada.

SMITH, M.L. Department of Psychology, EN8, Toronto General Hospital, 101 College Street, Toronto, Ontario MSG 1L7, Canada.

SOECHTING, J.F. University of Minnesota, Department of Physiology, Medical School, 6-255 Millard Hall, 435 Delaware Street S.E., Minneapolis, Minnesota 55455, U.S.A.

STAHL, J. Morris Brown College, 643 Martin Luther King Jr., Drive N.W., Atlanta, Georgia 30314, U.S.A.

THINUS-BLANC, C. Département de Psychologie Animale, Institut de Neurophysiologie et Psychophysiologie, CNRS-INP 9, 31 chemin Joseph Aiguier, 13402 Marseille cedex 9, France.

THOMSON, J.A. University of Strathclyde, Department of Psychology, Turnbull Building, 155 Georgia Street, Glasgow G1 1RD, G.B.

VANDERLINDEN, M. Université de Liège, Hôpital de Bavière, Clinique de Neurochirurgie, 66 Boulevard de la Constitution, 4020 Liège, Belgique.

VAUCLAIR, J. Département de Psychologie Animale, Institut de Neurophysiologie et Psychophysiologie, CNRS-INP 9, 31 chemin Joseph Aiguier, 13402 Marseille cedex 9, France.

WAGES, C. Department of Psychology, Georgia State University, University Plaza, Atlanta, Georgia 30303, U.S.A.

WILTSCHKO, R. Zoologisches Institut, Fachbereich Biologie, J.W. Goethe-Universitat, Siesmayerstrasse 70, Postfach 111932, D-6000 Frankfurt/Main, F.R.G.

WILTSCHKO, W. Zoologisches Institut, Fachbereich Biologie, J.W. Goethe-Universitat, Siesmayerstrasse 70, Postfach 111932, D-6000 Frankfurt/Main, F.R.G.

WINSON, J. The Rockefeller University, 1230 York Avenue, New York, N.Y. 10021, U.S.A.

THE COGNITIVE MAP CONCEPT AND ITS CONSEQUENCES.

Catherine THINUS-BLANC.

Department of Animal Psychology
Institute of Neurophysiology and Psychophysiology
CNRS - INP 9
31, chemin Joseph-Aiguier
13402 Marseille cedex 9
FRANCE.

The concept of cognitive maps (C.Ms) is central to studies on spatial orientation in animals and man. This does not mean, however, that there is a consensus as to its functional value. Simply, the amount of literature dealing with this concept, including several reviews and papers in which it has been extensively discussed and compared with other hypotheses, as well as numerous investigations on its neural basis, shows how much interest it has raised.

The so-called C.Ms were defined by Tolman in his 1948 article as follows : "we believe that in the course of learning, something like a field map of the environment gets established in the rat's brain ... the incoming impulses are usually worked over and elaborated into a tentative, cognitive-like map of the environment. And it is this tentative map, indicating routes and paths and environmental relationships which finally determines what responses, if any, the animal will finally release". (p. 192).

Since 1948, the C.M has become a well-known term. It has come to be used in many disciplines : Psychology of course, both animal and human, Ethology, Neurophysiology, Architecture, Town Planning, Geography, and Ethnology. Sometimes Tolman is not even quoted ; the term has become part of common usage in these fields, which shows that some kind of tacit acceptance has taken place.

Paradoxically, in spite of their successful career, C.Ms are still very puzzling, at least in Animal Psychology. They are often referred to when subjects' success in complex spatial tasks cannot be accounted for by simpler and better known mechanisms such as Stimulus-Response associations, sensory guidances, etc. The term is usually used rather loosely to mean "spatial representations" in a very broad sense. One mightjustifiably ask in fact whether this concept is really useful for understanding spatial processes.

In this paper, I shall discuss some points in connection with experimental Psychology research on spatial behavior in mammals. After looking back at Tolman's 1948 article, I propose to review the C.M concept in Animal Psychology (some of these points have been examined elsewhere, see Thinus-Blanc, 1984). Then, with the broad multidisciplinary outlook which inspired this ASI meeting, I shall present one example of a contribution made by cognitive Psychology to animal behavior studies.

The origin of C.Ms : Tolman, 1948

In support of his hypothesis, Tolman presented five classes of experimental data. The first concerns "latent learning", i.e. the fact that rats are able to learn where food and water are located in a maze, for example, even if they are neither hungry nor thirsty. This contrasts with the behaviorists' view that a need-reduction following immediately on a response contributed to learning by strengthening S-R associations.

The next two arguments were based upon observations of rats' behavior. For instance, at the very beginning of a discriminative learning task, or just before reaching the criterion, animals display hesitating behaviors (stopping at the choice point, looking in several directions, etc.), which have been called "Vicarious Trial and Error" (V.T.E.). Furthermore, during an avoidance conditioning with an electric shock associated with food, rats "... often seemed to look around after the shock to see what it was that had hit them". (p. 201). According to Tolman, these behaviors reflect active selection and searching for the stimulus and are not just passive responses to discrete stimuli.

The "hypothesis experiments" conducted by Krech (1932) provide the basis of the fourth class of arguments. Krech demonstrated that in the course of complex discriminative learning, rats display different successive strategies or hypotheses until that corresponding to the optimal level of reinforcement is found. For example, rats begin by choosing only right-hand doors, then only left-hand ones, and so on. Tolman's explanation is that each strategy may correspond to the use of a tentative cognitive map. The rats' non-perseverance with a fruitless strategy certainly reflects behavioral plasticity and adaptive abilities; but we have no proof that the strategies themselves reflect the systematic use of some kind of spatial representation such as a C.M.

"Spatial orientation experiments" were the last class of arguments put forward by Tolman in support of C.Ms. Besides the sunburst maze experiments which in fact, as pointed out by O'Keefe and Nadel (1978), clearly involved visually guided behavior, Tolman describes Ritchie's experiment showing that rats, when released from an unusual starting point in a maze, are able to choose the path pointing approximately to the side of the room where the goal is located. This ability to use various paths to get to a place had already been demonstrated however with other procedures long before by other authors using "reasoning" (Maier, 1938) and shortcut experiments (Shepard, 1931 ; 1933) for example.

In a nutshell, Tolman's arguments focus on the cognitive processing of spatial information, an active gathering of relevant cues. In the last class of experiments, the supremacy of environmental cues (those involved in place learning) over proprioceptive information (those involved in response learning) is stressed and the rat's ability to locate a place even when prevented from using the usual motor sequence is once again demonstrated. But Tolman delves no deeper and does not investigate the actual nature of C.Ms any further.

Given the weakness of Tolman's arguments, the present-day success of the CM concept seems in fact quite paradoxical, at least in experimental psychology.

After Tolman

There is a second paradox in the history of the C.M concept. This concept originated from animal Psychology, and was based on experimental data obtained with rats ; and yet, in the field from which it originated, it was only developed in the seventies. With the exception of the series of articles entitled "studies in spatial learning", which was completed in 1951 by Ritchie, Hay and Hare (Tolman's collaborators), nothing of importance was published on the subject of cognitive maps, or even spatial orientation in the field of animal psychology for twenty years.

Contrasting with this period of eclipse, the seventies marked the rapid growth of spatial orientation studies on animals (as well as human), which is still continuing today. This NATO book is the token of this lasting interest. In animal Psychology studies (Section 1, volume 1), the C.Ms have often been quoted in investigations on the nature of information structure supporting spatial knowledge. Nowadays, however, they are considered to be functional systems that are relevant to functional questions. The dynamic term "cognitive mapping" is used rather than C.M "... because it is not a noun and does not necessarily imply any internal object or entity that the animal in question must, figuratively speaking, scan in addition to looking at what it can perceive directly". (Menzel, 1984, p. 527).
As examples of the present functional approach, one might mention analyses of the strategies used to solve multiple solution problems (see in this volume : Buhot and Poucet ; Einon and Pacteau ; Fabrigoule) or to cope with classical situations such as detour-tests (Chapuis, this volume), or learning in cross-mazes (Poucet, this volume).

Closely related to this approach is the study of problem solving abilities with the Maier-3 table-task ; here heavy emphasis has been placed on the functional value of the spatial knowledge acquired during preliminary exploration phases (Ellen ; Stahl et al. ; Wages, this volume).

Research in Ethology (Section 2, Volume 1) has been dealing with spatial orientation for a very long time but explicit reference to the C.M has only been made very recently in this discipline (Peters, 1979 ; Fabrigoule and Maurel, 1982 ; Wiltschko and Wiltschko ; Bovet ; this volume). One cannot help wondering whether the "mosaic" or "navigational" maps claimed by Wiltschko and Wiltschko to be used by pigeons are actually the same as the C.Ms assumed by Tolman to be used by rats. But too little is known about C.Ms for it to be wise to carry this comparison any further.

Finally, the real thrust to the topic has been unquestionably given by studies on the neural basis of C.M, particularly the septo-hippocampal system subserving to a large extent spatial knowledge and memory (Section 1, Volume 2). In addition to data published directly in this field (cf. for instance Herrmann ; Gerbrandt ; Winson ; volume 2), it has provided, perhaps because of its analytical methods, some important distinctions which are also relevant to experimental Psychology, such as O'Keefe and Nadel's "locale system" based upon the use of C.Ms, versus a "taxon system" implying simpler sensorimotor mechanisms (O'Keefe and Nadel, 1978 ; O'Keefe, 1983 ; Morris, 1981 ; Bingman ; volume 2) ; and Olton's "reference" and "working" memories (Volume 2). Distinctions have also been made, within a given system of orientation, between different constitutive units or phases, such as "acquisition" and "retention" phases in a spatial learning task (Ammassari-Teule ; volume 2).

C.Ms : the trees hiding the forest

It may be useful to speculate on the reasons for the fluctuating interest in the C.M concept. One reason for the twenty years of silence needs hardly be explained : Psychology was dominated by Behaviorism until the sixties. Cognitive Psychology did not exist. Most of Tolman's arguments were <u>against</u> S-R associations. As emphazised by Bolles (1975), Tolman over-reacted in his forceful criticism of mechanistic models. He coined an expression that sounded rather good, but which is fundamentally self-contradictory and leads easily to misunderstandings. The term "cognitive" refers to an activity, a dynamic process. The word "map" is essentially static ; it suggests a static image of the real world. Tolman has said too much and not enough. Too much for his hypothesis to be forgotten. And not enough for his subtleties to outweigh his overstatements.

Another reason I see for the twenty-year eclipse of C.M concept is that the protagonists of the place versus response controversy debated about these two kinds of processes as if they were mutually exclusive. In a theoretical article, Restle (1957) attempted to solve the issue by suggesting that the probability of occurrence of either kind of learning depends on the availability and amount of relevant environmental cues. The presence of many easily perceptible landmarks would favour place learning whereas response learning would be more likely to occur in a barren environment. But instead of promoting new research in this field, these over conciliatory conclusions may well have contributed to the transitory disinterest in C.Ms. Of course the layout of the situation is one of the factors determining what means will be used to get around in space. But another determinant, which is at least as important, is the task requirement. Several mechanisms can be used efficiently to go from one place to another. For example, human beings do not systematically use spatial representations to perform complex oriented displacements. Many examples illustrating this question are to be found in everyday life. You are lost in a city you do not know very well. You have no map of the city and you want to go back to your starting point. Besides drastic solutions of less interest for our purposes, such as taking the underground or a taxi, other means may be available, such as the use of a distant landmark, or the position of the sun at different times of day. You can also try to walk back the way you came in order to get to a more familiar part of the city. Finally, you can ask someone your way. What will you be told ? "Take the first street on your right. When you reach a fountain, turn left", and so on. I have never seen someone trying to sketch a map : you will be given <u>instructions</u>, a list of procedures which defines a "route" according to O'Keefe and Nadel's acceptation (1978).

You will react to a sequence of stimuli by successively re-orientating your body, *i.e.* the trajectory of your displacements. Under quite different circumstances, while walking or driving in a familiar neighborhood or city where we are used to taking the same itinerary everyday, we behave in a very similar way : turning right at the corner, then left, etc., without paying much attention to these landmarks, at least, without having in our mind any image or representation of the topography of the place we are going through at any one moment.

Does this mean that we are unable to form and use spatial representations ? Cognitive maps ? No. It merely points to the fact that it is much easier and more economical not to.

Human beings, then, although gifted for the highest levels of conceptualization and abstraction, do not systematically use spatial representations to perform complex oriented displacements. *A fortiori*, why should animals not spontaneously adopt this "representational" economy principle ? In pointing out the distinction between "what one can do" and "what one would normally do", Downs (1981) brings up a point which is at the very heart of experimental Psychology. The problem is even more crucial when studying animals than man, since animal subjects cannot be given direct instructions. The experimenter tries to communicate with them by making "good" use of sometimes sophisticated apparatus, and trying to reach a subtle balance between the level of motivation and the amount of reinforcement. Solving a spatial problem, reaching the learning criterion in a cross-maze, for example, will be interpreted as the proof that C.Ms have been used ; whereas if a failure is observed, it is concluded that animals have failed to constitute or accurately use C.Ms. One may be wrong in both cases, because the possibility that other orienting systems might have been used is neglected. We formulate hypotheses, we design experiments according to our (human) way of coping with and thinking of space. How could we do otherwise ? This risk of mentalistic interpretation will be developed by Ellen (next chapter).

It is not impossible that under severe ecoethological constraints, on the one hand, and equipped with limited representational and cognitive mapping abilities on the other hand, some species during the course of evolution may have developed original orientation systems allowing them to get around in space and to survive. Bird migrations illustrate such species-specific orientation systems. Another possibility is that there may be non-specific orientation systems, existing also in man, which operate at the unconscious level. This appears to be the case with orientation based upon magneto-reception (Baker, this volume) or, on a smaller spatial scale, upon path integration and vestibular information (Etienne, this volume). Lastly, randomness can also play an important part in spatial behaviors (Bovet, 1979 ; Jamon, this volume). But other underlying systems of which we have not yet the least idea may participate in spatial orientation in both animals and man. The C.M concept is an outstanding example of trees hiding the forest.

To summarize, there are several reasons accounting for the lack of success of C.Ms for 20 years, namely the long lasting hegemony of behaviorism, a misunderstanding of the provocative, self-contradictory concept of cognitive maps, the fact that highest cognitive capacities may not be systematically involved in spontaneous but nevertheless adaptive behaviours, and finally the supremacy of cognitive maps in the mind of many psychologists who design the experiments but maybe not in that of the animals undergoing these experiments.

Only a little more time is necessary to consider the reasons for the relatively recent revival of interest in C.Ms. Such an analysis may actually not be very relevant. Lessons have been learned from the past, as the pluridisciplinary development of this topic shows. Nevertheless, one can briefly suggest a few reasons. First, cognitive Psychology came into being and flourished. Although dealing mostly with human functioning, this field has provided a general enough theoretical background to incorporate the animal studies. Additionnally, the part of cognitive Psychology devoted to environmental cognition in children and adults gave rise to concepts and theories that began to invoke the concept of C.M and provided a myriad of data (cf. the section on "human spatial orientation", this book, volume 2 ; "image and environment", Downs and Stea, 1975 ; etc.). Finally, the investigation of the neural basis of C.Ms (O'Keefe and Nadel, 1978) definitively dispelled the mentalistic connotation of the concept.

In such a context, it is not surprising that animal psychologists dusted off the C.M concept. Cognitive Psychology may have helped animal psychologists to have a new and different look at the C.M concept, in a heuristic way. It must be underlined that this process is the reverse to what usually happens ; for example, the animal models are extensively used in human Psychophysiology and clinical studies. The next part of this chapter provides an example of the contribution of cognitive Psychology to the study of C.Ms in animals.

The contribution of cognitive Psychology to animal studies : one example.

In his book "Cognition and reality", Neisser (1976) presents an original view of the C.M. conceived as an "orienting schema" that "... is an active information seeking structure. Instead of defining a C.M. as a kind of image, I will propose that spatial imagery itself is just an aspect of the functioning of orienting schemata. Like other schemata, they accept information and direct action. Just as I have an object schema that accepts information about my desk lamp and direct further exploration of it, I also have a cognitive map of my whole office and its setting to accept information about the office and to direct my movements within it. The lamp schema is a part of the larger orienting schema, just as the lamp itself is a part of the real environment. The perceptual cycle ... is embedded in a more inclusive cyle of exploration and information pick up that covers more ground and takes more time". (Neisser, 1976, p. 111 ; cf. figure 1).

To what extent the whole model is applicable to the cognitive functioning of animals living under particular laboratory conditions is difficult to assess. When a rat is passively transported in a box from the animal breeding room to an apparatus, then put in another cage during intertrial intervals, etc. it seems unlikely that the embedding will be accurate and continuous, except within each limited situation. But what is most interesting and new, is the conception of the dynamic role of these schemas and cognitive maps, which direct perceptual exploration, by

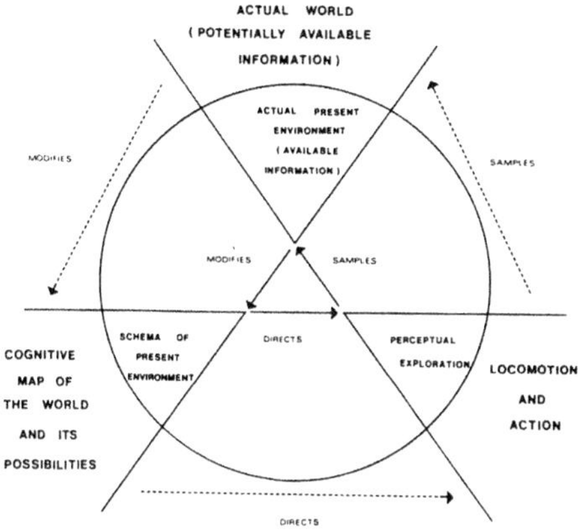

Figure 1. Schemata as embedded in cognitive maps. Redrawn from : Cognition and reality, 1976, by U. Neisser. (W.H. Freeman and Co.) Copyright 1986.

sampling the present environment. Among the three main elements composing this perceptual cycle, two are accessible to observation : first, the present environment, that is the experimental situation for example, and secondly, the overt manifestations of perceptual exploration.

In Rodents, many exploratory reactions can be easily observed. Even if animals can visually explore objects at a distance, they can sniff, mark, and generally make contact with any objects constituting a new environment. This activity, which is extremely intense at the onset of a confrontation with a new situation, decreases with time (habituation).

A few studies based upon behavioral observations have shown in Gerbils (Wilz and Boylton, 1971 ; Cheal, 1978 ; Thinus-Blanc and Ingle, 1985) that during exploration, animals learn not only the characteristics of the objects contained in the experimental device, but also their spatial relationships. If one or several familiar object(s) are displaced after habituation, renewal of exploratory reactions can be observed, showing that the change in the positions of familiar objects has been detected.

All these data fit Neisser's conception of C.Ms or orienting schemata and raise a major question : which among the spatial characteristics of a new environment are being attended to and integrated during exploration ?

In a series of experiments with hamsters, my colleagues (K. Chaix, N. Chapuis, M. Durup, B. Poucet) and I, we have obtained some data which bear on this question.

The subjects were allowed to individually explore a new situation : a circular open field containing four different objects and a conspicuous

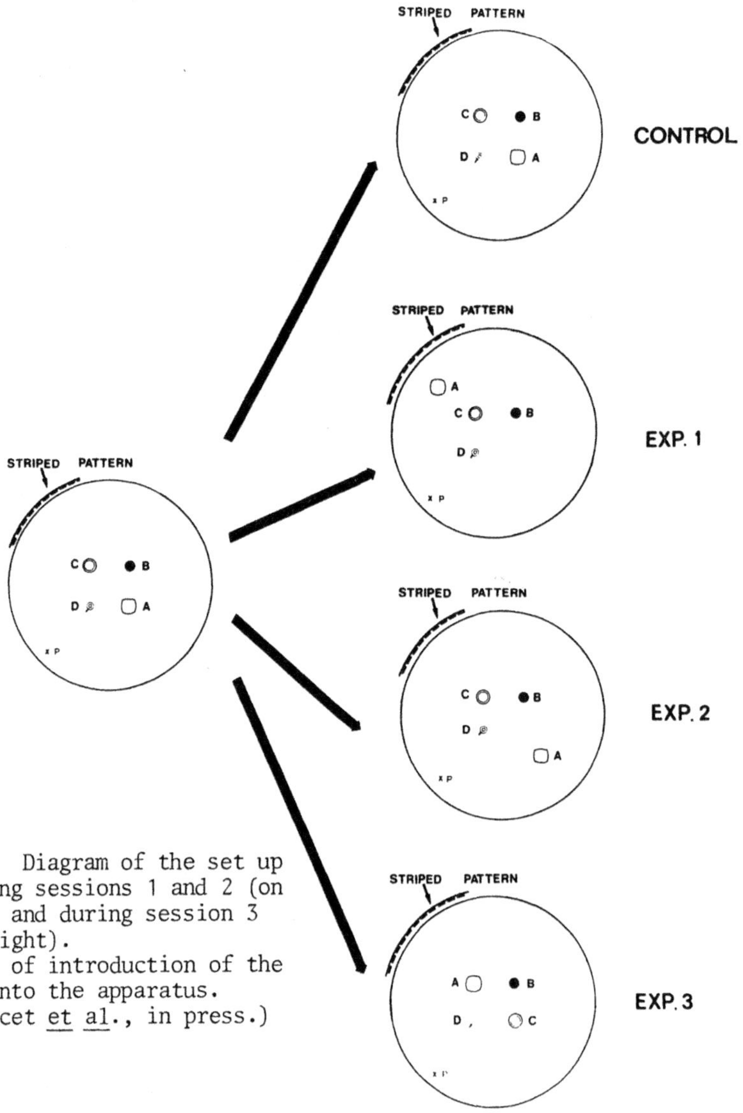

FIGURE 2. Diagram of the set up used during sessions 1 and 2 (on the left) and during session 3 (on the right).
P : point of introduction of the subject into the apparatus.
(From Poucet _et al_., in press.)

extra apparatus landmark. Two 15-min sessions, eight hours apart suffised for us to observe a marked decrease in exploratory activity. The latter was evaluated (via a video-system) from the number and duration of actual contacts made by animals with the objects. Then, during a third session, eight hours later, the position of one or several objects was changed with the experimental groups (but not for the control groups).

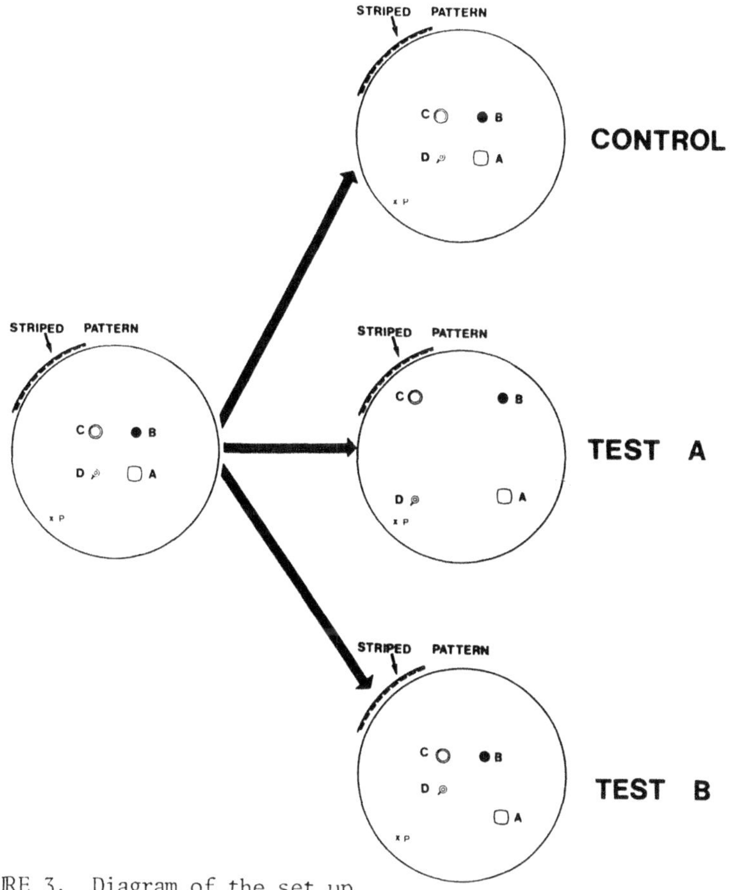

FIGURE 3. Diagram of the set up used in sessions 1 and 2 (on the left) and in session 3 (on the right).
(From Thinus-Blanc et al., in preparation.)

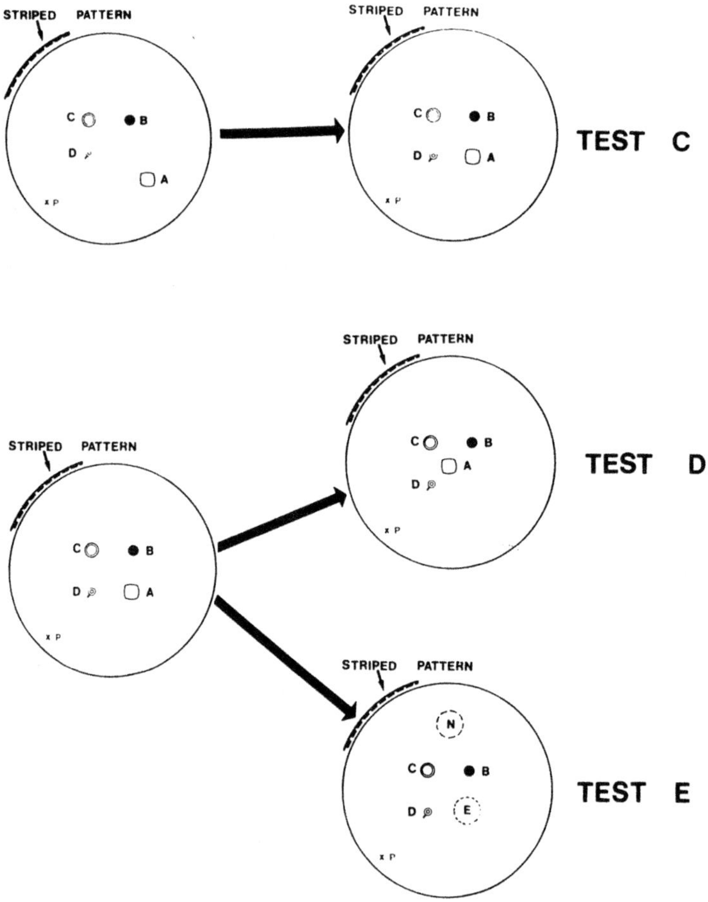

FIGURE 4. Diagram of the set up used in sessions 1 and 2 (on the left) and in session 3 (on the right). (From Thinus-Blanc et al., in preparation.)

This series of experiments was based upon a very simple idea : if, after habituation, the hamsters react to a change in the spatial relationships involved in the situation by displaying a renewal of exploratory activity, they must have detected this change by referring to an internal model, a C.M of the previous situation which no longer exists. Thus, by studying reactions to several kinds of spatial arrangements, it may be possible by comparing them to obtain some information about the characteristics of the internal model or C.M of very simple spatial configurations.

Figures 2 to 4 show some examples of the changes introduced at the third test session. Each test situation was presented to a new group of hamsters (containing 8 to 12 subjects). The displacement of one object can disrupt the overall configuration of the object set (figure 2, exp. 2; figure 3, test B ; figure 4, exp. C and D) and also (figure 2, exp. 1) their topological relationships. In exp. 3 (figure 2), the overall configuration was the same as the initial layout, but the topological relationships among the objects were modified. In exp. A (figure 3), the size of the square formed by the objects is increased. Exp. C (figure 3) was the reverse of exp. 2 (figure 2). In exp. E (figure 4), object A was removed. The behavior of the subjects at that place (E) was observed and compared with their behavior in a neutral zone (N). The details of the results have been given elsewhere (Poucet, Chapuis, Durup and Thinus-Blanc, 1986 ; Thinus-Blanc, Durup, Chaix, Poucet, in preparation). In summary, three categories of reactions to the change(s) are observed during the third session :

- Exploratory reactions were selectively directed towards the displaced object(s). Changes were correctly detected ; this precisely focused exploratory renewal might correspond to an <u>updating</u> of the situation.

- Exploratory reactions were directed to all the objects, whether or not they had been displaced, as if the situation were entirely new. In those cases, exploration might have a <u>mapping</u> function.

- No renewal of exploratory reactions was observed.

These data are difficult to interpret because they do not appear to be generally correlated with any particular class of spatial characteristics that was modified, namely the topological relationships of the four objects, their overall geometrical arrangement and its size, or the relationships (metrical and topological) between each object (or between the object set) and the external landmark. Nevertheless, there is one factor accounting, partly and <u>a posteriori</u>, for these results. This factor is related to modification of the area defined by the four objects, between the second and the third session. Figure 5 summarizes the data and the tests. When the change had either increased the area defined by the object set (exp 2 and B) or had not modified it at all (exp 3), the hamster's behavior was selectively directed towards the displaced objects. Conversely, animals displayed reactions towards non-displaced objects only when the change had caused a decrease in the area (exp D, E, 1, C). Lastly, a considerable enlargement of the area which was not accompanied by any change either in the shape of the object set or in the topological relationships, elicited no reaction.

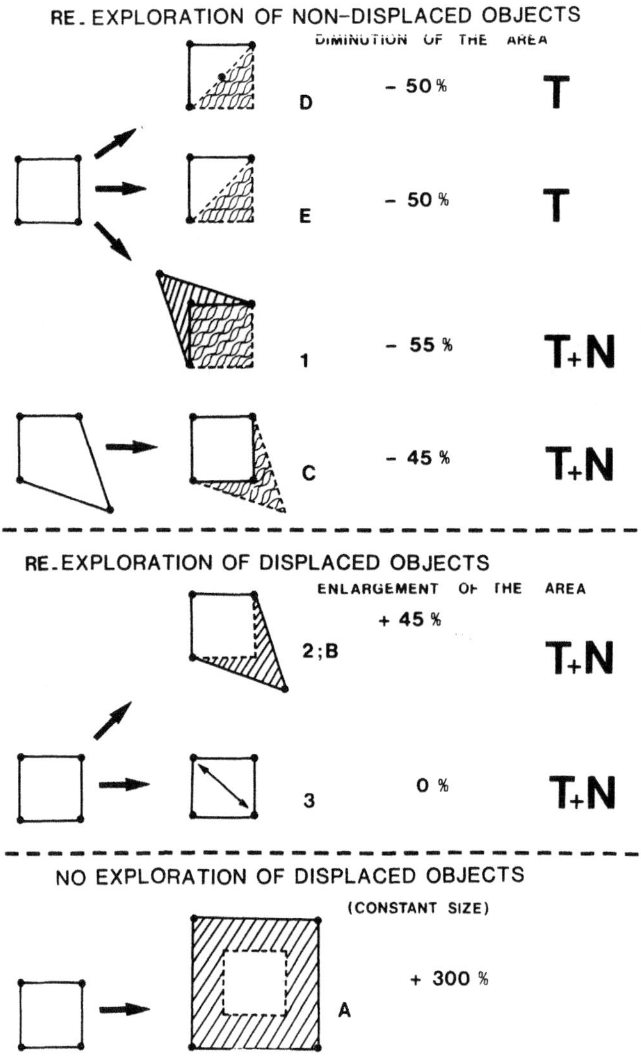

FIGURE 5. Diagram summarizing the results in terms of the variations in the area defined by the object set in session 3.
T : significant effect concerning the duration of contacts.
N : significant effect concerning the number of contacts.

These results call for several further experiments. But it can already be said that they seem to indicate that the area defined by the objects (inner area) might be a feature attended by the animals and given a different psychological "status" from the area exterior to the object set. The topological and geometrical relationships between the four objects appear to be other salient features for the animals, whereas neither the size (at least within certain limits) nor the external landmark appear to be prominent elements in the C.M of the situation as a whole.

It is possible to apply Neisser's cycle (cf. figure 1) to these experiments, considering several levels of functioning.

Let us consider a hamster placed into a totally new situation, E1 (the open field with the four objects for example). The schema of E1 does not exist yet (figure 6). The novel character of E1 can be detected only by referring to previous experience, previous schemata or C.Ms. I add to Neisser's cycle a systematic comparison test - which might be carried out by specific cells of the hippocampus (O'Keefe and Nadel, 1978).

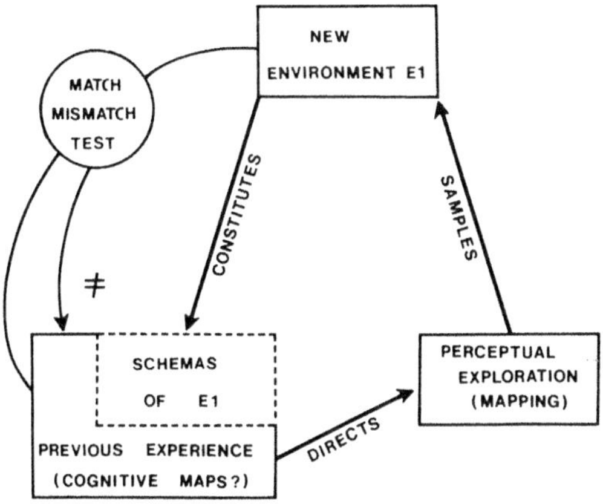

FIGURE 6. An adaptation of Neisser's cycle to "mapping exploration".

The match-mismatch test says : "no, there is nothing similar to what was previously sampled". If we assume that C.Ms are not only simple mental imagery, but generate acquired rules establishing spatial relationships by extracting spatial invariants, then these C.Ms are going to direct, to organize perceptual exploration which, in this case, will have a <u>mapping</u> function. We will observe a generalized and intense exploratory activity. Progressively, the schema of E1 will be constituted until the match-mismatch test does not detect any discrepancy. If we slightly modify the E1 environment (displacement of one object, figure 7), there are several possibilities : either the change is not perceptually detected, or, more importantly, the characteristic or the modality of the real situation affected by the change (the size of the object set for instance) is not mapped into the E1 schema. Another possibility is that the test indicates, according to the present state of E1 schemas, that some things are similar, and others different. Then, an <u>updating</u> perceptual exploration will be carried out, until the new information is integrated into the E1 schemas, until the response of the test is : "it is quite similar". This updating exploratory activity will mostly focus on the change. Finally, another possibility is that, still depending on what is contained in the E1 schema, the match-mismatch test may indicate that what is sampled from the real situation and the E1 schemas are quite different, so that the discrepancy elicits mapping exploratory activity.

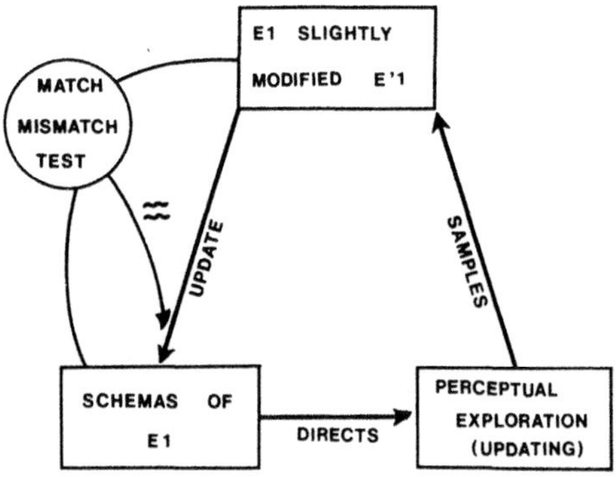

FIGURE 7. An adaptation of Neisser's cycle to "updating" exploration.

What this means, in connection with the experiments presented above, is that in the E1 schema, only relative spatial relationships may have been established within the set of objects, since the latter define a precise area, without any reference to external landmarks (the limits of the open field or the striped pattern). Otherwise, the displacement of an object within the circumscribed area would not have been perceived as disrupting all the spatial relationships as the renewal of overall exploration tends to indicate.

These data confirm the validity of one part of Neisser's cycle, namely the fact that perceptual exploration of a given environment leads to the constitution of a schematic internalized model of this environment. Further experiments are likely to yield more data about the features of these models. Nevertheless, although it seems intuitively acceptable, the hypothesis that orienting schemas guide and direct perceptual exploration has never been demonstrated. This point is crucial and needs to be investigated.

The first step in such a study might be to analyse the organization of exploratory activity. Would it be possible to find within a given species some steady patterns allowing one to characterize an apparently unorganized behavior ? More and more sophisticated recording techniques may help to answer this question, at least with rodents in which many exploratory reactions consist of actual contacts with objects ; in addition, because of the reduced ocular mobility of these species, easily observable head movements are displayed during exploration. Thus, video actographic methods which allow the recording and computing of the successive positions of the head axis (Lecas and Dutrieux, 1983, for example) appear likely to provide objective data about the behavioral manifestations of perceptual exploration. It is possible to know what part of the apparatus and what object the animal is facing, how long he does so, the sequences of places that are fixed, and the amplitude and frequency of head movements (cf. for example Thinus-Blanc and Lecas, 1985).

A second phase in this study would be to investigate with the same methods to what extent animals reared since birth in diverse spatial environments display differentiated exploratory patterns.

Such an approach, correlative to the study of the spatial relationships integrated during exploration (cf. above), should throw more light on questions more specifically related to the nature of the rules generated by orienting schemata and C.Ms.

DISCUSSION

From this survey, it can be concluded that the term "cognitive mapping" is more appropriate than "C.Ms" as it refers to a dynamic process. The function of this process is at least twofold : acquisition of environmental knowledge through exploration, and solution of spatial problems (in a wide sense, including spatial learning, homing, etc.).

Setting the problem in functional terms does not mean that maps, spatial representations, images of "the world in the head" do not exist. We will never be able to see what they are, but we already have some notion of how they work. Furthermore, according to the dynamic nature of the cognitive mapping system, maps should be subjected to continuous changes and hence difficult for investigators to grasp.

In a chapter entitled "Maps and mappings as metaphors for spatial representation", R. Downs (1981) interprets Tolman's C.Ms as follows : "In confronting the classic problem of the process-product dilemma, Tolman was forced to speak about maps as products. But these maps were only products in the sense of transitory stages in an ongoing process" (p. 149). This very loose interpretation of Tolman's ideas (if their meaning had been so limpid, many misunderstandings would have been avoided) appears up-to-date in the light of what we know today. More interestingly, this interpretation is that of a geographer. This convergence of interests strongly encourages a multidisciplinary approach to spatial orientation. Is such an approach without any risks, however ? Several questions should be kept in mind and invite us to be cautious : to what extent can the theoretical constructs and the related data obtained on man be used in research on animals ? And conversely, what is the validity of the animal model for human and clinical psychology ? Are cognitive processes of the same nature in animal and man, or is there a gap because of language ? In other words, is language a species-specific expression of a common cognitive functioning, which may be responsible in man as well as in primates, rodents, and other animals, for insight, problem solving, spatial orientation, etc. ? Premack (1979) gives some arguments in support of a common cognitive functioning by identifying in chimpanzees some factors which are shared by human intelligence and play a role in language, such as representations, causality, synonymy, etc. But as far as I know, nobody has found any trace of syntax in chimpanzees.

Still in favor of a common cognitive functioning is O'Keefe and Nadel's (1978) hypothesis that the cognitive mapping system could function as a deep structure of language. Semantic maps located in the left human hippocampus might have the same properties as those attributed to spatial maps : the ability to incorporate information regardless of the sequential order in which it is acquired, their plasticity, and so on. Perhaps the relatively new notion of modularity may shed some light on these problems (cf. Neisser, volume 2).

Honig (1978) does not object to applying concepts based on human experience to investigations on animal behavior and vice versa, provided that the <u>principle of biological continuity</u> among species is respected. This idea is attractive, but surely it is utopic. What is the principle of biological continuity ? Concerning brain structures, continuity, <u>i</u>.<u>e</u>. the fact that within vertebrates, for example, a structure is found at each level of the phylogenetic scale, without exception, is a criterion for deciding that there is homology concerning this structure. This is not an easy decision, and the results are often controversial. Do behavioral homologies exist ? Maybe, in some cases. Cineses and taxies in insects might be homologous to the visually guided behaviors governed by O'Keefe and Nadel's "Taxon system". The very name suggests this possibility, but what about more complex spatial behaviors ?

These issues will not be solved in a day. This does not mean however that we should be unaware of them or be crippled by the difficulties involved instead of facing the promising challenge of inter-disciplinary exchanges.

REFERENCES

Bolles, R.C., 1975. "Learning theory". Holt, Rinehart and Winston (Eds.), New York.

Bovet, P., 1979. La valeur adaptative des comportements aléatoires. Année Psychologique, 79 : 505-525.

Cheal, M.L., 1978. Stimulus-elicited investigation in the mongolian gerbil (<u>Meriones</u> <u>unguiculatus</u>). The Journal of Biological Psychology, 20 : 26-32.

Downs, R.M., 1981. Maps and mappings as metaphors for spatial representations. In "Spatial Representations and Behavior across the life Span. Theory and Application". L.S. Lyben, A.H. Patterson and N. Newcombe (Eds.), Academic Press, London.

Downs, R.M. and Stea, D., 1975. "Image and environment". Aldine Publishing Company, Chicago.

Fabrigoule, C. and Maurel, D., 1982. Radio-tracking study of foxes movements related to their home-range. A cognitive map hypothesis. The Quarterly Journal of Experimental Psychology, 34B : 195-210.

Honig, W.K., 1978. On the conceptual nature of cognitive terms : an initial essay. In "Cognitive Processes in Animal Behaviour". S.H. Hulse, H. Fowler and W.K. Honig (Eds.), Lawrence Erlbaum Ass., Hillsdale, New Jersey.

Krech, I., 1932. Hypotheses in rats. Psychological Review, 39 : 516-532.

Lecas, J.C. and Dutrieux, G., 1983. Head movements and actographic recordings in free-moving animals, using computer analysis of video-images. Journal of Neurosciences Methods, 9 : 357-366.

Maier, N.R.F., 1938. A further analysis of reasoning in rats. II. The integration of four separate experiences in problem solving. Comparative Physiology Monographs, 15 : 1-43.

Menzel, E.W., 1978. Cognitive mapping in chimpanzees. In "Cognitive Processes in Animal Behaviour." S.H. Hulse, H. Fowler and W.K. Honig (Eds.), Lawrence Erlbaum Ass., Hillsdale, New-Jersey.

Menzel, E.W. Jr., 1984. Spatial cognition and memory in captive chimpanzees. In "The Biology of Learning." P. Marler and H.S. Terrace (Eds.), Springer Verlag, Berlin.

Morris, R.G.M., 1981. Spatial localization does not require the presence of local cues. Learning and Motivation, 12 : 239-260.

Neisser, U., 1976. "Cognition and Reality." W.H. Freeman and company (Eds.), San Francisco.

O'Keefe, J., 1983. Spatial memory within and without the hippocampal system. In "Neurobiology of the Hippocampus." W. Seifert (Ed.), Academic Press, New York.

O'Keefe, J. and Nadel, L. , 1978. "The hippocampus as a cognitive map." Oxford University Press, London.

Peters, R., 1979. Mental maps in wolves. In "The Behavior and Ecology of Wolves." E. Klinghammer (Ed.), Garland STPM Press, New York, London.

Poucet, B., Chapuis, N., Durup, M. and Thinus-Blanc, C., 1986. A study of exploratory behavior as an index of spatial knowledge in hamsters. Animal Learning and Behaviour, in press.

Premack, D., 1979. Capacité de représentation et accessibilité du savoir. Le cas des chimpanzés. In "Théories du Langage, Théories de l'Apprentissage". M. Piatelli-Palmarini (Ed.), Editions du Seuil, Paris.

Restle, F., 1957. Discrimination of cues in mazes : a resolution of the place versus response question. Psychological Review, 64 : 217-228.

Ritchie, B.F., Hay, A. and Hare, R., 1951. Studies in spatial learning : IX. A dispositional analysis of response performance. Journal of Comparative and physiological Psychology, 44 : 442-449.

Shepard, J.F., 1931. More learning. Psychological Bulletin, 28 : 240-241.

Shepard, J.F., 1933. Higher processes in the behavior of rats. Proceedings of the National Academy of Sciences, 19 : 149-152.

Thinus-Blanc, C., 1984. A propos des cartes cognitives chez l'animal : examen critique de l'hypothèse de Tolman. Cahiers de Psychologie Cognitive, 4 : 537-558.

Thinus-Blanc, C., Durup, M., Chaix, K. and Poucet, B. An analysis of the role of exploratory activity in spatial knowledge. In preparation.

Thinus-Blanc, C. and Ingle ,D., 1985. Spatial behavior in gerbils (Meriones unguiculatus). Journal of Comparative Psychology, 99 : 311-315.

Thinus-Blanc, C. and Lecas, J.C., 1985. Effects of collicular lesions in the hamster during visual discrimination. An analysis from computer-video actograms. The Quarterly Journal of Experimental Psychology, 37B : 213-233.

Tolman, E.C., 1948. Cognitive mpas in rats and men. Psychological Review, 55 : 189-208.

Wilz, K.J. and Bolton, R.L., 1971. Exploratory behaviour in response to the spatial rearrangement of familiar stimuli. Psychonomic Science, 24 : 117-118.

COGNITIVE MECHANISMS IN ANIMAL PROBLEM-SOLVING

Paul Ellen

Georgia State University
Department of Psychology
Atlanta, GA 30303

In recent years there has been a burgeoning interest in the use of cognitive concepts to account for the performances of animals in complex test situations both in the laboratory and in the field. The recent volume stemming from the Guggenheim Conference on animal cognition (1984) published within six years of the major works by O'Keefe and Nadel (1978) and by Hulse, Fowler and Honig (1978) attest to the renewed interest in cognitive mechanisms in animal behavior. Animal psychologists have begun to reconsider the possibility that many of the behaviors exhibited by animals in laboratory as well as field test situations cannot be accounted for simply as chains of S-R associations strung together in complex patterns.

Modern-day theorists are now recognizing the necessity for invoking some sort of mediational concept to account for the fact that animals react to stimuli that are not physically present. However, the use of mediational concepts in animal psychology has tended to embroil theorists in controversy over such issues as "self-awareness, symbolic communication, intentionality, etc. Concern with animal consciousness has indeed dominated many of the major studies in modern animal cognition (Griffin, 1984). The recent conclusion of Putney (1985) that tool-using apes are "agents knowing what they are aiming at..."(p.60) exemplifies this concern.

Such controversy, while not being resolvable, does unmask the mentalistic and dualistic assumptions and biases underlying much of the current work in animal cognition. It reflects the fact that the concepts so disputed, were derived subjectively by analogy with human experience. As a consequence, these concepts lack objectivity and derive most of their support through anecdotal accounts of animal behavior.

While this type of theorizing has surfaced most noticeably in primate studies of tool-using and language, similar analogical inferences may be found in the writings of theorists concerned with studying the ability of animals, particularly rodents, to navigate in spatial situations. The attempts to clarify the mechanism become mentalistic and rely on an analogy with human map-reading processes as illustrated in the following account of how a rat navigates to a hidden platform in a pool of water when started from a variety of starting points:

> "At the beginning of a particular trial, the trained rat must determine its location among the configuration of distal objects and calculate a trajectory...in relation to at least some of these

> distal objects.... The rat calculates current position and trajectory for movement...No new problem-solving strategy emerges at novel starting points. It would not be inappropriate to claim that on each trial,...the rat infers the correct path"(Sutherland and Dyck 1984, p.329).

The making of inferences concerning the nature of animal mentation by analogy with subjective experience can only lead to a mentalistic type of theory and represents a regressive force in the attempt to study complex or higher processes in animals. Fifty years ago, Maier and Schneirla wrote,

> "...A system of animal psychology must be constructed from behavior data, and complex processes are no exception to this rule. It is therefore our task to develop an objective definition of, as well as objective criteria for, higher processes."(1935, p. 446).

This dictum takes on added urgency since, as Sutherland and Dyck also point out, the study of spatial ability has become a central concern of modern studies of animal cognition, cognitive and/or behavioral neuroscience, and ethology.

Although the animal learning literature contains reference to the varied ways rats can solve spatial problems, this literature does not distinguish adequately the conditions under which one or another behavioral strategy comes to expression, nor does it clarify or distinguish the psychological status of one or the other strategy. Sutherland and Dyck (1984) and O'Keefe and Nadel (1978) refer to mapping vs. taxis and guidance strategies or praxis and orientation strategies as potential mechanisms whereby rodents are capable of going from one place to another. Although mapping strategies are regarded as qualitatively different from taxis and praxis mechanisms since they allow animals greater flexibility and hence survivability in their adaptation to spatial problems, Sutherland and Dyck (1984) assert that there is little evidence for the presence of a hierarchy of navigational strategies. By implication the particular strategy which comes to expression at any given time simply depends then on the availability of environmental cues.

The apparent contradiction between the greater flexibility afforded by mapping strategies and the lack of any hierarchical relationship among the various strategies derives from a failure to specify the nature and properties of the mapping mechanism as opposed to that involved in praxis or taxis strategies. While the latter are fairly simple and can generally be subsumed within the rubrics of selective learning (Maier and Schneirla, 1942) or operant conditioning procedures, the psychological mechanism by which the mapping strategy confers greater adaptability is not as simple. It will be the thrust of this paper to demonstrate that the various behavioral mechanisms underlying adaptations to spatial problems are not only qualitatively different, but also, the kinds of adaptations resulting from the operation of such mechanisms differ with respect to their psychological status. Moreover, we will attempt to indicate the critical factors underlying the psychological status of a particular kind of adaptation. Finally we will demonstrate that intelligent adaptations to spatial problems can arise only when spatial information has been stored in a map-like format.

PROBLEMS IN MODERN ANIMAL COGNITION

One of the most vexing problems in animal psychology is the development of criteria which would denote intelligent behavior. Most generally, investigators create a task, and then argue that in order for the animal to be successful on the task, a certain process must be operational. Essentially, this procedure defines the process in terms either of the experimental variables being manipulated or the particular behaviors studied in a given task. Processes so defined involve nothing more than the reification of operationally defined variables and/or tasks.

Achievement- or performance-based criteria of the nature of psychological processes are prone to being highly subjective and/or highly specific either to the task or to the particular organism under consideration. Moreover, as Schneirla (1949) has pointed out, neither the complexity of behavior in learning problems nor the sudden improvement in efficiency which often occurs in such problems, can be used as evidence that the organism in question has any "insight" into the essential relations of the problem.

How then are we to proceed if we are not to utilize the achievement of the animal on a task as the basis for the inference of function? First, our interest must be in the psychological process itself rather than the particular behavior of an animal in a given situation. Its behavior is of interest only to the extent that it enables us to learn more about the process. As Nadel and O'Keefe noted, "the prediction and control of mind and behaviour are merely ancillary interests whose main role is in the verification of explanatory schemes" (1974, p.368). Accordingly, we must begin with some *a priori* notions of the characteristics and properties of the psychological processes in which we are interested and then create experiments which will test these notions. The achievement of the animal on a particular task will then not be the basis from which the function is inferred. Under such circumstances, each experiment will refer to a theoretically-derived process and the achievement of the animal in the experiment will be the objective measure with which to test the theory (Maier, 1938); thus, it will be the test of the theory rather than the data from which *ad hoc* mechanisms or processes are inferred.

TOWARDS A PROCESS-ORIENTED DEFINITION OF INTELLIGENCE

Few modern psychologists would disagree with the view that some sort of representational mechanism mediates between sensory inputs on the one hand and behavioral outputs on the other. Most are even willing to accept the idea that such representations can interact with each other, and that the behavioral output is the result of the interaction of the various representations rather than being merely the direct reflection of the inputs (Schwartz, 1984). To the extent that we admit the possibility of representations interacting with each other, the organism is freed from the constraints resulting from a view that an association between elements is simply the summation or chaining together of those elements. The notion that representations may interact provides a basis whereby the kind of organization resulting from the interaction is more than simply the summation of the individual representations themselves.

Behavior based on a chaining together of previously-acquired representations is merely evidence that the animal in question is able to learn. Maze learning studies in rats document quite fully the fact that different associations may be

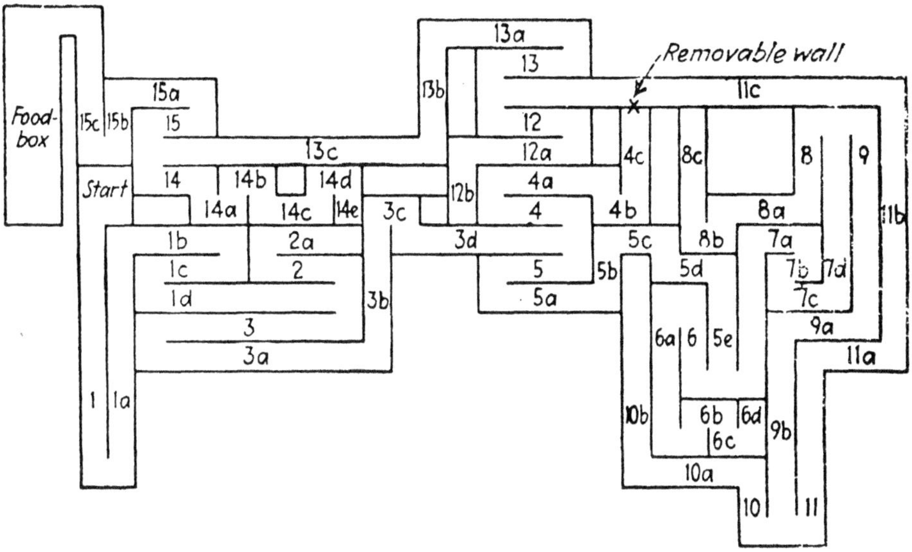

Figure 2.1. Maze used in John Shepard's unpublished experiments.

linked together in a particular order. However, not only can animals chain previously-acquired associations into a pattern, but also they are capable of spontaneously reorganizing them at the time of response expression.

One of the earliest demonstrations of this ability was shown in an old experiment by John F. Shepard of the University of Michigan (unpublished work cited in Maier and Schneirla, 1935). Using the maze shown in Figure (2.1), Shepard demonstrated the ability of rats to reorganize their previously-acquired representations in a quite dramatic manner. In this maze, the even-numbered alleys are blind alleys, while the odd-numbered alleys are the true paths. The letters refer to different sections of the alleys and are of use in tracking the animal through the maze. Mazes of this type are not used anymore; however, they are invaluable for uncovering the complexity of the rat's adaptation. This maze is readily learned. Following mastery, a change is made in the maze by removing the section of wall marked by the "X". This change is not reacted to by the animal since it is at the end of a blind alley. However, it can be discovered as the animal runs along the true path. Once detected, the animal explores the change for a short distance down the blind alley. The animal then returns to the true path and continues on to the goal box. What does the rat do on its very next trial in the maze? It choses blind alley 4 as a short-cut to the true path 11-c.

Results such as these suggest quite strongly that as an animal runs in a maze it acquires information about that maze, i.e. the spatial relationships existing in the maze independent of the particular location of the animal. Moreover, these results suggest that such information is stored in such a manner that it is readily accessible, readily fragmented, and capable of being linked together with other information from other representations which have been acquired at different times. Since the animal uses its past learning concerning the spatial relationships in a novel manner, it would seem that there has been a reorganization at the time of response expression of the stored cognitive structure.

This analysis suggests that if we are to demonstrate that a process more complex than that involved in learning is functional, then elements of previously-acquired representations must be integrated with elements of other representations or with new learning, at the time of response expression. Problem-solving based on a mechanism which involves nothing more nor less than the mere chaining together of one association or representation with another reflects nothing more than the sequential retrieval from memory of previously-stored representations in their entirety. However, it is quite a different matter when problem-solving requires that part of one representation and part of another be integrated.

The spontaneous adjustment to a problem situation, involving as it does the reorganization of traces of past experiences, places the problem-solving process then within the context of a perceptual experience rather than within the domain of learning and memory phenomena. While learning and memory phenomena may provide the raw materials with which the problem-solving process operates, that is all such phenomena do. They do not provide the solutions to the problems. Support for this view comes from the fact that scores made on maze learning tasks tend to correlate negatively with scores made in tests of the ability to reorganize past experience (Maier and Schneirla, 1935).

From this perspective, solutions to problems come about as the result of a perceptual experience over which the organism has no control, in much the same manner as a figure-ground reversal is experienced. At each instant there is a sudden perception with a change in the meaning of the elements. Elements at one moment are seen as one unity; at the next moment, another unity appears with the same elements.

Our emphasis on the reorganization of past experiences as being critical in distinguishing intelligent adaptations from those reflecting the mere retrieval from memory of previously acquired information does not make the organism or the task and situation a central element. Rather it characterizes a mechanism which can be tested in a variety of different situations or tasks. By being process-oriented rather than species- or task-oriented, it has generality across species and tasks thereby allowing us to determine whether various organisms are capable of such reorganizations of past experience. Organisms can be classified in terms of their performance on process-defined tasks rather than in terms of experimenter-defined or species-defined characteristics. According to this view, the organism is merely the carrier of the process. Parko's work with school age children (see Volume 2) shows clearly that although the specific task used to demonstrate how independently-acquired past experiences may be combined to solve a problem may vary from that used with rats, nonetheless it is clear that the same type of psychological process is involved.

There is a major advantage resulting from this type of approach. If we assume that the appearance of spontaneous adjustments in a problem situation reflects the operation of a psychological process which is similar to creativeness and reasoning in man and is fundamental to the formation of new patterns of behavior, then we have a basis for believing in the notion that it is not necessary to utilize mentalistic concepts in the analysis of human reasoning and complex functioning. To the extent that the process is similar to that which is engaged in by humans when they are engaged in drawing inferences, we can avoid an unncessary separation in animal and human work.

LEVELS OF ADAPTATION TO SPATIAL PROBLEMS

Taxes(guidances) and Praxes(orientations)

Most organisms have a number of behavioral mechanisms available to them in their attempts to move and behave adaptively with respect to objects located in space. At the simplest level are those mechanisms that either allow animals to move towards or away from some specific stimulus object without regard for the particular patterns of movement involved or equally simply, enable animals to approach or withdraw from an object with a particular pattern of movements such as turn right or turn left at the choice point. These mechanisms has been variously called taxes and praxes (Sutherland and Dyck, 1984) or guidances and orientations (O'Keefe and Nadel, 1978). I make the point that these are relatively simple behavioral mechanisms since they invariably are available to brain-damaged animals (Ellen and Wages, 1984), and are the means whereby such animals can function quite adaptively. Herrmann (Volume 2) has described in detail such adaptations. It is the case however, that these adaptations are quite limited and specific to the particular situation. In general, one would not assert that a cognitive function is involved in these kinds of adaptation since in each instance, the stimuli controlling the behavior are present at the time of response expression.

Rule-governed Behavior

Organisms can also adapt to spatial problems by performing in terms of some rule or consistent food searching strategy. For example, Olton and Schlosberg (1978) and Kamil (1984) have reported that rats and birds are predisposed to follow win-shift strategies in foraging situations. Kamil noted, while recording the visits of color-banded territorial birds to a set of mapped flowers, that resident birds avoided revisits to flowers which had just been emptied, while intruder birds displayed no such tendency.

Let me note however at this juncture that an adaptation to a spatial problem based upon the expression of some consistent strategy such as win-shift reflects a more complex level of psychological function than an adaptation based simply upon the expression of a body turn or upon the approach to or withdrawal from a specific stimulus without regard for the pattern of movements involved. As I have already indicated, animals with brain damage have no difficulty adapting by taxes or praxes. However, they are unable to adapt to spatial problems by means of rules or strategies such as win-shift or win-stay (Ellen and Weston, 1983) unless they have had experience with the task prior to the brain-damage (Ellen, 1980).

A clue as to the basis of the failure of septally-damaged rats to utilize strategies in solving spatial problems comes from a recently-completed study (Ellen, Taylor and Wages, 1984) utilizing the Maier 3-table task. On the 3-table task, the animals are generally first given a preliminary exploratory experience over three tables and their interconnecting runways. Subsequently, they are given a feeding experience on one of the tables. They are then tested by being placed on one of the two remaining tables. Their task can then either be to return directly to the table on which they were just fed or to go to the table that was not baited during the feeding experience, depending upon the particular experiment in progress. Each day the table on which they are fed and the start table is varied in order to prevent the animals learning either a specific location of the food or a specific turn towards or away from the food. In a typical application of the task, the animals are given one test trial each each day. Since the problem varies daily, the animals are required to reorganize part of their past experiences to conform to the new requirements of either the daily change in the locus of food table or the start table (i.e., the place where they were fed on the preceding day may or may not be the place where the food is currently available).

In the Ellen, Taylor, and Wages experiment the animals were given 3 trials per day. Under these circumstances, only the first trial each day can be considered as reflecting the ability to reorganize experience. The other two daily trials are learning trials in which the animal can locate the food table. Since the problem changes from day to day, the animals have no opportunity to learn a specific solution that will be effective from day to day, i.e., that food is in a particular location or that a particular turn will always be successful. Rather the animals can learn a strategy that, regardless of the outcome of the first trial, will always be successful on the second and third trial each day. That is, if the animal is correct on Trial 1, it should learn on Trials 2 and 3 that a win-stay strategy is in order. If however it fails on Trial 1, then it should learn that a lose-shift strategy is in order.

The animals were given 10 days of 3 trials a day testing on the task. During this time the animals had an opportunity to learn the spatial relations among the tables since they were given daily exploratory experience. Following this testing, the animals were injected either with atropine sulfate which crosses the blood-brain barrier or atropine methyl nitrate which doesn't. The drugs were injected 30 min prior to daily testing. Following 10 days of additional testing under drugs, the drug injections were discontinued for 2 days and then resumed for an additional 10 days. Now however, those animals which received atropine sulfate received methyl atropine and vice versa. Trial 1 performance was markedly impaired with atropine sulfate but not with atropine methyl nitrate. More important, was the fact that performance on the second and third trials of each day was not impaired with atropine sulfate (see Figure 2.2).

If we assume that in order to follow some rule such as lose-shift, win-stay from Trial 1 to Trial 2 and/or 3, the animals must remember on the second and third trial each day whether they received food or not following a particular response choice on the previous trial, then it would appear that the central cholinergic blockade did not affect the response strategy which was acquired prior to drug treatment. In contrast, since the Trial 1 performance was impaired it would appear that the memory as to the locus of food, i.e. a spatial memory, acquired each day was affected by the drug treatment.

Thus animals can learn a response rule such as win-stay, lose-shift as a means of adapting to a spatial problem, even though the mechanism critical for the reorganization and integration of spatial experiences, i.e. spatial memory, is not functional. It would follow then that rule-governed behavior represents the operation of a simpler mechanism than that involved in behavior which reflects the reorganization of experience. However, to the extent that the animals must remember whether the response they made on the previous trial led to food or not, then the expression of response rules such as a win-stay, lose-shift would be somewhat more complicated than behavior representing the operation of a taxis or a praxis mechanism.

Finally, to the extent that animals with brain-damage can perform successfully on the 2nd and 3rd trials of this task provided they have had

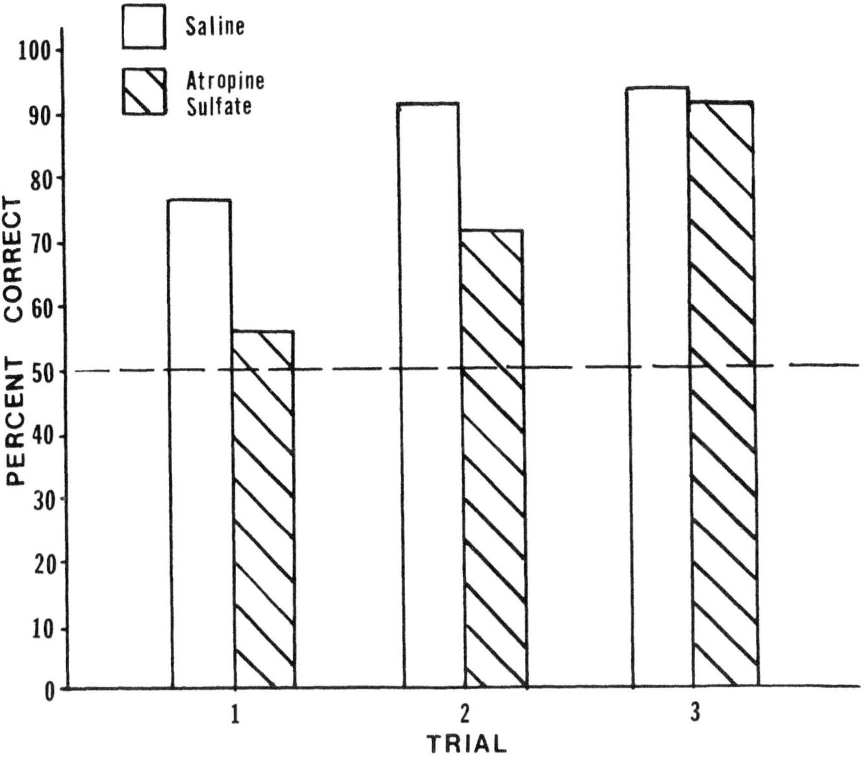

Figure 2.2 Percent correct following cholinergic blockage on either Trials 1, 2, or 3 of 3-Table task.

experience with the task prior to the brain damage (Ellen, 1980) it would appear that the memory which is spared by either the drugs or brain-damage (long-term or reference memory) is not selective with respect to the nature of its content. In contrast, the memory of the daily locus of food (working memory) is spatial in character.

Intelligent Spatial Problem Solving

Turning now to the most complex level of adaptation to spatial problems, Herrmann, Bahr, Bremner and I (1982) showed that when animals are already familiar with the spatial relationships of the various locations in the environment, then regardless of whether the problem on the Maier 3-table task requires them to return to the baited table or to avoid the baited table on the first daily test trial, the animals have no difficulty in locating the food source (Figure 2.3). Moreover, these data indicate that not only are the animals expressing their knowledge of the various feeding locations relative to their current location, but also their behavior is sensitive to the state of the food supply at a given location. When there is no food left after the feeding experience on the 3-table task, rats perform poorly when required to return to the previously-baited table; however, when food remains there then they have no difficulty in returning to the food on the test

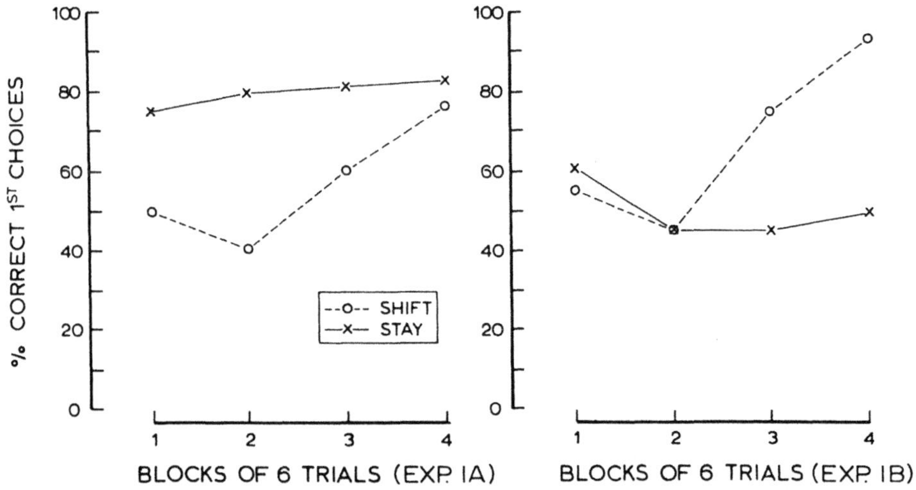

Figure 2.3. Performance of rats tested on either a stay or shift problem: A. Following partial feeding. B. Following complete feeding. (Reprinted from Herrmann et al. 1982 copyright 1982 Psychonomic Society, Inc., with permission)

trial. In contrast, rats have some difficulty in expressing the win-shift tendency when food is left after the feeding experience. These findings would suggest that in food-seeking problems, it is the knowledge of the availability of food at a given location which is integrated with the knowledge of the spatial relations existing among various locations that is a critical determinant of the behavior. More importantly, these data imply that even in a behavior as biologically important as food-searching, animals do more than simply express some species-typical food-searching response rule that may have been selected for by the evolutionary process.

THE COGNITIVE MAP AS A BASIS OF INTELLIGENT ADAPTATION

What is required in order for an animal to be able to reorganize elements of its past experience in order to solve a spatial problem such as that posed by the 3-table task? I have previously indicated that the information stored in a cognitive representation must be readily accessible, readily fragmented and capable of being linked together with information from other representations acquired at different times. Inasmuch as organisms acquire information about a region by virtue of their exploratory activity and inasmuch as exploratory behavior involves a series of successive experiences, it is the case that such successively-acquired information must be transformed into a cognitive structure (a simultaneous pattern) in which the distance and direction between the various successively-experienced objects are indicated in order for spatial information to result. Otherwise, such successive experiences will simply remain as a series of temporally-ordered experiences. Ellen, Soteres, and Wages (1984) demonstrated that although the animals were allowed to explore different combinations of two tables and their interconnecting runways on successive days, nonetheless they

GROUP	EXPLORATION	FEEDING	TESTING
0	none	daily	daily
1		every fourth day	every fourth day
2			
3			
TRADITIONAL		daily	daily

Figure 2.4. Procedure used to demonstrate the integration of separate exploratory experiences.

could be successful on the test trial which was given after the feeding experience on the day after all possible combinations of two tables and their interconnections were explored. The design of the experiment is illustrated in Figure 2.4. Different groups of rats were allowed to explore either one, two or all three tables on successive days. On the fourth day of each cycle, the animals were merely fed on one of the tables and given a test trial. It can be seen from Fig. 2.5 that all of the animals allowed to explore three tables each day were successful while of none of the animals allowed to explore only one table and its runway each day were successful. However, six of the 10 animals which explored two tables and their interconnecting runways were able to perform successfully providing clear evidence that rats can link together into a unified cognitive representation information acquired on separate occasions (see Fig. 2.5). Moreover, it is clear that the feeding experience and test trial on the fourth day led to a reorganization of these separate experiences since the animals were correct on that test trial regardless of whether or not the route to the food on the day of the test trial corresponded to the route explored the previous day (see Figure 2.6). In contrast, those animals which were unable to be consistently successful following the piecemeal exploratory experience, were successful only when the test route corresponded to the route explored the previous day. This latter finding suggests that for these animals at least, the information concerning the spatial relations was not stored in a map-type of storage whereby any part could be accessed.

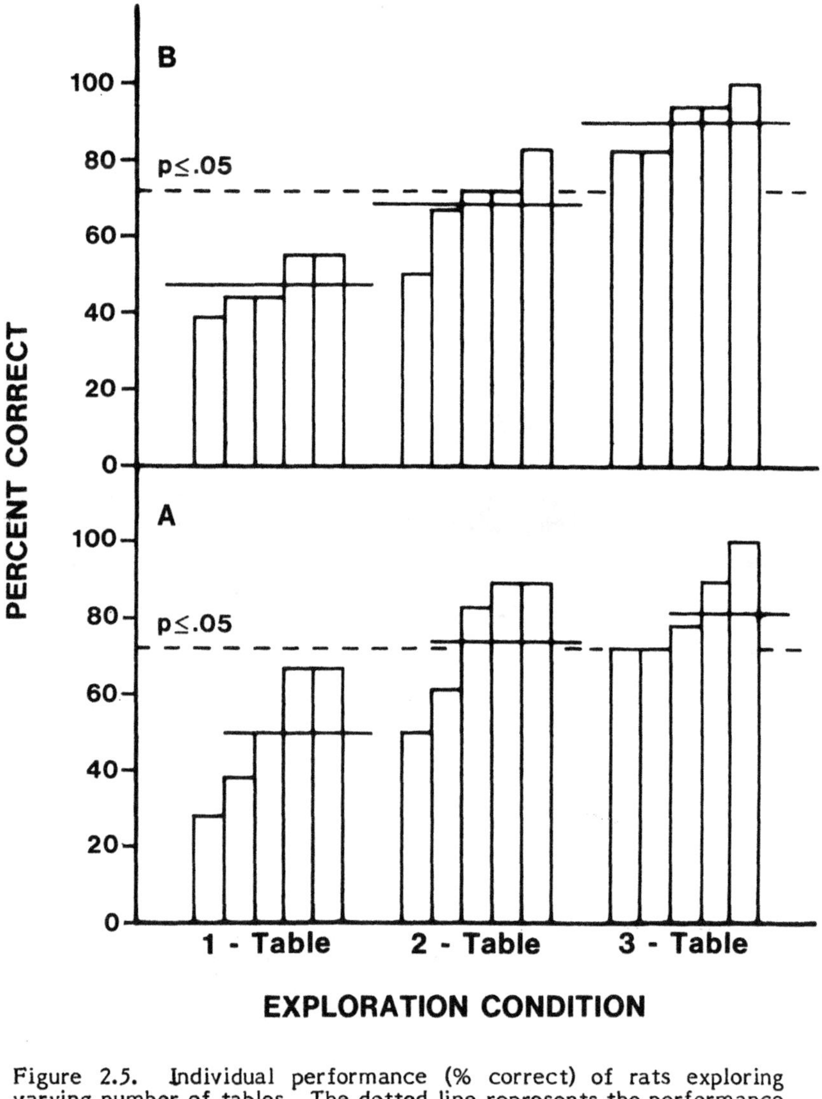

Figure 2.5. Individual performance (% correct) of rats exploring varying number of tables. The dotted line represents the performance level that could occur by chance 5% of the time or less. The solid line represents group means. (Reprinted from Ellen et al. (1984) copyright 1984 Psychonomic Society, Inc., with permission).

Rather, for the animals which failed, we have the distinct impression that the storage is in the form of some list or register since the accessibility of information is determined by the recency with which it was entered into the register. In contrast, it would appear that the storage of the spatial information was in map-form for those animals which were successful since their success did not depend on whether the test route corresponded to the route just explored the day or several days previous. In short, this study would seem to lend support to

Correct Performance as a Function of When The Test Route Was Explored During The Exploratory Cycle

Group	Total Percent Correct	Percent Correct When Test Tables Matched: First Pair Explored	Second Pair Explored	Third Pair Explored
Animals scoring above 72% N = 6	81.1	83.2	76.2	83.8
Animals scoring below 72% N = 3*	58.7	38.7	64.3	73.0

* One animal which showed a turn tendency (15/18 turns to the same side) was omitted from this analysis.

Figure 2.6. Correct performance as a function of when the test route was explored during the exploratory cycle. (Reprinted from Ellen et al (1984) copyright 1984, Psychonomic Society, with permission)

the notion that one of the necessary conditions for the appearance of a reorganization of experience is that the various experiences of the organism be stored in a map rather than list format.

ADAPTIVE ADVANTAGES OF INTELLIGENT PROBLEM SOLVING

As I indicated earlier, the ability to reorganize elements of past experience confers an adaptive advantage on organisms so endowed. Such organisms are capable of taking short-cuts in spatial problems, of chosing the shorter of several alternative routes to a goal, etc. These intelligent behavioral achievements have their root in the cognitive mapping mechanism. Once animals have a cognitive representation or map of a problem space, then not only can they "react at a distance", but also they can recognize those relationships in the problem space such as "shorter than", "more direct than", etc., which if utilized would confer an adaptive advantage.

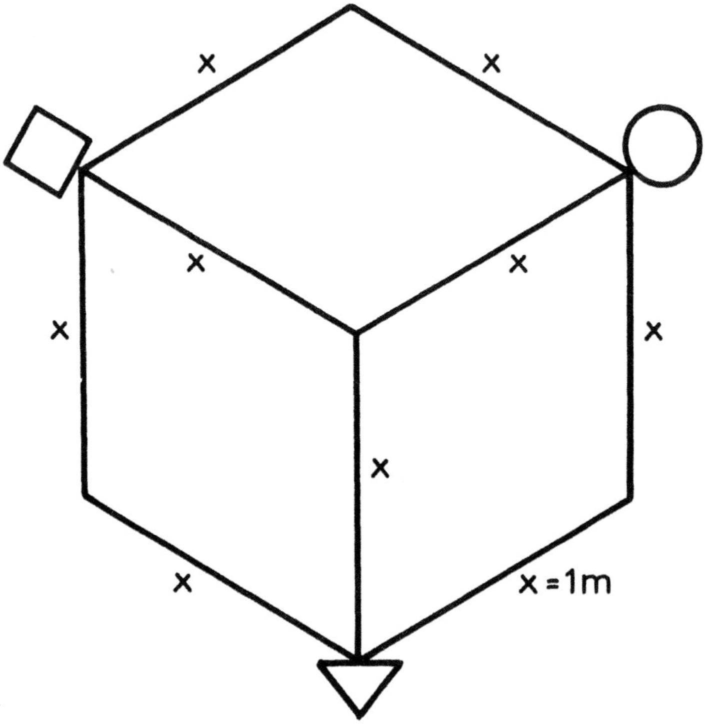

Figure 2.7. Diagram of apparatus used in Herrmann et al. (1985) copyright 1985 American Psychological Assn., with permission)

Recently, we have had the opportunity to see how the existence of a cognitive map of the spatial relations existing amongst tables of the 3-table problem allowed animals to take advantage of another similar type of relationship which existed in the problem space. Herrmann, Poucet, and Ellen (1985), demonstrated that rats prefer to delay making choices if the delay allows them to make the choice closer to the goal. Using the apparatus illustrated in Figure 2.7 normal and septal rats were tested in the same manner as is used on the traditional 3-table task--that is, the animals were given a 15 min exploratory experience, followed by a brief feeding on one of the tables, and then the test trial. Normal animals had no difficulty with this task, while septals failed. While this difference is of interest, the qualitative character of the animals' behavior during the exploratory period and test trial is of particular interest. Both normals and septals explored the outer pathways of the apparatus twice as frequently as the inner pathways. However, during the test trial, while the septal animals continued to prefer to run on the outer pathways, 8 of 10 normals preferred to run on the inner pathways. This shift in choice of pathways by the normal animals from the exploration period to the test trial is of particular significance. The only difference between inner and outer pathways is that once the animal runs down the outer pathway it is committed to its choice of table. However, if the animal choses to run on the inner pathway, it does not have to make a choice of table until it is almost a meter closer to the tables. In short, it would appear from the shift in pathway preference, that normal animals prefer to delay making choices in adapting to a spatial problem, whereas animals with septal damage are constrained to express on the test trial their initial response preferences the outer pathways. I would like to suggest that the basis for this difference in pathway preference during the test trial stems from the fact that the normal animals have a representation of the spatial properties of the problem, while the brain-damaged animals do not. More importantly, this representation allows the normal animals to recognize the adaptive advantages of the inner routes.

Support for this hypothesis was found in other data of the experiment. In this experiment the animals were allowed to self-correct an incorrect first choice. When this occurred, an animal could self-correct by immediately leaving the incorrect table via the same route or pathway that it entered the table or by taking a different route or path. Thirty-five percent of the animals in both groups self-corrected by chosing a pathway different from the one they took on their first choice. Thus, if they took an outer path on the first choice, they took the inner path when they self-corrected, and vice versa. Normals, under these circumstances, reached the food table on 75% of their second choices. Septals choosing an alternate route on the self-correction opportunity were successful only 32% of the time. That is, the septal animals, when they shifted from an outer path to an inner path and vice versa on the self-correction opportunity, found themselves back on the start table of the trial. In contrast, normal animals, when they shifted routes, were able to get to the food table. Thus it would seem that the normals, although they made an error on their first choice, had some awareness of the spatial relations existing among the tables - they could get to the food table via an alternate route. Septals simply could not. This finding would lend much support to the importance of the mapping mechanism for the storage of the spatial relationships among the tables and for enabling animals to recognize the adaptive advantages of the various relationships existing between objects represented in the map.

CONCLUDING COMMENTS

It would appear that there are several levels of complexity in the adaptation of animals to a spatial problem. At the simplest level, exemplified by the performance of animals with brain damage, behavior is directed by the most proximal, salient stimuli available. There is no evidence in the performance of such animals that memorial mechanisms have any role in the adaptation of such animals to spatial problems. These animals successfully adapt to such problems by means of either taxis mechanisms, if specific local cues are consistently associated with success, or by praxis mechanisms as long as the starting point and goal remain in a constant relation to one another.

However, when long term or reference memorial mechanisms are functional, then despite brain-damage or drugs, animals can adapt to spatial problems simply by following some sort of rule such as win-stay, lose shift. Having a stored rule or strategem to follow frees animals from the constraint of having to rely on the most salient stimuli available for their adaptation to a spatial problem. This type of adaptation, involving as it does a memorial mechanism, would seem to be of a higher psychological status than those involving a direct response to stimuli, either external or response-produced.

While adaptations based on rules or on taxis and praxis mechanisms give rise to behavior that appears to be "psychologically smart" (Rumbaugh, 1985), such adaptations do not reflect the operation of an intelligence or complex psychological process. In either instance the animals are constrained to behave simply in terms of their past learning. While adaptations of this sort are adaptive to a degree, they fail when past learning does not provide the solution to the problem. It is only when animals reorganize their past experiences which have been stored in a map-like format, that we can see the operation of the complex intelligence process.

References

Ellen, P., 1980. Cognitive maps and the hippocampus. Physiological Psychology, 82:168-174.

Ellen, P. and Wages, C., 1984. The use of intramaze stimuli in the attenuation of the problem-solving deficit of septal animals. Physiological Psychology, 12:133-140.

Ellen, P. and Weston, S., 1983. Problem solving in the rat: Septal lesion effects on habituation and perseverative tendencies. Physiological Psychology, 112:112-118.

Ellen, P., Soteres, B. J. and Wages, C., 1984. Problem-solving in the rat. Animal Learning and Behavior, 12:232-237.

Ellen, P., Taylor, H. S., and Wages, C., 1984. Atropine sulfate affects Spatial integration, but not cue discrimination performance. Paper presented at the Twenty-fifth Annual Meeting of the Psychonomic Society, San Antonio, Texas.

Griffin, D. R., 1984. "Animal Thinking." Harvard University Press, Cambridge, Mass.

Herrmann,T., Bahr, E., Bremner, B., and Ellen, P., 1982. Problem-solving in the rat: Stay vs. shift solutions on the three-table task. Animal Learning and Behavior, 10:39-45.

Herrmann, T., Poucet, B. and Ellen, P., 1985. Spatial problem solving in a dual runway task by normal and septal rats. Behavioral Neuroscience, 99:631-637.

Hulse, S. H., Fowler, H. and Honig, W. K. (Eds.), 1978. "Cognitive Processes in Animal Behavior." Erlbaum, Hillsdale, N.J.

Kamil, A. C., 1984. Adaptation and Cognition: Knowing what comes naturally. in "Animal Cognition". Roitblat, H. L., Bever, T. G. and Terrace, H. S. (Eds.), Erlbaum, Hillsdale, N.J.

Maier, N. R. F., 1938. A further analysis of reasoning in rats.II. The integration of four separate experiences in problem solving. Comparative Psychology Monographs, 15:1-43.

Maier, N. R. F. and Schneirla, T. C., 1935. "Principles of Animal Psychology," McGraw-Hill, New York.

Maier, N. R. F. and Schneirla, T. C., 1942. Mechanisms in conditioning. Psychological Review, 49:117-134.

Nadel, L. and O'Keefe, J., 1974. The hippocampus in pieces and patches: an essay on modes of explanation in physiological psychology. in "Essays on the Nervous System: A Festscrift for Professor J. Z. Young." Bellairs, R. and Gray, E. G. (Eds.) Clarendon Press, Oxford.

Olton, D. S. and Schlosberg, P., 1978. Food searching strategies in young rats Win-shift predominates over win-stay. Journal of Comparative and Physiological Psychology, 92: 609-618.

O'Keefe, J. and Nadel, L.. 1978. "The Hippocampus as a Cognitive Map." Clarendon Press, Oxford.

Putney, R. T. P., 1985. Do willful apes know what they are aiming at? The Psychological Record, 35: 49-62.

Roitblat, H., Bever, T., and Terrace, H., 1984. "Animal Cognition." Erlbaum, Hillsdale, New Jersey.

Rumbaugh, D. M., 1985. Comparative Pyschology: Patterns in Adaptation. The G. Stanley Hall Lecture Series, Vol.5, pp.7-53. Rogers, A. M. and Scheirer, C. J. (Eds.), American Psychological Association, Washington, D.C.

Schneirla, T. C., 1949. Levels in the Psychological Capacities of Animals. in: Philosophy for the Future: The Quest of Modern Materialism. Sellars, R. W., McGill, V. J., and Farber, M. (Eds.), The Macmillan Company, New York.

Schwartz, B., 1984. "Psychology of Learning and Behavior." W. W. Norton & Co., W. W. Norton Co. New York.

Sutherland, R. J. and Dyck, R. H., 1984. Place navigation by rats in a swimming pool. Canadian Journal of Psychology, 382:322-344.

Acknowledgement

I should like to take this opportunity to thank Duane M. Rumbaugh, James L. Pate and Charlene Wages for reading and commenting on previous versions of this manuscript. Their input encouraged and facilitated its preparation.

SECTION I. EXPERIMENTAL ANIMAL PSYCHOLOGY

BEHAVIOR IN RELATION TO OBJECTS IN SPACE: SOME HISTORICAL PERSPECTIVES

Michael E. Rashotte

Florida State University

Some of the best minds in learning theory circa 1930-1950 dealt with problems involving spatial orientation in animals. In the ensuing Skinner era of the '50s and '60s, spatial problems were mostly ignored while attention was focused on reinforcement schedules and repetitive manipulative behavior. Late in the 1960s, however, learning researchers began returning to spatial problems, as did others whose interests lay in memory, reasoning and the neural mechanisms of behavior. Today, the ambience in parts of the field is strongly evocative of the 1930s. Problems, apparatuses and theoretical languages that were in vogue half a century ago appear in the current literature once again, and spatial behavior figures prominently in this trend.

In this paper, I will review the two major theoretical approaches that were applied to spatial behavior in the 1930-1950 era, cognitive theory and stimulus-response (S-R) theory. The developments in cognitive theory during that era provide the foundation for much of the cognitive theory that predominates in today's work on spatial behavior. However, unlike today, cognitive theory experienced vigorous competition in the 1930s. S-R theory was the competitor which offered a non-cognitive approach to spatial behavior. In some ways, the issues raised in this early competition persist today in the form of questions about the relationship between cognitive and non-cognitive (e.g., neural) approaches to spatial behavior.

The main cognitive approaches of the 1930-1950 era were proffered by the Gestaltist, Kurt Lewin, and by the Sign-Gestaltist, Edward Tolman. In fact, Lewin is widely credited with bringing spatial problems to the attention of mainstream learning theory (Hull, 1938; Tolman, 1932a). Tolman himself commented early in his influential book, Purposive Behavior In Man And Animals, that much of what he was to say "seems to bear a close relationship to the doctrine of Lewin, as the latter is to be gleaned from his writings and those of his students" (Tolman, 1932a, p 37).

The main theoretical competition for the cognitive approaches came from a rather sophisticated version of S-R theory that must be distinguished from Watsonian or Skinnerian approaches. This competitor approach to spatial behavior was developed by Clark Hull and was amplified by Neal Miller. In today's cognitive literature, the Hull approach is still noted, but usually in passing and as a caricature that seems designed for easy dismissal in favor of cognitive theory. Hull's theory is often misrepresented as a melange of S-R and operant ideas that bear little relation to what might be called "the real thing." Once presented in this fashion, its inadequacies are noted (often having to do with alleged failures surrounding the treatment of the stimulus that controls behavior), and the whole approach is characterized as a theoretical venture that has been eclipsed by the progressive new emphases of today's cognitive approaches (which often turn out to be very similar to the cognitive approaches of the 1930s).

This style of dealing with its past competitor is undoubtedly good for what might be called the business side of cognitivism. The introduction to many a cognitive paper is enlivened by such discussions of S-R theory, which often imply that a Kuhnian scientific revolution is at hand (or is already completed), thereby imparting an undeniable air of progress to the whole enterprise. If the historical record is examined, however, we find that although the Hullian approach to spatial behavior was easy to distinguish from the cognitive approaches on the basis of theoretical language and style, it was often hard to distinguish on the basis of ability to make predictions about behavior in spatial situations. In some instances, Hull and Miller even seemed to be ahead in predictive capability. Be that as it may, I think there is value in revisiting these two theoretical approaches with an eye for accurate historical detail, not as a way of reviving Hullian theory, but as a way of improving the perspective on today's cognitive theory. The following passage from the philosopher of science, Paul Feyerabend, captures the spirit in which the present paper is written.

> Theories cannot be justified and their excellence cannot be shown without reference to other theories. We may explain the success of a theory by reference to a more comprehensive theory (we may explain the success of Newton's theory by using the more general theory of relativity); and we may explain our preference for it by comparing it with other theories. Such a comparison does not establish the intrinsic excellence of the theory we have chosen. As a matter of fact, the theory we have chosen may be pretty lousy....The rejected alternatives...serve as correctives (after all, we may have made the wrong choice) and they also explain the content of the preferred views (we understand relatively better when we understand the structure of its competitors...)" (Feyerabend, 1975, p. 5).

I will illustrate the cognitive and non-cognitive theoretical approaches of the 1930-1940 era by considering how Lewin, Tolman and Hull conceptualized the classic detour problem. This particular problem of spatial behavior is still of interest to contemporary researchers, as chapters elsewhere in these two volumes will indicate.

THE DETOUR PROBLEM

Two instances of the detour problem are shown from an overhead perspective in Figure 1. The one on the left represents a case in which an individual is located at "S" inside a U-shaped barrier. Outside, at G, an attractive object is visible to the organism. The most direct route to that object is the one marked A, which involves locomotion towards G. However, that pathway is blocked by the barrier in such a way that the only successful route to the object is a route of the sort marked B, which involves an initial phase of locomotion away from G.

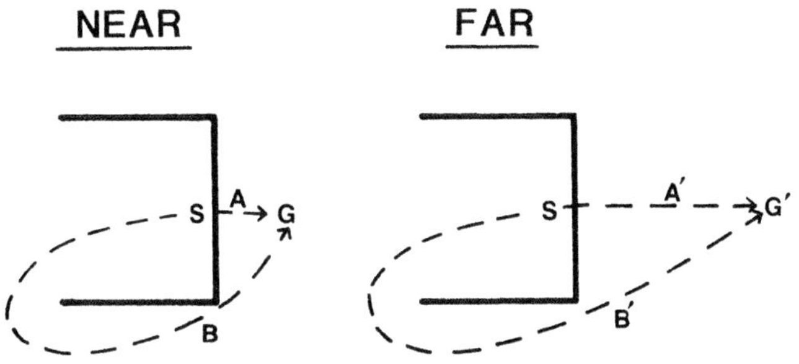

Figure 1. Overhead perspective on the detour problem when the attractive goal object (G) is near the barrier or far from it. (After Hull, 1938).

The left- and right-hand sides of this figure illustrate detour problems that differ only in how near the attractive object is to the barrier. The theorists of the 1930s were attracted to the performances that this small procedural difference yields. In particular, when G is near, naive organisms have considerable difficulty solving the problem because they persist in trying to reach G along the blocked path, A. In contrast, when G is distant from the barrier the naive organism is able to solve the problem more quickly. Experienced organisms, such as older children, find it relatively easy to solve either of these two detour problems (Lewin, 1935). It is the theoretical analyses of these aspects of performance on the detour problem that will be the focus here.

LEWIN'S ANALYSIS

Kurt Lewin's analysis of the detour problem was made in terms of his field theory (Lewin, 1933, 1935). According to this theory the main determinants of behavior are psychological forces that operate within the confines of an organism's cognitive representation of the problem. The way a detour problem is represented in the cognitive structures in any given instance is determined by the organism's interpretation of the problem which, in turn, is influenced by the organism's current internal state and its accumulated past history of experiences with similar problems. The organism's actual behavior on the problem is viewed as being strictly determined by the array of forces operating on the entities in the representation.

Figure 2 summarizes Lewin's conceptualization of the entities in a child's representation of the detour problem. The left-hand panel shows the case for a child that encounters the problem for the first time. Here, the problem is represented as one in which the child (C) is surrounded by a U-shaped barrier that blocks any direct pathway to a piece of chocolate

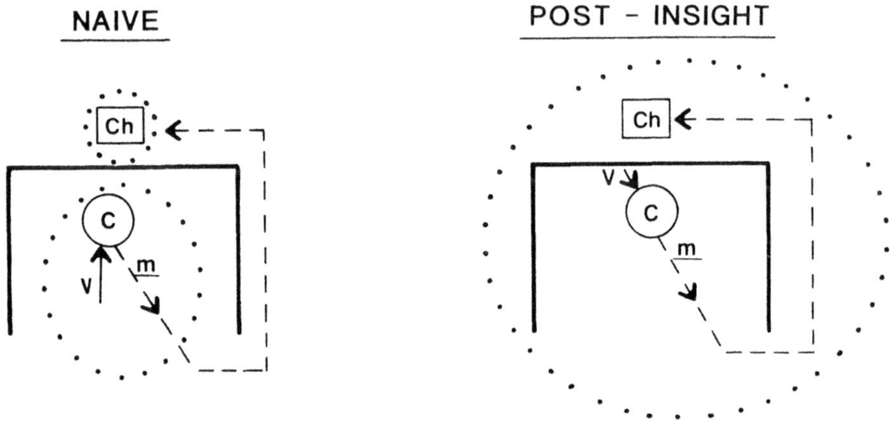

Figure 2. Cognitive representation of the detour problem by a naive child and by a child for which insight has occurred. The arrow marked V shows the direction of the force responsible for locomotion in the child's representation of the problem. (Based on Lewin, 1933, 1935). See text for details.

(Ch), the attractive object on the other side of the barrier. A successful route to the chocolate requires moving in a direction away from the chocolate. One such route is shown by the dashed line labeled m. The child and the chocolate are each encircled by a dotted line to indicate that they are unconnected regions in the child's representation so far as locomotion is concerned. Because of this, any movement in the direction marked m will be interpreted by the child as a movement away from the chocolate. The positive valence of the chocolate exerts a force on the child, indicated by the vector (V). The direction of the vector is towards the chocolate, and it is the action of this vector which causes the naive child to move directly (and unsuccessfully) towards the chocolate.

By Lewin's account, a child who represents the detour problem in the way shown for the naive child in Figure 2 must experience difficulty in solving it. Furthermore, Lewin assumed that the strength of the vector (V) increases in direct proportion to the nearness of the chocolate to the barrier. Hence, he identified one factor that makes the detour problem more difficult to solve in the "near" case: as the strength of the vector increases, the child is impelled even more strongly along the direct route to the chocolate.

How does the naive child eventually come to solve the detour problem? And, why does an experienced child have little difficulty with it? Lewin's answer involves a new factor: the organism's cognitive representation of the detour problem can be restructured through a process of insight. According to Lewin, the solution to the detour problem results from rearrangement of the forces within such a restructured representation. The idea is illustrated in the right-hand panel of Figure 2. After there has been an insight, of the sort Kohler and other Gestalt psychologists favored, all the elements of the problem are seen as being of a single piece (illustrated in Figure 2 by having all the entities in the representation enclosed in one large circle), so that, as Lewin put it,

> After the changing of the structure of the field, the direction m is psychologically no longer a direction away from the goal but toward it. Therefore, the direction of the psychological force to the goal is then to be characterized as the direction of m, and the child is moving in the direction of this force to the goal in accordance with our basic theory about the connection between locomotion and forces (Lewin 1933, p 338).

He went on to add that, in reality, the spatial connections between child and chocolate shown in the figure are further complicated by visual connections so that "a ratio of two fields in which the direction and distances are to be defined through different psychological functions" would have to be dealt with (Lewin, 1933, p 338).

Lewin (1935, p 84) proposed that insight comes about when the organism achieves "relative detachment and inward retirement from" the vector that operates in the original representation of the problem. This detachment-insight process was supposed to be impeded if the vector was particularly strong, but, apparently, the process would be hastened if the barrier acquired negative valence through the organism's repeated unsuccessful encounters with it along the blocked pathway. In the end, the relatively great difficulty in solving the detour problem when the chocolate is close to the barrier is attributed to two main factors: strong pressure from the force vector to approach the chocolate directly, and slowness in achieving the detachment that allows insight and a realignment of the force vectors to occur. The experienced child would have relatively little trouble in any detour problem because it would have the post-insight representation of the problem shown in Figure 2.

It goes without saying that this theoretical approach to spatial behavior raises several basic questions pertaining to the way representations are formed and modified, the way the strength of vectors is to be quantified, and so on. This is not the place to discuss these questions, but it may be noted that the thrust of some of them can be directed at today's cognitive approaches.

TOLMAN'S ANALYSIS

Let us turn now to Tolman's analysis of the detour problem (1932a, b). As background, let me remind you that Tolman's "cognitive maps" were also known as "means-end-fields" which he viewed as being "practically identical with" the representational fields of Lewin that we have been discussing (Tolman, 1932b). In his own words,

> My "means-end-field", like Lewin's "environmental field", is thus the environmental milieu (insofar as the organism is sensitive to it) at the center of which he finds himself and relative to which he behaves. The milieu radiates out from

> the organism as a center. And the objects along its radii are in varying degrees demanded-for and demanded-against (loaded with positive or negative valences). All such objects have, relative to one another and to the organism, all sorts of complicated interrelations of "distance" and "direction." (Tolman, 1932b, p 4).

The parenthetical comment that the means-end-field is equivalent to the actual environmental milieu "insofar as the organism is sensitive to it" is Tolman's way of endorsing Lewin's position that the organism interprets its current environment. Such interpretations yield the idiosyncratic cognitive representations that are the basis for behavior. In Tolman's (1932a, b) system, behavior in relation to objects in space is determined by the motivational significance of the objects represented in the means-end field. The demand-value of the objects (to use Tolman's term for their motivational significance) is functionally identical to the force vectors of Lewin's theory.

In the detour problem, the organism is faced with obtaining an object that is highly demanded, but which requires the taking of an indirect pathway in order to reach it. Tolman's view was that the various pathways (even parts of pathways) that exist in an organism's cognitive representation of the problem have subordinate demand-values that are compared by the organism. The pathway with the highest relative demand value will be taken. Figure 3 shows an example of pathways that might be found in the cognitive representation of an organism (S) faced with obtaining an attractive goal object (G) located at a distance from the barrier. Tolman's theoretical analysis of such a problem was this:

> ...other things being equal, a direct route always gets more subordinate demand-value distributed onto it than does an indirect route. The demand-value of the goal lends subordinate demand-value or attractingness to all routes but, to the degree of the organism's cognitions of the relative directnesses of these routes, the goal lends more subordinate attractingness to direct routes than to indirect ones. When, however, the direct routes are barred, then these latter, by virtue of a further cognitive sensitiveness to their "barred character", will lose some of their relative attractingness and the indirect ones will become relatively more attracting. (Tolman, 1932b, p 8).

In Tolman's theory, then, a solution to the detour problem arises from a reassignment of demand-value among the various pathways in the means-end field, much the same idea as the reassignment of forces in Lewin's theory. The Gestalt process of insight is not explicitly invoked by Tolman, but the idea that the reassignment of demand-values occurs when there is a "cognitive sensitiveness" to the fact that some pathways are blocked is functionally the same as insight.

With respect to the greater difficulty naive organisms experience in solving the problem when the goal is near as opposed to far from the barrier, Tolman's account is in terms of the relative demand value of all the path segments that point towards and away from the goal. When the goal is near, the ratio of demand-values for all the path segments pointing towards the goal to path segments pointing away from the goal will be

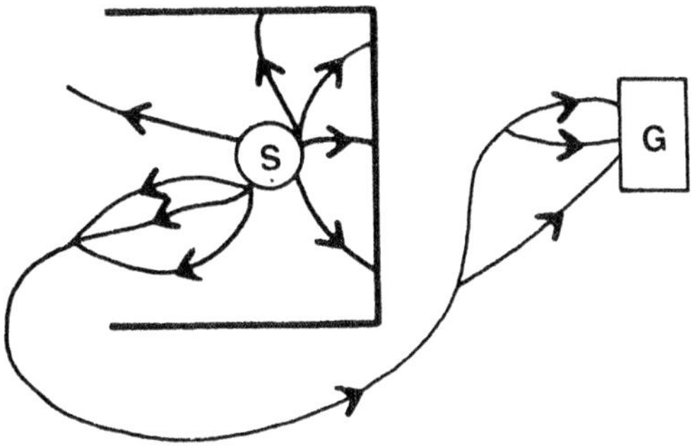

Figure 3. Some possible pathways in the cognitive representation of a detour problem by an organism (S) blocked from an attractive goal object (G) by a U-shaped barrier. (After Tolman, 1932a, p. 177).

large. Because the organism must move in the relatively most-demanded direction (the unsuccessful direct route), the problem is difficult to solve. When the goal is far from the barrier, on the other hand, the ratio of direct to indirect path-segment values is smaller and the organism solves the problem more readily (Tolman, 1932b, p 10). One contributing factor to the solution that should be noted is that the barrier can acquire a negative valence as a result of the organism encountering it on the blocked path, and this helps reduce the attractiveness of paths in the blocked direction. Finally, as for the ease with which older children solve the detour problem, Tolman's theoretical account was very similar to Lewin's. In his own words,

> The difference between the younger and the older child is then, according to me, simply one of differences in knowledge possibilities as regards "direction" and "distance" and "barred-characters" [of the paths with barriers] and the fact that these knowledge-possibilities are greater in the older child than in the younger. (Tolman, 1932b, p 8).

Many features of Tolman's approach are similar to Lewin's, although the language is different. At the most general level, behavior in the problem situation is viewed as being determined by processes that operate within each organism's cognitive representation of the problem. Experience can alter both the content of the representation and the valences of entities in the representation, and such alterations will affect the configuration of determining processes. An insight-like process is important in restructuring the representation in a way that allows the problem to be solved. Lewin and Tolman quibbled over some details, such as whether certain motivational and cognitive processes operate behind the field independendently of the processes in the field itself (Lewin, 1933; Tolman, 1932b), but these were not significant issues. The important point is that these approaches showed how spatial problems could be conceptualized in cognitive terms that were, for the most part, rather distant from the molecular processes of physiology and reflexology that characterized the S-R psychology of Thorndike and Pavlov. It turned out, however, that roughly the same range of spatial problems could be tackled with considerable success at the relatively molecular S-R level. Let us now turn to Hull's S-R approach.

HULL'S ANALYSIS

An initial background comment or two may be useful here. Today, Hull is mostly remembered as a psychologist who wrote a theoretical book in 1943, Principles of Behavior, that is usually viewed as an attempt to combine hypothetico-deductive logic with S-R learning theory. A conventional thumbnail historical sketch is that Hull's Principles was very influential in psychology during the 1940s, but overambitious, and that by the 1950s the limitations of the whole Hullian approach had become so painfully obvious that its centrality to learning theory was displaced by Skinner's approach. Elsewhere, Abram Amsel & I have urged a broader historical view of Hull's contribution to learning theory, particularly as it is represented in 21 theoretical papers published in the Psychological Review between 1929 and 1950 (Amsel & Rashotte, 1984). With respect to the present topic, I simply want to point out that Hull's theoretical papers include a sophisticated S-R analysis of several instances of behavior in spatial situations. More generally, these papers include many attempts to explore the extent to which S-R mechanisms might account for behavior in a variety of complex situations where cognitive processes seem to operate. The relative merits of Gestalt theory, Sign-Gestalt theory and sophisticated S-R theory as ways of dealing with a wide array of behaviors were aired in these papers and, in the process, many new ideas were introduced into S-R theory. It is probably not known to most people that in Hull's book, A Behavior System, published in 1952 just after his death, the longest chapter was devoted to problems involving spatial behavior. That book was cast in the hypothetico-deductive style of the earlier Principles, and the chapter to which I refer includes about 50 theoretical deductions concerning behavior in a variety of spatial situations. (The title of Hull's chapter appears as part of the title of the present paper: "Behavior in Relation to Objects In Space"). In what follows, I will use material mostly from a theoretical paper Hull wrote in 1934 (published in 1938) which attempted to analyse many of Lewin's spatial problems, including the detour problem, in S-R terms (Hull, 1938).

The particular commitment of any S-R theory is, as far as possible, to reduce the causes of behavior to stimulus-response mechanisms. On the whole, the mechanisms of S-R theory in the 1930-1950 era, were formulated at a far more molecular level, and gave the appearance of being more immediately related to a physiological substrate, than were the force-vector and demand-value mechanisms in the representational stimulus fields of Lewin's and Tolman's theories. As we will see, it appeared possible to account for many instances of spatial behavior by reference to theoretical mechanisms at these very different levels. In this respect, the earlier era mirrors present-day cases described elsewhere in these volumes where cognitive accounts and neural accounts of behavior in spatial situations are available.

In Hull's formal system, behavior is determined by a single theoretical construct, excitatory potential, whose value reflects the influence of a variety of factors that determine the strength of tendency to make a particular response in the presence of some stimulus. The behavior that actually occurrs in a given situation is determined by competition among the various reaction tendencies that are activated by stimuli in the situation. The stimulus whose excitatory potential is momentarily greatest in the situation determines the behavioral response. The strengths and limitations of this feature of Hull's approach are discussed elsewhere (e.g. Koch, 1954; Logan, 1959). Here, I simply want to make two points. One is that, in Hull's theory, excitatory potential functions as the main determiner of behavior in a way that is analogous to the force-vector in Lewin's theory or the demand-value in Tolman's (Lewin, 1933, p 329). The other point is that Hull's emphasis on excitatory potential as being attached to stimuli raises questions about how he conceptualized stimuli that impinge on an organism in spatial situations.

Briefly, at each point in time, Hull saw the organism as imposing control on the kind of stimulation that reaches the nervous system by the manner in which it orients its sensory receptors in the current environment ("receptor adjustment acts", Hull, 1952, p 93). Once stimulated, the receptors provide the organism's nervous system with an array of neural activity whose individual components are altered by other neural activity in the system ("afferent neural interaction", Hull, 1943, 1945). Finally, stimulus conditions originating from three main sources within the organism entered the analysis as important factors: 1) movement-produced feedback stimulation arising from proprioception, 2) interoceptive stimulation arising from the chronic state of the organism (e.g. deprivation), and 3) stimulation from the learned anticipation of goal events (Hull, 1930, 1931).

Hull's conceptualization of "the stimulus" is far richer than that found in other versions of S-R learning theory. In fact, it provided Hull with the conceptual tools for analysing many of the complex behavioral phenomena that seemed to require cognitive processes. I do not wish to imply that Hull's approach was without its problems in this regard (see Amsel & Rashotte, 1984), but in my opinion these problems were no greater than those posed by the conceptualizations found in Lewin's and in Tolman's theories, and in many of the current cognitive theories. It seems to me that a common mistake in much of today's portrayal of S-R theory is to assume that in such a theory the functional stimulus must be isomorphic with the stimulating conditions that are defined by the physical properties of an organism's current environment. This was certainly not the case in Hull's version of S-R theory. In fact, he provided S-R theory with a very sophisticated analysis of the sources of stimulation that it should take into account (see Amsel & Rashotte, 1984, p. 64-80 for a summary). For present purposes, it may be said that Hull's analysis of the stimulating conditions operating in spatial situations had many of the functional properties of Lewin's and Tolman's analyses, including considerable latitude for individual differences in the "representation" of a problem.

Let us return now to the detour problem where we can see some aspects of Hull's approach applied to a specific case. The key to his analysis of the "near" and "far" detour problems presented in Figure 1 was to recast them as problems in choice between long and short pathways to a visible goal object. There are 3 steps in this analysis when naive organisms are considered.

First, the excitatory values of stimulus conditions that the organism encounters at the beginning of the direct and of the indirect paths to the goal are estimated for each of the two detour problems shown in Figure 1. Hull illustrated this process, as he often did, by making some quantitative assumptions about the underlying processes in the situation. In this case, a decaying gradient of excitation radiating out from the goal was assumed to operate, the so-called goal-gradient principle which Hull used widely in his theorizing. Table 1 shows the values of excitatory tendencies that would operate at different distances from the goal objects. At one unit from the goal, the tendency has a maximum value of 10, and it falls off in a logarithmic function as the organism's distance from the goal increases to 50 units. For illustrative purposes, Hull proposed that when the goal is close to the barrier in the detour problem, the direct path between organism and the goal is 5 units long and the indirect path is 29 units long. When the goal is far from the barrier, the direct path is 12 units and the indirect path is 35 units. When these pathway-lengths are read from the table, the excitatory strengths of the direct vs. indirect pathways in this example are 6 units <u>vs</u> 1.63 units in the near-goal case, and 3.82 <u>vs</u> 1.16 units in the far-goal case. Because the pathway with the higher excitatory strength must be taken, the analysis requires that an organism would initially choose the direct pathway in both of the problems shown.

The second aspect of Hull's analysis concerns the mechanism by which the organism overcomes this initial tendency to take the unsuccessful direct route. In terms of the theory, what must be achieved here is a reduction in the excitatory strength of the direct path to a value below that of the indirect path. At that point, the stronger excitatory potential of the indirect path will control behavior and the organism will solve the problem. The general theoretical idea here is similar to that in Lewin's and in Tolman's theories where the problem is solved when forces are rearranged or demand-values are reassigned among pathways. In Hull's theory, this is accomplished by the mechanism of experimental extinction of the tendency to move along the direct pathway. He proposed that repeated failures to reach the goal directly would reduce the excitatory strength of the direct pathway through extinction. When the goal is near the barrier more failures would be required to reduce the direct path's relatively high excitatory strength below the value of the indirect pathway. In this way, Hull provided an S-R theory account of why it would be more difficult for the organism to solve the detour problem when the goal was near as opposed to far from the barrier. I pointed out earlier that in their theories both Lewin and Tolman specified that unsuccessful attempts along the direct pathway contributed to the solution. In Hull's theory, that factor played a much more central role.

The third factor in Hull's analysis concerned the way that an organism's prior experience in other spatial situations would influence its initial behavior in the detour problem. Figure 4 summarizes the idea. Hull presumed that in the normal course of events, organisms (S) would be faced with the problem of obtaining visible goal objects (G) in relatively unconstrained spatial situations. In these situations, organisms are presumed to learn through random locomotion that the angle of

TABLE I

This table shows the hypothetical strength of the conditioning of stimuli received at different distances from the goal to reactions occurring simultaneously. This is on the assumption that the excitatory tendency one unit from the goal acquires a strength of ten units and that the strength of the excitatory tendency acquired at each of the remaining distances diminishes with their remoteness according to the equation

$$E = a - b \log D,$$

where a has a value of 10, b has a value of 4, and the logarithms are taken with a base of 5.

Units Distant from Goal (*D*)	Strength of Excitatory Tendency (*E*)	Units Distant from Goal (*D*)	Strength of Excitatory Tendency (*E*)
1	10.000	26	1.903
2	8.277	27	1.809
3	7.270	28	1.718
4	6.554	29	1.631
5	6.000	30	1.547
6	5.547	31	1.465
7	5.164	32	1.386
8	4.831	33	1.310
9	4.539	34	1.236
10	4.277	35	1.164
11	4.040	36	1.094
12	3.824	37	1.026
13	3.625	38	.959
14	3.441	39	.895
15	3.270	40	.832
16	3.109	41	.771
17	2.958	42	.711
18	2.816	43	.652
19	2.682	44	.595
20	2.554	45	.539
21	2.433	46	.484
22	2.317	47	.431
23	2.207	48	.379
24	2.101	49	.327
25	2.000	50	.277

Note: This table actually shows the excitatory tendency of conditioned stimuli at various points along the pathway to a goal. Hull (1952, p. 271) commented that he abandoned the logarithmic decay function in his theorizing between the time he wrote the paper from which this table is taken (1934) and the time the paper was published (1938). The numbers in the table should be viewed as being illustrative of the approach he took. (From Hull, 1938).

the pathway with which they begin to approach the goal object is differentially correlated with distance to the goal. Direct paths are always the shortest, and the distance increases as the starting angle diverges from the direct path. This kind of prior learning provides the organism with what amounts to a hierarchy of response tendencies that can influence its behavior when it first encounters a detour problem (Hull's habit-family hierarchy, 1934a, b). It is because of this prior learning that the naive organism will have a relatively strong tendency to take the direct path to a visible goal. In part, the kind of response hierarchy shown here is S-R theory's analogue of the array of pathways that make up Tolman's cognitive maps (see Figure 3).

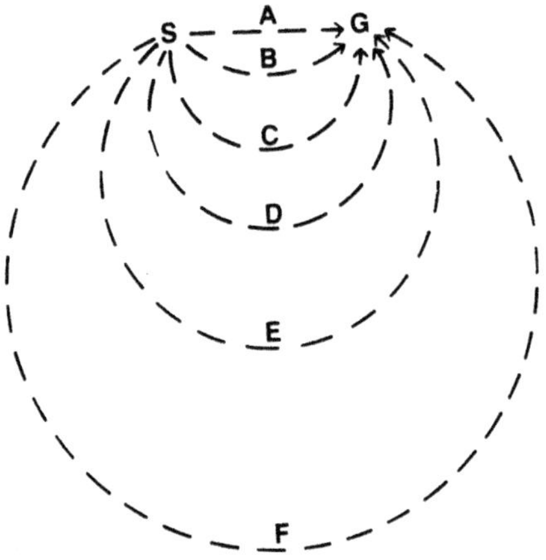

Figure 4. Illustration of the idea that prior learning by animals in unconstrained spaces will result in a hierarchy of response tendencies with respect to the initial angle of the pathways that lead to a visible goal (G). Path A results in the shortest delay to reach G. Paths B through F result in increasingly long delays. (From Hull, 1938).

Why is the detour problem easy to solve for organisms that have had experience with it? Hull's answer was not fully developed, but the idea was clear. He noted in his book chapter on spatial behavior that the principle of patterned discrimination learning would enter in this case (Hull, 1952, p 270). That is, organisms would come to discriminate situations with barriers from those without barriers, a kind of conditional discrimination learning (Hull, 1945) that would allow the experienced organism to take appropriate action from the outset in a detour problem.

Finally, I want to note that Hull often used his theoretical analyses to make rather explicit predictions about behavior in new situations. Figure 5 gives one example that draws on the ideas just discussed about the angle of different pathways leading to a goal. This figure represents a naive organism (S) faced with a detour blocking its direct path (A) to the visible goal (G). The detour in this case is asymetrically shaped so that there is a long and a short path around the barrier. Hull's analysis predicts that the organism should choose the physically longer pathway to the goal that runs along the dotted line marked C, rather than the shorter path marked B. The prediction is based on the logic just reviewed: the excitatory strength of a starting pathway with the angle of C should be greater than one starting with the angle of B (compare the routes along which the angles of those pathways usually lead, C' vs B'). Recent work with cats and with dogs in a situation very much like that with which Hull's 1938 prediction was concerned has yielded data consistent with the prediction (Chapuis, Thinus-Blanc, & Poucet, 1983; Poucet, Thinus-Blanc, & Chapuis, 1983). Interestingly, that research also showed that when the goal object was not visible in the detour problem, the length of the

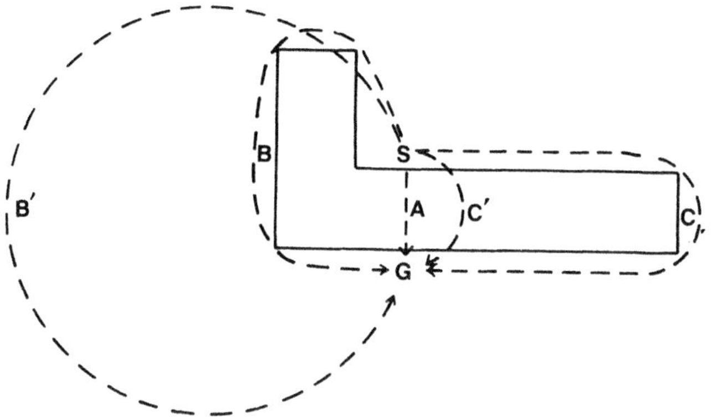

Figure 5. Hull's analysis of a detour problem in which the barrier is asymetrical. (From Hull, 1938).

pathways to the goal became the principal controlling variable. That particular part of the result is not anticipated by Hull. The point of describing this prediction from Hull's theory is to illustrate the kind of implications to which that theory often led (Hull, 1938, 1952). Certainly, S-R theory was far from being hamstrung in dealing with behavior in spatial situations. When the true scope of S-R theory is acknowledged in these cases, at least, it is not so easily dismissed as a competitor to cognitive theory.

According to Hull's analysis, spatial behavior could be understood in terms of processes operating at the sensory, motor, and central levels of the nervous system. S-R psychology provided the general framework, and Thorndikian and Pavlovian research provided many of the specific ideas about the role that experience plays in spatial situations. Taking the detour problem as an example, Hull's approach contrasts sharply with the more relaxed approach taken by Lewin and Tolman who, as we have seen, freely appealed to such concepts as "insight," "inward retirement from vectors that operate in representations," "cognitive sensitiveness to the barred character of routes," and "all sorts of complicated interrelations of distance and direction." Hull encouraged the rigorous comparison of theoretical approaches that relied on different concepts (e.g. Hull, 1935), but few such comparisons were carried out. However, many questions raised by Hull about the cognitive theories of the 1930-1950 era seem worth considering when the cognitive theories of today are evaluated. I hasten to add that Hull's analysis of complex instances of behavior also revealed some of the difficulties that are, perhaps, inherent in a rigorous account of such behaviors that appeals to molecular processes (see Amsel & Rashotte, 1984).

CONCLUSION

In the foregoing historical review, I revisited the time when cognitive approaches to spatial behavior were being formulated and when a non-cognitive approach was being vigorously advocated by S-R theorists. What can be learned from this exercise?

One thing is that spatial behavior can be effectively conceptualized in a non-cognitive framework. This point is worth making today when cognitive theory predominates in the study of spatial behavior. It is clear from the historical review that Hull's S-R analysis of the detour problem was at least as illuminating as the cognitive analyses of Lewin and Tolman. Hull's non-cognitive framework was also capable of making important predictions about performance in novel detour problems, as illustrated in the case of the asymetrical barrier problem.

I note that the detour problem is not the only instance of spatial behavior in which S-R theory provided an effective non-cognitive account of spatial behavior. The classic conflict situations involving objects with positive and/or negative valences provide a very excellent example of another instance. I think it is fair to say that Neal Miller's (e.g. 1944) extension of Hull's (1938) analysis of these situations went far beyond Lewin's (e.g. 1935) original conceptualization and observations (see also, Hull, 1952). As another example, Hullian theory competed favorably with Tolman's theory in accounting for the elimination of entrances into goalward-pointing blinds in maze learning (e.g. Hull, 1932, 1934a, b; Tolman, 1932a, Chapter VIII). A more extensive historical review would provide other examples as well. For present purposes, however, it is sufficient to note that the role played by Hull's theory in the 1930s as an alternative to cognitive theory is in a sense, being recreated today by those approaches to spatial behavior that appeal to neural mechanisms. The latter approaches are amply documented elsewhere in these volumes.

A problem that arises when there are several theoretical approaches to a problem concerns the relationship between the approaches. In the 1930s, Lewin's and Tolman's cognitive theories were clearly cast at a more molar level than Hull's non-cognitive theory which appealed heavily to the relatively molecular sensory and reflex processes known at that time. If the molar theory is simply reducible to the molecular mechanisms, it would obviously be profitable to undertake the translation between levels. Some of today's attempts to link cognitive theory to neural mechanisms seem to be undertaken in this spirit. It is probable, however, that behavior in at least some spatial situations will turn out to be jointly controlled by cognitive and by non-cognitive processes acting as separate entities. Most present-day cognitive accounts of behavior lack reference to reflex or motivational mechanisms that are known to influence behavior in many situations. It may be necessary to develop an approach in which these different levels of processes are combined, rather than one level being reduced to the other (see Winson's chapter elsewhere in these volumes).

The current tendency in parts of psychology is to view the vast majority of behavioral phenomena as being mediated by cognitive processes, even in phylogenetically diverse cases where relatively simple instances of behavior are under consideration (e.g., Dickinson, 1980; Mackintosh, 1983).

One consequence of the lack of theoretical alternatives is that a consensus has arisen in some quarters that animals have rather extensive cognitive capabilities. It is not without irony that this "fact" of cognitive abilities in animals is now cited by the animal rights movement as an important reason for terminating all animal research (Singer, 1985, p. 5). There are likely to be many benefits from placing the cognitive approach in better perspective with non-cognitive accounts of behavioral phenomena, especially as it becomes clearer that relationships between cognitive and non-cognitive determinants of behavior need to be explicated (e.g., Searle, 1984).

I thank Paul Ellen, Catherine Thinus-Blanc and Darryl Bruce for helpful comments on an earlier version of this paper.

References

Amsel, A., and Rashotte, M. E. (Eds), 1984. "Mechanisms of Adaptive Behavior. Clark L. Hull's Theoretical Papers, With Commentary". Columbia University Press, New York.

Chapuis, N., Thinus-Blanc, C. and Poucet, B., 1983. Dissociation of mechanisms involved in dogs' oriented displacements. Quarterly Journal of Experimental Psychology, 358:213-219.

Dickinson, A., 1980. "Contemporary Animal Learning Theory". Cambridge University Press, Cambridge.

Feyerabend, P., 1975. How to defend society against science. Radical Philosophy, 2:4-8.

Hull, C. L., 1930. Knowledge and purpose as habit mechanisms. Psychological Review, 37:511-525.

Hull, C. L., 1931. Goal attraction and directing ideas conceived as habit phenomena. Psychological Review, 38:487-506.

Hull, C. L., 1932. The goal gradient hypothesis and maze learning. Psychological Review, 39:25-43.

Hull, C. L., 1934a. The concept of the habit-family hierarchy and maze learning: Part 1. Psychological Review, 41:33-54.

Hull, C. L., 1934b. The concept of the habit-family hierarchy and maze learning: Part 2. Psychological Review, 41:134-152.

Hull, C. L., 1935. The conflicting psychologies of learning - A way out. Psychological Review 42:491-516.

Hull, C. L., 1938. The goal-gradient hypothesis applied to some "field-force" problems in the behavior of young children. Psychological Review, 45:271-299.

Hull, C. L., 1943. "Principles of behavior: An Introduction to Behavior Theory". Appleton-Century-Crofts, New York.

Hull, C. L., 1945. The discrimination of stimulus configurations and the hypothesis of afferent neural interaction. Psychological Review, 52:133-142.

Hull, C. L., 1952. "A Behavior System. An Introduction to Behavior Theory Concerning the Individual Organism". Yale University Press, New Haven.

Koch, S., 1954. Clark L. Hull. in "Modern Learning Theory". W. K. Estes, S. Koch, K. MacCorquodale, P. E. Meehl, C. G. Mueller, W. N. Schoenfeld and W. S. Verplanck (Eds.), Appleton-Century-Crofts, New York.

Lewin, K., 1933. Vectors, cognitive processes, and Mr. Tolman's criticism. Journal of General Psychology, 8:318-345.

Lewin, K., 1935. "A Dynamic Theory of Personality: Selected Papers". McGraw-Hill, New York.

Logan, F. A., 1959. The Hull-Spence approach. in "Psychology: A Study of a Science, Vol. 2". S. Koch (Ed.), McGraw-Hill, New York.

Mackintosh, N. J., 1983. "Conditioning and Associative Learning". Oxford University Press, Oxford.

Miller, N. E., 1944. Experimental studies in conflict. in "Personality and the Behavior Disorders: Vol. 1". J. McV. Hunt (Ed.), Ronald Press, New York.

Poucet, B., Thinus-Blanc, C. and Chapuis, N., 1983. Route planning in cats, in relation to the visibility of the goal. Animal Behavior, 31:514-599.

Searle, J., 1984. "Minds, Brain and Science". British Brodcasting Corporation, London.

Singer, P., 1985. Prologue: Ethics and the new animal liberation movement. in "In Defense of Animals". P. Singer (Ed.), Basil Blackwell Publishers, Oxford.

Tolman, E. C., 1932a. "Purposive Behavior in Animals and Men". Appleton-Century-Crofts, New York.

Tolman, E. C., 1932b. Lewin's concept of vectors. Journal of General Psychology, 7:3-15.

BEHAVIOR AS A LOCATIONIST VIEWS IT

E. W. Menzel, Jr.

Department of Psychology
State University of New York at Stony Brook
Stony Brook NY 11794 USA

One of the most fundamental beliefs of behaviorists and of ethologists that "All that can ever actually be observed in fellow human beings and in the lower animals is behavior" (Tolman, 1932)-- statements about another being's mentality being mere inferences that we draw from our behavioral observations. This belief is still widespread today even among those who call themselves students of cognition; the principal change is that, like Tolman himself, these investigators would delete the word "mere," and say that such inferences are scientifically warranted and valuable. In my opinion, however, the belief is false-- unless, perchance, "behavior" is stretched to cover not only anything which we as observers record and use as our dependent variables, but also anything else that we can see, could possibly see, or believe that one might conceivably be able to see, assuming that the appropriate equipment were available and one knew what to look for. (And such an extended definition of behavior renders Tolman's statement tautological, if not indistinguishable from phenomenology.)

In saying this, I am not advocating a return to turn-of-the-century mentalism or a literalistic belief in mind reading. On the contrary. My point is that I do not know precisely where observables leave off and inferences begin; and neither, I suspect, does anyone else. To be more specific, we do not know with any precision what we see; and we often know more than we can tell (Polanyi, 1958). In Gibson's words, "The most decisive test of reality is whether you can discover new features and details by the act of scrutiny. Can you obtain new stimulation and extract new information from it? Is the information inexhaustible? Is there more to be seen? The imaginary scrutiny of an imaginary entity cannot pass this test" (1979, p. 257).

I will grant that to say one can directly see a monkey's intent, or that it is an intelligent being rather than a stupid one, is a more controversial statement than saying that one can see the same animal is approaching a banana, or is walking rather than standing still, or is located, at this instant, exactly 10 meters due North of that big tree. I'll grant, too, that neither "an intent" nor "intelligence" (not to mention any other analogous terms) is a directly tangible object or entity, and that to avoid the risk that such terms might be reified I should not have used nouns here, but rather some adjectival or adverbial expression instead. Nevertheless, I still believe that the differences between these statements are matters of degree rather than of kind. That, at least, is the central thesis of the present paper.

Insofar as possible, I shall develop and expand on this thesis by means of a series of concrete examples. These examples will all entail nonhuman primates. For the most part they will be drawn from my own research. I call my approach locationistic for two reasons: (a) Its basic methods, questions and concepts are neither behavioristic nor mentalistic but spatio-mechanical. (b) Surely no one today would be so naive as to believe that all one can really see are locations (motion, those forms of motion we call "real behavior," and cognition all being mere inferences we make from changes in locations). To be more precise, my major interest over the past 20 years has been to specify where monkeys and apes are located at any given moment; to discover why they are there rather than elsewhere; and to predict where they will go next. My hard core data amount to nothing more than a series of maps or snapshots showing where an animal, or each member of a small social group, is in an indoor room or an outdoor field is every X sec. As someone else, or a camera, records these data, I sit back and watch. Without such direct experience, and some "feel" for the particular animals in question, I would have no confidence in the "hard core data"-- or vice versa.

BEHAVIOR AND THE UNCERTAINTY PRINCIPLE

In quantum mechanics the Heisenberg uncertainty principle holds that one cannot simultaneously specify both the position and the momentum (mass times velocity) of the same particle, to any arbitrary degree of exactitude. A precisely analogous situation exists in psychology (Lewin, 1951; Menzel, 1979); but I confess that I did not recognize it until confronted with a straightforward case in point, in my own research.

My co-workers and I were studying the reactions of infant wild-born chimpanzees to their being released into an unfamiliar indoor room. The job of the observer was to state where the animals were located in the room and what they were doing, every 10th sec-- at the moment that an electrical timer sounded a click. Locations were to be recorded to the nearest 2 ft (.6 m) square; the behavior categories included a variety of different activities. Frequent reliability checks were made, in which two or more observers scored the same sessions independently.

Even after observers had made every effort to use precisely the same criteria in arriving at their judgments, and even after they were highly practiced at their task, it was impossible for them to simultaneously reach better than 90 per cent agreement with one another as to which cell in the room an animal was in, and to achieve a similar level of agreement as to what the same animal had been doing at that same instant. Indeed, the more exactly they tried to narrow down that portion of the timer "click" at which their locational judgment should be made (so that they would both place even a running animal in the same cell of the room), the less confident they became as to whether the animal was (for example) "really" in the process of running or vocalizing or scratching itself, or merely momentarily positioned in such a way that more casual observers might make such an inference, based on what they had perceived just before or on what they expected to perceive in the immediate future. Conversely, if observers allowed themselves a long enough interval of time to be reasonably confident that the animal was (say) walking rather than running or standing still, their agreements with one another as to where the animal had been located at the same moment dropped precipitously.

The situation was not improved when we took films of the chimpanzees and made our judgments from the film record rather than by direct observation. As one assistant put it, "I have no idea any more as to what is actually going on out there, and all criteria as to where one behavior category leaves off and another begins seem to me almost totally arbitrary in the final analysis. To get good agreement with another observer, I simply try to guess how he is going to vote, and then vote the same way myself." This assistant, coincidentally, soon quit, and abandoned his aspirations to become a research psychologist for a job as a salesman.

Such an exercise does not, of course, destroy the scientific validity of behavioral concepts any more than quantum mechanics can be said to have destroyed Newtonian mechanics. In either case, the "old view" remains valid from the point of view of an everyday or commonsensical observer who makes no pretext of having fathomed the ultimate nature of reality or having arrived at a "purely objective" account of its nature. Stated otherwise, almost every behavioral event in which a psychologist or an ethologist might be interested has some appreciable duration in time; different sorts of events have very different durations; whether or not any given event can be directly perceived depends in a very fundamental way upon the interval of time that we allow observers to make their judgments; and our choice of this time interval, which is to a great degree arbitrary, might immediately make it difficult, if not impossible, to simultaneously perceive events whose temporal limits are much greater or much smaller than the interval chosen. It is, however, only if one looks at behavioral descriptors such as "running" under a locationistic microscope that they appear too inference-laden and subjective to use in scientific discourse; and behaviorists and ethologists ordinarily solve this problem by simply not using too powerful a microscope on their own behavioral categories.

THE PROBLEM OF INSIGHT REVISITED

What if we were to adopt the same ("telescopic") perspective regarding other, much more macroscopic and controversial, terms? Might this not enable us better to see what the proponents of these terms claim to be able to see?

One such macroscopic term is assuredly "insight." I myself have heretofore deliberately avoided using this term; for this very reason I shall pick it for scrutiny now. Is it possible, at least in some cases, to almost directly see that an animal is acting insightfully, and to be virtually as confident of one's judgment here as one might be that the animal one is looking at is a "chimpanzee" or that it is "walking" or "climbing"? (Actually, no proponent of insight has gone quite that far. I exaggerate the position in order to clarify it.) I pose this question not as a mentalist, but as a locationist.

Figure 1 is as complete, straightforward and objective a record of one of my observations as can reasonably be reproduced here. The several snapshots are presented (from left to right, top to bottom) in the exact sequence in which they were taken; the total length of time they encompassed was a few minutes. This was only the second time I had ever seen the same overall event take place in the same locus, and a comparable set of snapshots was also obtained on the previous occasion. A little earlier, on arriving at this outdoor enclosure, I had discovered the chimpanzees up in a tree. Obviously, they had somehow or other gotten past the electric shock wires that were wrapped around each tree. How did they do it? Did this feat entail any performances that one might characterize as being insightful? Readers may make their own judgments. I shall simply recount, in as neutral a language as possible, some details that were relevant in my own case.

A long, denuded tree branch just so happened to be lying at the base of the tree in which the chimpanzees were located, and it had not been there the day before. This, plus what we had previously seen the same animals do with branches, was enough to make us suspicious. However, my assistant and I acted as if nothing special had happened, and we simply lured the chimpanzees down from the tree and commenced our "official" experiments, which had nothing to do with tool using. (We had not, coincidentally, deliberately given the animals any training of any sort on any variety of stick-using; nor were they food-deprived or given any reinforcers other than those that they discovered on their own.)

When I saw one of the chimpanzees (Rock, the main actor in these snapshots) glance toward the tree tops and then at the misplaced long branch, I started to shoot with a still camera,

Figure 1. One chimpanzee (Rock) sets up a "ladder" for getting up into the trees and averting the electric shock wires on the trees, while another one (Bandit) watches him. The "runway" on which the animals sit was elevated about 2.4 m above the ground and about 15 cm wide. Some of the photos are from Menzel, 1972.

taking another single frame every time I anticipated any significant change in the scene. As any professional photographer will tell you, human reaction time is too slow to capture "live action" sequences in snapshots unless you in effect know what is going to happen before it happens. And the film record alone is sufficient to demonstrate that it was not merely the molecular changes in the positions of the chimpanzees and of the tree branch that were anticipated. In brief, as soon as Rock glanced at the branch on the ground, I could well have said, "I can directly see that he is commencing the process of getting up into the trees with a 'ladder'." Indeed, a description in terms of one's inferences rather than direct percepts might reflect little more than one's relative degree of confidence in one's judgments.

Rock was, coincidentally, for a few weeks the only member of the group of eight chimpanzees who was capable of this particular variant on 'ladder using.' The others were at a loss for one simple reason: they focussed on placing the top of the branch against the tree, but did not first look down and secure its base on the substrate.

How representative is Rock's behavior of what chimpanzees in general can do? Field studies of nonhuman primates have yet to report any systematic use of portable branches or logs as ladders or bridges. Here, however, is an exception.

Sugiyama and Komans (1979) had followed a group of chimpanzees to a fig tree. The trunk of this tree was too large for the chimpanzees to scale, and only one limb of an adjacent tree (a thorny kapok) permitted possible access. While the rest of the group watched from the ground, two adult males (alternately) stood on this kapok limb and hit at the fig tree's nearest branch with sticks which they had broken from the kapok tree and stripped of their bark and thorns. Many different sticks were tried and then discarded, and they varied in their length, weight and curvature, and precisely how they were used.

Each time either animal managed to catch his stick on the fig branch or hit it almost within his reach, group members on the ground made "hooting and booming" noises. One male bounced the limb up and down and rose bipedally when the limb was at its peak; but he still could not reach the fig branch. He went back to using sticks until he was displaced for a time by the other male. When he got another turn he commenced to break off most of the branches from the limb (on which he sat). This time he did not strip the branches as before, and simply dropped them to the ground. The limb became lighter and started to rise. Then he rose to all-fours, bounced the limb again, rose bipedally, and managed this time to get hold of the fig branch. He climbed into the fig tree but did not eat for several minutes-- instead racing around the tree, uttering high pitched calls, as everyone else hooted and boomed. It had taken him almost an hour to get into the tree.

A third male now climbed into the kapok and attempted the same technique but without success. Then, however, he used a very long stick while standing bipedally, and managed to thus secure the fig branch and get into the tree (and the chimps on the ground again went wild). With his weight on the fig branch, the branch hung down lower than usual, and the other male in the kapok raced up and grabbed it. Many other animals now ascended the kapok, pulled the branch lower, and also got access to the ripe figs.

Ethologists might note that all of the so-called response elements here, including branch-breaking, limb-bouncing and "hitting at," are commonplace primate behaviors, especially in frustrating situations. If, however, there was nothing novel from a more macroscopic point of view, why is it that not only the human observers but also the entire chimp group got so excited? (Surely all had eaten figs before, too.) The authors themselves say, "This type of excitement has rarely been observed among wild chimpanzees, even by the senior author. It is apparently based on human-like emotions within a thinking mind" (p. 523). They note the similarities between their observations and those of Kohler; they stress the inventiveness of the animals, especially in using sticks in varied ways; but at the same time they note the animals' failure to select sticks with good hooks for raking in the fig branch, or to utilize sticks that had fallen to the ground. This last is notable, for the same animals did transport their tools in other situations. On subsequent days, the kapok tree was eventually depleted of usable sticks, and the chimps were no longer able to get into the same fig tree.

Inasmuch as various observers' criteria of insightful behavior will surely be much more variable than would be the case for terms such as "running," let me now spell out more formally what criteria the original proponents of this term used. First of all, the terms that Kohler used were the German words "Einsicht" and "Einsichtig," for which there are no exact English equivalents. "Intelligent" would come as close as the English words "insight" or "insightful". Secondly, the term was intended principally as a commonsense description of observable behavior, not as an explanation. (To be sure, this rule was sometimes violated.) Thirdly, the central question is whether the animal shows some trace of comprehending what it is doing-- some sign of apprehending the relationships between novel presented facts and guiding its actions appropriately. (A sudden or "one-trial" change in the organization of its behavior is one such sign; but it is not a sufficient sign, or otherwise taste aversion learning would qualify as insight.) How the animal came to be capable of organizing its performance in such a fashion as to convey such an impression to us is a different, and logically independent, question-- except to the degree that one must have some objective basis for believing that the degree of novelty in the test situation and in the animal's behavior is clearly greater than zero. In other words, it is not necessary that every element of the performance be "totally novel" (whatever that means), or that chance factors, the animal's species-characteristic motor patterns, and its prior experiences with other situations account for none of the variance of the performance. Viewed in this light, the recent articles by Robert Epstein and his colleagues (Epstein, 1984; Epstein et al, 1984) may more accurately be described as a reiteration or paraphrasing of Kohler's position than as a critique, for these articles themselves characterize the performances of the subjects (which were pigeons) as non-Thorndikean and intelligent, and stress the fact that the organization of these performances entailed "genuine novelty."

THE PSYCHOLOGY OF STICKS AND ROCKS, AND THE METRICS OF SPACE

Working in West Africa, Boesch and Boesch (1984) and Sugiyama and Komans (1979) repeatedly found pieces of hard nut shells lying on or immediately next to a hard root or a rock, while a short, heavy stick or another rock also lay nearby, with fragments of nuts on it. What could account for the relative positions of such objects, and the absence of the meat of the nuts? Some highly manipulative animal that eats nuts would be anyone's first suspect, even if one had not read Dennett's (1982) recent article on the use of the Sherlock Holmes method in cognitive science. Here, as in the domains of archaeology and operant conditioning, one does not necessarily even have to see one's animals directly, let alone see their bodily movements directly, to study many aspects of their psychology. Boesch and Boesch's analysis of the day-today movement patterns of rocks (which they had good reason to believe were energized by the movement patterns of chimpanzees) are a case in point.

The chimpanzees that Boesch and Boesch studied were too shy to follow continuously or at short range, and it was seldom possible to see more than 20 m in the dense forest; but fortunately there were relatively few available rocks in the forest. By marking each rock with an identification number, and tracking its movements as closely as was possible, the authors produced quite strong evidence for "planning ahead" and "cognitive mapping" in their animals.

The various species of nuts that the chimpanzees ate differed in hardness. For cracking the very hard Panda nut the chimpanzees used rocks rather than the softer clubs they used on Coula nuts, and they usually selected heavier and harder types of rocks and carried them for greater distances. They seemed to remember the locations of these rocks from one day to the next and to choose rocks that were close to the nut tree they were out to exploit, over rocks that were farther away. Both weight of the stone and distance seemed to be taken into account simultaneously, but "least-distance" was a favored strategy, probably because in carrying very heavy rocks the chimpanzees had to walk tripedally. Given that it was impossible for the chimpanzees to directly see from one tree or rock to the next, and of course the highly nonrandom movements of the rocks, the authors conclude that the following mental operations seem necessary to explain the locational variance of their data: (a) measurement and conservation of distance; (b) comparison of several distances; (c) permutation of objects in this "map;" and (d) permutation of points of reference, such that one can measure the distance from any momentary starting point to any destination. And all this, they believe, adds up to some sort of concept of Euclidean space.

This last sentence-- which one may also find in many other studies of cognitive mapping (e.g., Cheng, this volume)-- should, in my opinion, be taken with a grain of salt. As Einstein would have put it, space and time are not conditions in which animals live; they are concepts which humans have invented to describe the conditions in which they live. The concept of Euclidean space assumes that space is best viewed as an empty container (ala Newton and Kant) rather than in terms of the relationships between real objects (ala Descartes and Leibnitz). It also assumes that 10 meters is 10 meters regardless of where and when you make your measurement and even in which direction you measure from a given starting point. These assumptions greatly simplify human mathematics and might be useful as a sort of "null hypothesis" about how the world is actually organized; but they are empirically questionable assumptions even in modern physics (Shapiro et al, 1985) and surely false in the domain of "mental geometry."

A BIT MORE ON COGNITIVE MAPPING

I take it as an article of faith that where any primate will go next is highly correlated with where it has been in the past, and what objects (or classes of objects) and experiences it encountered in each particular locality. The problem is not to demonstrate this, but to spell out more of its details. Formal experiments are an aid, not a necessity. Thus, for example, as one of my chimpanzees (Belle, then about three years old) was riding on my back one day, I happened to walk past a small blackberry bush. She spotted it and made a soft "food grunt," but (for reasons I can no longer remember) I did not stop, but continued on my path for 10 m or so. The bush was not in sight from this point. When she climbed down from my back about 15 sec later, she did not retrace my exact path but ran almost directly to the bush and began to eat the berries. Would you expect that a wild infant, riding on the back of its own mother, and going through a similar experience, would be any less capable of doing the same thing? Wouldn't you in fact think that this feat could be observed every day in the field, if someone were to look for it?

Many if not most of my experiments were based on such fortuitous observations of what the animals had been doing "naturally." Procedurally, I tried to make the least amount of change that was necessary to convincingly demonstrate the phenomenon in question. The phenomenon in question here, as in the case of insight, is how chimpanzees organize their travel route and their more molecular behaviors. Short-cutting or the use of novel paths is one index of of this organization; ability to get to a particular location after having passed it while being passively transported is another; and Belle seemed to have done both. To test this possibility I used the following procedure.

One chimpanzee, from a social group of six, was carried about a .4 hectare enclosure by one experimenter and permitted to watch as another experimenter hid some food in each of 18 randomly selected locations. We did not walk in a randomly-determined path from one place to the next largely because the chimps were heavy and often wiggly; such paths would have been several hundred meters long; and we were lazy. After having shown the chimpanzee the food, we returned it to its sleeping cage, in which its five group-mates had been previously locked up for the interim. Then we left the enclosure, ascended an observation tower, and pulled a cable and turned all of the animals loose. Why not test each animal individually? First of all, chimpanzees ordinarily live and travel and forage in groups. Secondly, even if we had wanted to work with isolated individuals, it would have taken weeks of effort, for these chimpanzees were wild-born and strongly attached to one another. Two of them would never let us carry them more than ten meters from the group holding-cage; and even the boldest of them would not have gone for food by itself if no one else followed, unless the food were very close. (With age and experience, of course, this changed.) Thirdly, there was information to be gained by using a group-testing procedure. The five animals that were "uninformed" obviously may, if we choose, be viewed as yoked-controls for all of the cues that we did not control for-- odors, the trails we might have left through the grass, so-called Clever Hans errors, and so forth. Thus, statistically significant results might in principle be obtainable with even a single trial per informed animal.

Such data are shown in Figure 2. The chimpanzees had never before been tested with multiple hidden objects, so instead of conducting a single trial per animal we conducted four (in the first such experiment), and selected for display the one on which they recovered the largest number of foods. The sheer number of foods recovered is of no particular interest; the crucial question is how the animals organized their move. (Judging from prior tests of the same animals' ability to detect almost any novel items or changes in their environment, I would not hazard any guess as to the upper limit of the number of objects or places they could remember. For a more extended discussion of this issue, see Neisser, this volume.)

Obviously the chimpanzees did not follow the path on which we had carried them, and used a path that was far more efficient than one would expect by chance. By Monte Carlo

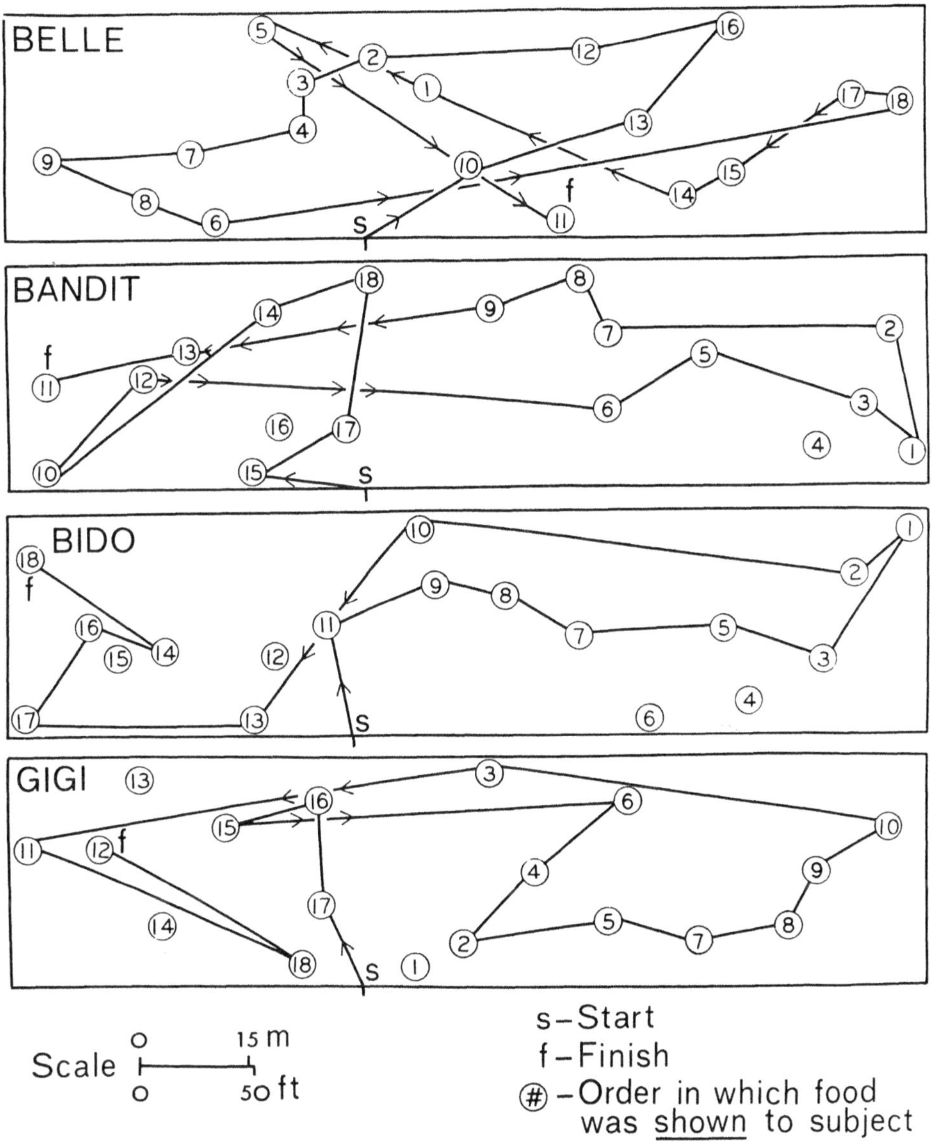

Figure 2. Maps showing each test animal's performance on the trial (out of four) on which it found the largest number of hidden foods. The connecting lines show the order in which the various places were searched and give a rough idea of travel routes. From Menzel, 1973.

techniques, the probability of paths this short-- even assuming that the animals could see the food directly-- was less than one out of a thousand for each and every one of the 16 trials, considered separately. The individual that had actually seen the food being hidden made almost all of the finds; its companions scored primarily by searching wherever it did. Especially since a novel set of locations was used on every trial, and the informed animals often ran directly to hiding places that were not visible from the point from which their run commenced, the term "cognitive mapping" seems as good a descriptor as any.

ARE MARMOSETS INTELLIGENT?

The problem of animal intelligence may be approached in at least five different ways:

(a) Ecologically speaking, the question is how the particular species or individual one is studying goes about making its living in the world at large. Any genotype that can make its own living in this day and age is, I would think, scarcely unintelligent. Its particular brand of intelligence is whatever it does, especially insofar as these actions give it an advantage over its competitors.

(b) Philosophically speaking, the issue is whether animals comprehend what they are doing. One necessary requirement here (before one answers in the affirmative) is that their actions be "significantly different" from that which one might expect of some being, hypothetical or actual, whose performances are predictable purely by the laws of Brownian motion or Newtonian physics. (I know of no way to empirically prove that sticks and rocks are totally devoid of intelligence. I assume it.) Any living thing can, in some respects, pass this minimal test. To this date, none of the additional hurdles that have been proposed, such as (non)instinctive behavior or (non)conditioned behavior or ability to reason, have laws that are clearly formulated enough or stable enough from one decade to the next to furnish an objective, quantitative standard of what an intelligent or unintelligent being "should" do; so for the duration at least we are stuck with judging this issue qualitatively, and possibly subjectively.

(c) Psychometrically speaking, the question is how their performance ranks, in comparison with other species, on "standardized" tests which we believe (for whatever reasons) might constitute reliable tests of intelligence (however one chooses to define it).

(d) Anthropocentrically, the question is whether animals can do what we ourselves do. (I assume, too, that "we" are intelligent.)

(e) Technologically speaking, the issue is whether one

could build a robot, program a computer, or "shape" some other, presumably unintelligent, animal to perform actions that appear "fundamentally the same" as those of the animal with whom we are concerned. (If so, the latter animal might be less of a genius than we once thought.) What does "fundamentally the same" mean here? To students of artificial intelligence, if not to most behaviorists, it seems to mean only that an otherwise uninformed human judge, who cannot see the actors in question, and has access only to their formal data transcripts, cannot tell whose transcript is whose.

A sixth issue is also implicit; its importance was particularly clear to me in the studies I have done over the past five years with marmoset monkeys (Saguinus fuscicollis). Do the animals look intelligent; and, if so, why? This is obviously a question in folk-taxonomy. But so too in their own way are most of the other approaches listed above.

Within the first few minutes that I laid eyes on these animals I was struck by how intelligent they looked. When I said as much, Dr. Gisela Epple, a zoologist, immediately remarked, "But of course! They are primates, aren't they?" Their taxonomic status was not, however, as far as I am aware, the basis for my judgment. To me, they looked very much like squirrels, except for their heads. In fact, many taxonomists have called them squirrel-like too, as well as referring to them as among the most primitive of true simians. As best as I could judge, my inference was based on how alert, visually curious, and reactive to me and my actions (and indeed to almost any strange noises or inanimate objects) they appeared. This observation formed the primary basis for the tests that my co-workers and I (Menzel and Juno, 1982, 1985; Menzel and Menzel, 1979) subsequently devised. As in my chimpanzee studies, we worked with social groups rather than isolated individuals; but here we studied natural family groups (a breeding pair, together with all of their offspring). Insofar as possible, we avoided all physical handling of the animals (they are very easily stressed) and tried to adapt our tests to them, rather than vice versa. The basic data were, once more, locational data. Each of our experiments amounted to little more than presenting the animals (in their home environment, or in an adjacent room, to which they were allowed access for about 15 min each day) some additional minor novelty or change in their environment.

I would frankly be suprised if chimpanzees, or indeed any other animal, could surpass the marmosets on some of the tasks that we used-- in the same test rooms, and with nothing more by way of pretraining or prior instructions. Right from the start of our testing the marmosets could, for example, detect a single novel object from among 30 test objects, and on the next trial show evidence of remembering its visual appearance, location in the room, orientation, whether or not it had contained food, where on the object the food was located, whether it was a preferred or nonpreferred food, or a large or small quantity, and (if the object was moved between trials) where in the room food had been found. I should add that the test trials here were 24 hr or more apart. Tests that amount to delayed matching-to-sample, non-matching, nonpositional delayed response (once thought to be extremely difficult for chimpanzees), and others besides all seemed to solved without any obvious slow, gradual learning. To be sure, we did not test their tool using ability (they are not very manipulative), language acquistion, or other behaviors at which humans excel. But why should one-- unless, of course, one is interested in the anthropocentric question? Even more to the point than these studies, however, is this fact. Since we commenced our studies, papers have appeared from zoologists which severely criticize the characterization of marmosets as squirrel-like in their habits, or as "primitive" primates (for a review, see Sussman and Kinzey, 1984); and ecologists (Terborgh, 1983) have obtained field data on the foraging habits of the species we worked with, which mesh very closely with our own data.

What scientific stock should one put in one's initial impressions as to whether or not any given animal is intelligent? Not much, of course. But, on the other hand, these impresssions probably furnish an aspiring student of animal intelligence as good a source of research ideas as any.

CONCLUSIONS: LOCATIONISM AND THE CONCEPT OF BEHAVIOR

Some years ago I was of the opinion that all students of animal behavior, whether they call themselves geographers, ecologists, sociologists, ethologists, behaviorists or cognitive psychologists, face a common problem, if not a single problem: Simply to account for the variance in animals locations and changes in location as such. Call this view "locationism pure and simple." I found it attractive for many reasons. Almost any overt actions of any animal in any situation may be described in locational terms, if one so chooses. Inasmuch as most quantitative methods were derived in the first place from locational problems (the formulas for most parametric statistics being derivable from the generalized Pythagorean formula for distance; probability theory stemming from statistical mechanics), the possibility of developing systematic, quantitative models for one's problems seems obvious. By the same token, various different levels of analysis (populational, group, individual, intra-individual) can, at least in principle, be dealt with in a common, systematic framework. And finally and most importantly, there is no other approach whose questions, methods and concepts so fully utilize the complexity of the real world in their very formulation, and yet are so simple, generic and intelligible to members of any discipline, regardless of the differences in their more technical and specialized vocabularies.

The prime difficulty that I had in trying to remain a locationist pure and simple was one of communication. No audience to whom I have talked ever took me seriously if I commenced by talking about the locations and motions of particles in space, and then showed them a picture of a monkey or a chimpanzee in an outdoor field. Until I showed such a picture they yawned; and as soon as I did show it they asked, What are you really getting at? What is your real problem? What does your study mean, so far as chimpanzee (or psychology) is concerned? Slowly but gradually I learned that probably no one is actually interested merely in accounting for the variance of locations and motions as such, or of behavior as such, so long as one is talking about living beings so akin to ourselves. There is, in brief, more than "a single problem" to consider. Furthermore, there is not just one common language of scientists-- the 'language' of space and time-- but at least two, the other being that of "common sense."

Kant formulated the problem in even more general terms. According to him, all knowledge that is derived from observation may be characterized in either of three ways: (a) Logically, or in terms of logical categories, such as when various plants and animals are classified into species regardless of where or when they occur; (b) Geographically, or in terms of where in space the phenomena in question occur; and (c) Historically, or when in time the phenomena occur. In the vocabulary of that branch of artificial intelligence called computer vision (if not in the vocabulary of today's common sense), these are "What" questions and "Where" (in space or time) questions. To them we should of course add "Why" questions-- Kant's problem of (d) Causality.

Needless to say, locationism pure and simple is concerned primarily with "Where" questions. But at least to this date, no one has succeeded in reducing "What" or "Why" questions into spatio-temporal terms alone; and the prospects for such an accomplishment in the domain of animal psychology are not promising. Such a reduction remains the aspiration of mechanistically-inclined sciences, but not only today's common sense but also the theoretical descriptions and explanations that are used in most sciences remain Kantian, in the sense that they invoke four different and largely independent questions rather than just one.

Most ethologists would, indeed, say that there is not just one "Why" question, but at least several independent ones. To say why an animal is doing something, we must consider not only immediate causal factors, but also the adaptive function of the behavior, its ontogenetic or developmental history and its evolutionary history. I concur, and would furthermore be inclined to agree with cognitive ethologists and psychologists, who would add to this list questions regarding "knowing and wanting" (Griffin, 1984; Mason, 1980). There is nothing in locationism per se that "forces" this; quite simply, it reflects my perception of kinship with the subjects that I study-- just as Descartes' "human versus animal" and "mental versus mechanistic" dichotomies reflected his perception of kinship and his beliefs regarding the origin of species.

Where does this leave the concept of behavior? It seems to me that the difficulties with this concept are even more serious than the difficulties with being a locationist pure and simple. The central problem is that it is by no means clear whether "behavior" is a "what" question, a "where" question or a "why" question. Examine any ethogram or attempt to formulate a taxonomy of behavior that you choose, and you will find that all of these questions are in effect invoked, if not frequently confused with one another. To pick only one example, ethologists' "fixed action patterns" do not merely describe the motions of an animal's body parts relative to one another (a spatio-temporal problem); the concept of fixed action pattern also stipulates that these movements be species-typical or innate-- which is, at least implicitly a "Why" question (usually involving both developmental and phyletic problems). And this, of course, is the major reason that ethology was once so hotly contested by learning theorists and developmental psychologists. To quote Wilson, the classification of behavior "is a straightforward taxonomic exercise limited by the built-in arbitrariness in the definition of unit categories and clustering procedures. The difficulty is exacerbated by the fact that ... (some) behavior is very far from the genotype and is unusually genetically labile... To collect behaviors of different species in the same categories is largely a matter of judging analogy rather than homology, a largely subjective procedure" (Wilson, 1975, p. 217).

Even if the behavioral uncertainty principle (see above) could be ignored, and even if perchance a given behavior could be defined in purely spatial terms (e.g., captured in a single "instantaneous" snapshot, and without either implicitly or explicitly invoking anything that happened before or after that instant, or any consideration of "Why" questions), that behavioral descriptor must, in brief, be considered an hypothesis regarding the organization of nature. More likely than not this hypothesis will prove problematical on closer analysis. Whether one calls behavior a directly observable or an inference is a matter of opinion; but the differences between behavioristic and mentalistic descriptors seem matters of degree rather than of kind.

Konrad Lorenz has on more than one occasion said that the concept of the fixed action pattern is the "Archimidean fixed point" from which all of ethology must be constructed. Theorists of other persuasions have not been as explicit as this, but their claims for their own primordial concepts boil down to much the same sentiment. I am more inclined to agree with the grandfather of locationism, Rene Descartes: "There are no Archimidean fixed points except insofar as we choose to view them as fixed by our.own thought."

In sum: If behavioral (or mental) concepts enable me better to see what is "out there," I shall by all means use them. If, however, anyone tells me that all anyone can really see is what they themselves see, and that no one else should look for anything else, I shall trust my own eyes and my own intuitions rather than theirs. Nor do I believe that any really good students of animal psychology have ever really done anything different, regardless of their theoretical persuasion.

REFERENCES

Boesch, C. and Boesch, H., 1984. Mental map in wild chimpanzees: An analysis of hammer transport for nut cracking. Primates, 25:160-170.

Dennett, D.C., 1983. Intentional systems in cognitive ethology: The "Panglossian" paradigm defended. The Behavioral and Brain Sciences, 6:343-390.

Epstein, R., 1984. Pigeons, canaries and problem solving. Nature, 312:313.

Epstein, R., Kirshnit, C.E., Lanza, R.P. and Rubin, L.C., 1984. 'Insight' in the pigeon: Antecedents and determinants of an intelligent performance. Nature, 308:61-62.

Gibson, J.J., 1979. "The Ecological Approach to Visual Perception". Houghton-Mifflin, Boston.

Griffin, D.R., 1984. "Animal Thinking". Harvard University Press, Cambridge.

Lewin, K., 1951. "Field Theory in Social Science". Harper and Row, New York.

Mason, W.A., 1980. Minding our business. American Psychologist, 35:964-967.

Menzel, E.W., 1972. Spontaneous invention of ladders in a group of young chimpanzees. Folia primatologica, 17:87-106.

Menzel, E.W., 1973. Chimpanzee spatial memory organization. Science, 182:943-945.

Menzel, E.W., 1979. General discussion of the methodological problems involved in the study of social interactions. In "Social Interaction Analysis: Methodological Issues". M. Lamb, S. Suomi and G.R. Stephenson (Eds.), University of Wisconsin Press, Madison.

Menzel, E.W. and Juno, C., 1982. Marmosets (Saguinus fuscicollis): Are learning sets learned? Science, 217:750-752.

Menzel, E.W. and Juno, C., 1985. Social foraging in marmoset monkeys and the question of intelligence. Philosophical Transactions of the Royal Society, London, B308:145-158.

Menzel, E.W. and Menzel, C.R., 1979. Cognitive, developmental and social aspects of responsiveness to novel objects in a family group of marmosets (Saguinus fuscicollis), Behaviour, 70:251-278.

Polanyi, M., 1958. "Personal Knowledge". University of Chicago Press, Chicago.

Shapiro, S.L., Stark, R.F. and Teukolsky, S.A., 1985. The search for gravitational waves. Scientific American, 73:248-257.

Sugiyama, Y. and Komans, 1979. Tool-using and tool-making in wild chimpanzees at Bossou, Guinea. Primates, 20:513-524.

Sussman, R.W. and Kinzey, W.G., 1984. The ecological role of the Callitrichidae: A review. American Journal of Physical Anthropology, 64:419-449.

Terborgh, J.W., 1983. "Five New World Primates: A Study in Comparative Ecology". Princeton University Press, Princeton.

Tolman, E.C., 1932. "Purposive Behavior in Animals and Man". Appleton-Century-Crofts, New York.

Wilson, E.O., 1975. "Sociobiology". Harvard University Press, Cambridge.

LOCAL CUES AND DISTAL ARRAYS IN THE CONTROL OF SPATIAL BEHAVIOR

W. K. Honig

Dalhousie University

1. LOCAL CUES AND DISTAL ARRAYS

In the psychology of animal learning and animal behavior, psychologists have generally divided stimuli into two classes: Those that direct or control or initiate behavior, and those that provide the background, or setting for the behavior. The directive stimuli (as I shall call them) include the conditioned stimuli in Pavlovian conditioning, discriminative stimuli in instrumental learning, and sign stimuli in ethology. In contrast, the background, or environmental stimuli provide a favorable setting in which the control of behavior can be observed and even exercised by the experimenter. Directive stimuli are most often punctate and local: their location is restricted and consistent, and they are small enough so that the subject's response to them, including approaches and withdrawals, are easily identified. In many procedures they are presented briefly, especially in classical conditioning and in discrete-trial instrumental learning. Acquisition of behavior is facilitated if the local cue is small in relation to its background, and brief in relation to its absence (Hearst and Jenkins, 1974). If the controlling stimuli can be held constant or made to vary along specific dimensions, this facilitates the study of habituation, discrimination, generalization, and other processes central to the area of animal learning.

For these cogent reasons, and perhaps others, the vast bulk of work on animal learning in experimental environments involves such punctate stimuli, or local cues. Such cues signal both the availability of the reward and the location for the correct response. The rest of the environment, or "background", is generally kept as constant as possible, again with good reason. Animals are more easily adapted to respond in an environment that offers few surprises, and changes in the environment acquire eliciting or discriminative properties if they happen to occur together with appetitive or aversive events. Background stimuli become "neutralized" if they are present both when such events begin and when they end.

The prevalent concern with local or punctate stimuli no doubt derives in part from the physiological roots of classical conditioning. Much of traditional neurophysiology was based on an input-output model, the input being a specific stimulus, and the output a discrete muscular, glandular, or neural reaction. The area of classical conditioning remains largely in that tradition, although the importance of contextual cues is becoming more widely recognized (Balsam and Tomie, 1985). However, these background cues do not usually support spatial discriminations.

Instrumental discriminations of spatial locations usually involve an approach to a particular, marked location, or responding directed at a localized object. Responses to the wrong stimulus in the wrong place suffer extinction.

In recent years, research on the learning about spatial locations, and memory for them, has forced us to revise this traditional conceptualization of the stimulus. Behavior is readily controlled by distal cues that are well separated from the local cue to which the subject responds. Aggregations of such cues were called distal arrays by Gibson (1950, 1979), who argued that perception is in large part determined by such arrays. In particular, Gibson emphasized the orderly nature of such arrays, which underlie the perception of spatial orientation and location. Such distal arrays are also important for spatial learning and memory in animals, as I will try to show in this chapter.

Three paradigmatic procedures illustrate the control of behavior by such arrays. Menzel (1973, 1978) studied spatial memory in a naturalistic setting that involved both distal and local cues. He carried chimps around in a field to watch food being hidden in various locations marked by local cues. When the chimps were allowed to search for the food on their own, their pattern of travel was shorter and more direct than the route used for hiding the food, and more efficient than a random progression from one location to another. As the local cues were not readily visible from a distance, the chimps were presumably guided to food sites by the arrays of distal stimuli.

The radial arm maze of Olton and Samuelson (1976) provides an abstracted version of a foraging environment. The arms of the maze restrict the directions of travel, and lead the subject directly to feeding locations at the ends of the arms. The memory of arms visited depends on the stimulus array surrounding the maze, not on local cues. The rat does not have to discover the feeding locations, but it does have to remember which ones it has visited, and that depends on distal cues.

Discriminations without local cues have been studied rarely, but a recent example is the "milk bath" or "swimming pool" for rats, popularized by Morris (1981). The subject is put into a tank filled with opaque water, and has to discover a platform under the surface on which it can stand until rescued. The hidden platform provides no local cue, and the rat depends entirely on a distal array. The interesting aspect of this procedure is precisely that the platform which would normally guide the animal to safety is not detectable; the rat depends upon other cues that are separated from the one stimulus of interest to him.

In these procedures, then, the subject is guided or directed by stimulus arrays removed from the locations that it must find and remember. The processes that enable efficient performance in such situations have not been analyzed adequately, either conceptually or experimentally. Experimenters have often been more interested in the memory functions that affect behavior in these kinds of situations than in the way that the available distal stimuli direct the behavior. The controlling stimuli are casually described, and their functional effectiveness is not often studied experimentally. With an emphasis on spatial memory, we may have been putting the cart before the horse. The subject has to discriminate a location before it can remember the same. This kind of discrimination is the main concern of this chapter.

2. PERCEPTUAL FIELDS, PERCEPTUAL WORLDS, AND DESTINATIONS

Behavior in space takes place in a perceptual world. A portion of the world is perceived at any one time as a perceptual field. These terms are adapted from Gibson (1950) who, in line with his primary interest, wrote about the visual field and the visual world. The perceptual field refers to the information mediated by the proximal stimuli, that changing and evanescent pattern of sensations provided by the receptor surfaces. The perceptual world is the stable psychological environment which is in some way abstracted from a succession of perceptual fields. As the animal moves about, it generates a succession of perceptual fields associated with different locations. One kind of location is of particular interest both to the animal and to the psychologist, namely the destination. A destination is a location at which an approach is terminated, and at which other behavior, often consummatory, is initiated. Like any other location, a destination is identified by a set of perceptual fields for the subject. But the destination often also provides local cues of particular importance for the subject, namely those that are associated with reinforcement.

It is reasonable, therefore, to suggest that a perceptual world is integrated from the information provided by a succession of perceptual fields. Differences in these fields reflect differences in locations, and they are generally produced by the subject's movements. The manner in which these difference are integrated into a perceptual world is a mystery. For the moment, let it suffice to suggest that the subjects remember particular fields and thus particular locations in their perceptual worlds, especially when these are marked by events of importance, such as feeding or danger.

Take, for example, the rat in Morris's milk bath. At first it swims around, producing a constantly changing array of perceptual fields. At one point in its more or less random movements, it comes upon the platform and rests there. Now the aspects of the environment at that location -- that particular perceptual field -- are registered as a destination, because of reinforcing value of that location as a place of safety, and possibly also because of the marked change in behavior from swimming to standing. When the rat is put back in the milk bath for further trials, it may move again in a random fashion to begin with, but these movements will result in an approximation to the destination associated with safety. In the course of training, the destination is reached with great efficiency. This process is similar to the successive approximations by which animals learn to make particular skilled responses in a constant location in order to gain a reward. The difference is that a destination is approximated rather than a particular response.

The search for food by Menzel's chimps is more complex, but not very different conceptually. The chimp needs to encode and remember a "list" of destinations, each of them comprised of a set of perceptual fields in a world with which it is familiar. The chimp does not reach the next point in its route by some random wandering about, and we know that it does not usually memorize and follow the path taken initially by the experimenter. It sets out in a specific direction, which suggests that it anticipates the destination; the movement is ballistic in the sense that a decision of where to go is made at the outset, and a direction of travel is established. Thus, the chimp must have some process for determining in advance what route will permit it to attain the proper set of perceptual fields. The same process in a simpler form presumably enables a rat to swim directly to the hidden platform in the milk bath.

In the radial arm maze, the routes from the center of the maze to each destination are circumscribed. The critical event during each run appears to be the "registration" of each destination, with its particulr perceptual field, at the end of each visited arm. The memory for that field prevents a return to that location, while other, unvisited locations are available. Now rats do not generally learn the problem by running out to the ends of already visited arms and then, failing to find food, run out for shorter and shorter distances in some slow process of extinction. They either run to the end of the arm or they don't. Thus the subject presumably makes a decision at the central platform, anticipating the destination at the end of the arm. The memory of locations at which the subject has already been fed contributes to the decision, and this results in "ballistic" behavior in another sense, namely the inhibition of movement in a particular direction.

This preliminary analysis draws attention to three problems that do not generally arise when behavior is controlled only by punctate stimuli or local cues. The first is the process by which a stable perceptual world is generated from a set of fields, or "looks", obtained sequentially from various locations. The second problem is the identification of particular locations as destinations on the basis of distal cues, as in the milk bath or on the radial arm maze. The third problem, closely related to the second, is the origin of "ballistic" spatial behavior, both when that involves a particular direction and distance of travel, and when it results in the suppression of movement toward particular places.

3. CONTROL BY DISTAL ARRAYS

Perceptual fields are comprised in large part of arrays of distal stimuli, which, together with local cues, control movements among different locations. Local cues vary in their prominence and salience; in the limiting case, there are no local cues at all. In this section, I review evidence that perceptual fields generated by distal stimuli control the discrimination among specific locations and memory for them.

In the area of discrimination learning there is a strong tradition which supports the notion that learning is facilitated when discriminative stimuli are close to the reinforcer, both in space and in time. Therefore, spatial discriminations have most frequently involved local cues that identify particular locations for the subject. The best evidence for discriminative control by distal arrays came at first from studies of spatial memory rather than spatial learning.

Spatial Discrimination in the Radial Arm Maze

Rats and other foraging animals tend not to return to places where they have found food. Therefore they discriminate and remember them. The radial arm maze provides a set of locations, but within the maze, local cues at the ends of the arms are presumed to be non-differential. We may assume that rats use an array of distal cues to identify the locations and remember them, but this requires experimental support. In several studies, variations in the location of the maze, or in the arrangement of distal cues around the maze, have shown that such cues control spatially governed behavior.

Roberts (1981) studied the role of extramaze cues through retroactive interference in the spatial memory of rats. Following the general method of Beatty and Shavalia (1980), he forced rats to run into four pre-selected arms of an eight-arm radial maze, and then interpolated experiences that might interfere with the memory for these arms during the rest of the trial. In one study, he ran rats in a second maze which was located in the same room, and only 30 cm removed from the first maze. He forced them either to enter arms pointing in the same direction as the four arms on the first maze, into the arms pointing into the four remaining directions, or into a randomly chosen four arms. A control group received no interpolated training. He then allowed his subjects to finish the original trial by running them on the maze in its original location.

The first of the interpolated treatments should result in rather little interference, as the experience would be redundant with the initial entries into the four arms of the original maze. The second treatment (entering arms into four "new" directions) should result in the most interference, and the effects of the third, "random" treatment should fall in between. To his surprise, Roberts found no evidence of interference from any intervening training, when performances were compared to the control condition. He replicated the study by placing the second maze 60 cm directly above the first. He ran the corresponding conditions, and again obtained no reliable interference. In a third study, Roberts used only one maze. After four choices, he placed the rats either at the ends of the entered arms, the unentered arms, or a random set of arms, to eat a pellet of food. In this case, he did find some interference, as one might expect, when the rat was fed at the ends of arms that he had not entered previously; but even so, the treatment was not devastating. These findings imply that the rats could readily distinguish locations between visual fields at the ends of the arms, separated by 30 cm horizontally or 60 cm vertically. The differences between these fields must have been rather small, and it is reasonable to suppose that the rats discriminated the locations through arrays or configurations of distal cues, rather than attending to particular, isolated stimuli.

This conclusion is supported by another experiment by Roberts (1979), on the so-called hierarchical maze. This maze is composed of eight primary arms, each with three secondary "branches". This produces 24 feeding locations, spaced fairly close together (see Fig. 1A). In different phases of the work, Roberts either left all branches open during a trial, or he closed them off selectively so that the rat could obtain food from only one or two branches upon entering each arm. When three branches were left open, the rat would collect food from all three before going to the next arm. This is hardly surprising and would not involve a severe memory requirement for different branches. But the subjects also did well when given the opportunity to visit only one or two branches at a time on each arm. They remembered which branch(es) they entered on the first run down an arm, and then avoided them on later visits.

These studies suggest that rats base discrimination and memory for spatial locations on rather subtle differences among spatial arrays. To do so, they would require a fairly wide view of the environment. This is supported by an analytic study carried out Mazmanian and Roberts (1983). First they let a single group of rats learn to collect food from the ends of a four-arm radial maze with no restrictions on the field of view. Then they pre-fed the rats on two of the four arms on each trial. The experimental treatment was the width of the field of view permitted on the pre-feeding arms. This was either a tunnel, which greatly reduced the field, a 180-degree restriction of the field, or a full view (see Fig. 1B). Then they allowed the rats a free choice of arms with the restrictions removed. Accuracy of choice of the "unfed" arms was greatly reduced on test trials following "tunnel vision" during pre-feeding, and moderately reduced following pre-feeding with the 180-degree field. This finding was replicated with independent groups in a second experiment. Rats which were pre-fed in the "tunnel" condition failed to improve their performance over 18 trials, while rats in the other groups improved.

Perceptual fields are composed of distal stimuli in particular arrangements or configurations. It is important to determine whether the configurational aspect is important for the control of spatial behavior. While the work of Roberts and his associates suggests that this is likely to be the case, it is conceivable that when marked individual cues are available, the subject will attend to them in isolation. This question was addressed by Suzuki, Augerinos, and Black (1980). They ran rats in an eight-arm radial maze surrounded by a curtain, on which were hung seven distinctive, prominent decorations, each of them near the end of one of the arms. The eighth arm was marked by absence of a cue. When the rats got very good at choosing eight arms without repeating any, trials were interrupted after three choices, and three experimental test conditions were imposed. (1) The positions of the decorations were left unchanged for the fourth to eighth choices in the control condition. (2) The relative positions of the decorations were maintained, but the entire array was relocated by rotating it around the maze by 180 degrees. (3) The array was rearranged. The rats made few errors after the interruption in the control or the relocation conditions, but they made numerous errors following the rearrangement of the stimuli.

The authors concluded that the subjects encoded locations not in terms of a single stimulus close to each arm, but in terms of the relative positions of the stimuli. It would be interesting to determine whether rats could learn to run a radial maze at all in an environment where a similar set of cues was frequently rearranged between trials -- that is, whether they could learn to ignore the spatial relationships and attend only to specific local cues. Even if this were possible, I think it likely that the problem would be mastered much more quickly if the relations are kept constant.

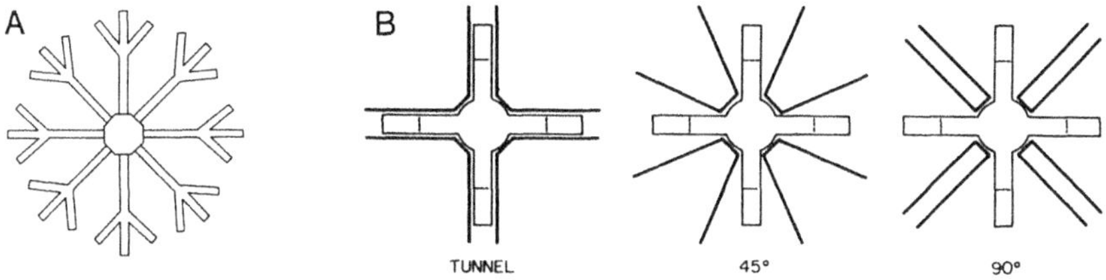

FIGURE 1. (A) Diagram of the radial maze used by Roberts (1979) to provide 24 different food locations for rats. These locations are almost equally spaced on a circle, and are separated by about 15 degrees, providing rather similar views of the environment from adjacent locations. (B) Diagram of the viewing conditions provided by Mazmanian and Roberts (1983) in a four-arm radial maze. The heavy lines represent partitions that restricted the views of the environment to various degrees.

Spatial Discrimination in the Arena

The radial arm maze provides a set of fixed destinations and of restricted, circuitous paths for going from one destination to another. The local cues at the ends of the arms, which mark food locations, do not differ, while distal arrays are used to identify locations already visited. In natural settings, animals take more direct routes from one possible source of reinforcement to another. The locations providing reinforcement may or may not be marked by a distinctive local cue. The open field in which Menzel and his associates have done much of their work represents this kind of setting. A more restricted setting may be called an arena--an enclosed area in which the subject can move freely, and into which it may be admitted from various locations. Experiments in arenas can provide some approximations to foraging in a natural setting, while incorporating experimental control over distal and local cues.

Spetch and Edwards (unpublished research) have studied spatial memory in pigeons in an arena, taking advantage of their propensity for feeding on the ground. Eight 2-litre milk cartons were placed in a circle on the floor of a small room and weighted with grit. A few grains were hidden behind an opening cut in one side of each carton. The pigeons walked from carton to carton and looked for the grain, which they could not see until they put their heads into the opening. They learned this procedure readily, and made few errors within the first eight choices. Both the locations and the appearances of the cartons can be varied within this setting. Various sorts of milk products come in cartons of different colors. At first, three different colors were associated with particular locations. Eventually Spetch and Edwards replaced some cartons so that they were all of the same color, but performance was not affected. Then the cartons were rearranged from a circle of eight into two straight rows of four each. Again there was little effect on performance, even when the rows were quite close together. The numbers of different cartons chosen in the first eight approaches are shown in Fig. 2 for various phases of the experiment. Clearly, the transitions from the training to various test procedures caused little disruption of the behavior.

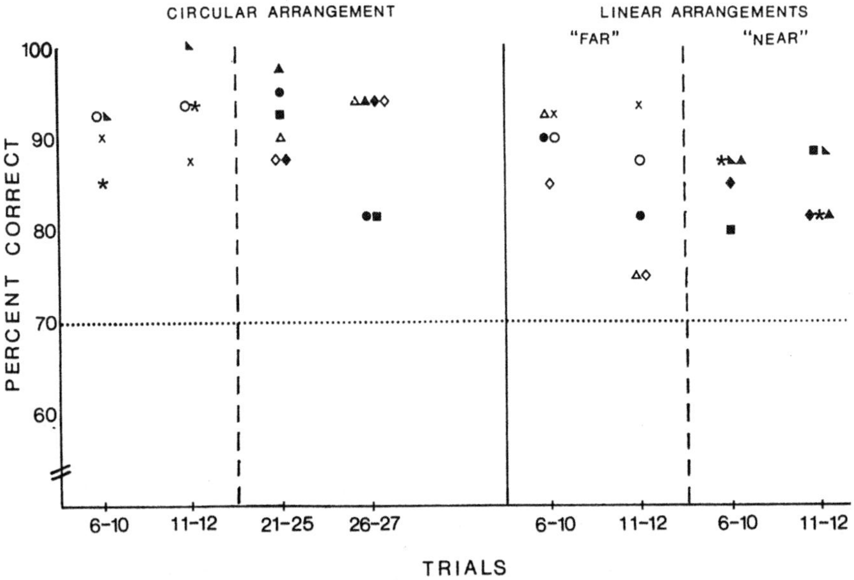

FIGURE 2. Performance on the "walking memory" task by individual pigeons in the task designed by Spetch and Edwards (submitted). The symbols represent different subjects. The score is the percentage of different food locations visited on a trial, the maximum being eight. Seventy per cent correct is the level expected from random performance. Under the "circular arrangement", the left panel shows data from four birds who had prior experience on a related task; the right panel represents naive birds. Under the "linear arrangements", independent groups of birds were trained with the food sources placed far apart (left) or close together (right). In each panel, the data at the left are taken from five trials with milk cartons of different colors; those at the right from two test trials with black and white cartons only. The change from the more discriminable to less discriminable cartons had little effect on performance.

This method of feeding pigeons (or other animals) on the floor permits flexibility in the location of food sources, and it will be of interest to locate cartons in different and more or less random positions on each trial, so that the pigeon would never learn to identify a restricted number of feeding locations. It will be important to determine whether the stability of such locations is at all critical for spatial working memory.

The arena is useful for the study of working spatial memory, but it can be used more directly to determine the degree to which distal cues control the "absolute" discrimination of spatial locations. Olton and his associates have carried out several studies of this sort (see Olton, 1982, for a review). In one procedure, two blocks were placed inside an arena 1 m square. The blocks differed in appearance from trial to trial, but their locations did not vary. Only one location (actually the top of one block) provided food on each trial. The arena was surrounded by a narrow runway, which contained three openings on each of its four sides to admit the rat. Access was controlled by the experimenter.

During training trials, the rat entered the arena from the center door on each of three sides. After the subjects mastered the discrimination of the correct location, side doors on each of the sides were opened on test trials to admit the subject. The rats showed excellent transfer. A more impressive test was carried out next, when the rats were admitted from a door in the fourth side, which had up to that time not provided access to the arena. Transfer was again very good; the rats went directly "from the opening in the wall to the correct location with no signs of hesitation" (Olton, 1982, p. 348). (Olton's chapter in this book also provides a more detailed discussion of "location learning".) Since the local cues provided by the two possible sources of food varied at random, the rat had to use the distal array to make the appropriate discrimination. It is surprising (and regrettable) that this rather simple research method has not been exploited to a greater degree.

As indicated in the introductory section, Morris (1981) extended this general strategy by providing rats with an arena in the form of a tank of water, in which there are no local cues to mark the location of reinforcement. In his original work, Morris trained the rats until they readily found the platform. Then he moved it, and the rats swam to the original location and searched there before they finally found it elsewhere in the pool. Clearly, their behavior was controlled by the spatial array of stimuli outside the pool.

This research is particularly important because the destination is not marked by any local cue at all. Yet the subjects do not, apparently, memorize a route by coordinating their movements with specific changes in the distal arrays. Morris showed that the rats would adjust their route immediately to reach the platform if they were placed into the water at a location other than that used in original training. The desired perceptual field, which identified the destination, could be attained from various starting points.

Sutherland and Lynggard (1982) went one step further with a latent learning procedure. They pretrained rats by placing them directly several times on a hidden platform before they required them to swim to it. For the positive transfer group, this hidden platform stayed at the same location when regular training began. For the negative transfer group, the platform was moved to the quadrant of the pool diagonally opposite the location used for pretraining. A third group was given pretraining with a tank and platform in a different room. When regular training began, the positive transfer group showed considerable savings in time and path length to find the platform in comparison to the other two groups. The negative transfer group spent more time in the quadrant used for pretraining, searching for the platform, than did the group pretrained in a different room. This study strongly suggests that the subjects learn the safe destination even without having to approach it in the first instance, and then move to attain the appropriate perceptual field when placed elsewhere. There is no reason why this method could not be adapted for other species and other situations; for example, food could be hidden in a specific location under sawdust or sand in an arena.

Sutherland and Dyck (1984) carried out several further studies with the hidden platform. In the most interesting, they divided the swim tank into four equal quadrants with the use of white vertical dividers crossing in the center. A modest space of 12 cm was left between the rim of the pool and the outer edge of each divider, so that the rats could enter all of the quadrants. The visual fields were restricted in extent, but differential cues were available outside the pool at each quadrant. The rat was placed at one interior corner created by the intersection of the dividers, and had to find a hidden platform located at one of the other interior corners. Subjects in the "map" group had to swim to a particular corner from any of the other corners. Rats in the "response" group had to swim from the starting corner in any quadrant to the quadrant to their left in order to find the platform.

The "map" group learned the problem somewhat more quickly, but a greater difference was found when the dividers were removed, and the rats were placed in the water at the edge of the pool. The rats in the "map" group learned quickly to swim to the platform, and some did so almost immediately, integrating visual information from portions of the visual fields which had not previously been seen at the same time. The rats in the "response" group swam in a random manner.

In their final experiment, Sutherland and Dyck trained a single group of rats (with no dividers) in a "map" condition, in which visual cues were available, and also in a so-called "praxis" condition, in which black curtains obscured the cues outside the tank. After some training the rats received alternating trials, and they used the search strategies typical for these conditions -- the direct approach in the visual condition, and a circular search in the praxis condition. When the platform was then moved, the control by visual cues hindered the animals from finding the new location in the "map" condition; with visual cues lacking, however, the rats used the so-called praxis strategy to good advantage, and found the platform more quickly. Apparently, the distal visual cues so dominated the rats that they did not revert to the "praxis" strategy when they failed to find the platform in the accustomed location. Results obtained from one rat under the "map" and the "praxis" conditions are shown in Fig. 3.

The Role of Landmarks in the Discrimination of Locations

Research with the arena is valuable in demonstrating the importance of distal cues in the discrimination of specific locations. When such cues are removed, the subject reverts to a "praxis" that can narrow the search. When they are available, the subject presumably uses the landmarks to identify the location, or at least to narrow the search. There has been little study of this process. Suzuki et al. did show that landmarks are not perceived in isolation, but that their relative positions guide the subject in the radial arm maze. But the role of the number and position of landmarks in locating a destination not marked by local cues has received little attention.

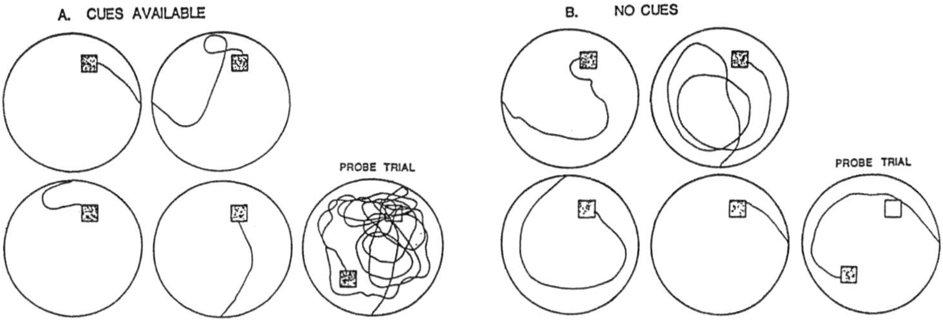

FIGURE 3. Routes taken by representative rats in the milk bath toward a hidden platform (shaded box), when distal cues were available (A) and when they were not (B). Tracings from the former condition show that the rat swam toward the platform rather directly from any starting location. When the platform was moved, the rat did not revert to an efficient "praxis" strategy, but searched for the platform near its original location (open box). The praxis strategy is illustrated in the tracings at the right. The rat easily found the relocated platform, which was placed at the same distance from the side of the pool (Sutherland and Dyck, 1984).

If the subject is trying to find a particular location, a single landmark provides information only with respect to distance, if it is large enough to subtend a reasonable visual angle. (This assumes that the landmark does not possess distinctive features providing a cue to direction, and that the subject does not possess a compass.) Two landmarks subtend a visual angle between them, and this provides a circular arc of possible locations at which a goal might be located. Three cues are sufficient for a "fix" because they subtend two separate angles, and arcs based on them will intersect only at one point. From a geometric point of view, further cues would be redundant.

Little systematic work on the question of localization in an open field with precise control of landmarks has been carried out with subhuman vertebrates. Sutherland and Dyck (1984) hung a couple of "landmarks" over the "swimming pool" for their rats, which was otherwise surrounded by a black curtain. Their description is scanty, and specific results are not reported. In this volume, K. Cheng describes interesting research in which he used walls of a rectangular arena as extended features, and concluded that for rats, the relationship between adjacent walls is important in guiding their behavior. However, the method does not provide flexibility for the study of discrete landmarks and their placement in the control of localization without local cues.

Collett and Cartwright (1982) carried out ingenious experiments with bees in an arena in which an unmarked source of food was available. They provided the subjects with a landmark in the form of a black cylinder placed upright on a floor, 50 cm from the food source. The bees could estimate the general direction from the food, as some room cues were also available, and they searched rather widely at an appropriate distance, as the cylinder subtended a reasonable angle. When the size of the cylinder was reduced, the bees moved closer to it for their search; when it was enlarged, they moved away. This shows that they were using the visual angle of the object as a cue. In further research, three cylinders were placed in the field to form an obtuse angle, and the bees markedly reduced the range of their search. It was concentrated at the center of an arc on which the three landmarks were placed (see Fig. 4). The experimenters then carried out various changes in the size of the landmarks, and their distance from the food source, while maintaining the bearing of each landmark from that souce. With these variations, the bees still searched in the same place, in spite of the changes in the visual angles subtended by the individual objects. The relative positions of the objects (or the angles subtended by the separations between them) controlled the location of the search. This brief account does not do justice to the interesting and ingenious work carried out by these authors on their bees. It indicates a general strategy which should be adapted for the study of localization in vertebrates.

4. PROBLEMS, PROCESSES, AND PROSPECTS

I have reviewed a fair body of evidence to support the control of behavior in space by distal stimuli. But this evidence is sketchy, and it does more to show that such control is important than it does for analyzing its nature. Several theoretical problems remain to be resolved, and I can only touch upon them here.

The first problem is the integration of the spatial fields, provided by successive "looks", into a stable perceived location in the perceptual world. The <u>spatial field</u> is the evanescent "projection" of the environment upon the sensory systems of the animal. It has the same relation to the perception of location as Gibson's (1950) <u>visual field</u> has to his <u>visual world</u>. Gibson suggested that the detailed analysis of a visual field may not be necessary for understanding the visual world because the important psychophysical relation is between the external stimulus order and the perception of it. Likewise, we can analyze the manner in which distal stimuli govern spatial behavior without worrying too much about the sensory "appearance" of these stimuli to the organism. Nonetheless, the analysis of spatial fields may provide insights into the nature of spatial perception. For example, the role of the number and redundancy of spatial cues needs to be analyzed. The work of Cartwright and Collett (1983) is only a beginning. Furthermore, attention to spatial fields will make us more sensitive to the geometric organization of the environment.

A second problem is the "ballistic" nature of the spatial behavior. In the situations of interest here, the subject approaches a destination that is identified by distal cues. In the "purest" examples, the subject moves freely toward a location with no distinguishing local features. Generally, such an approach is direct, and does not involve a series of approximations to the correct location. Even if the route is circumscribed as in the radial maze, the subject generally acts in a "ballistic" manner; if it enters an arm at all, it tends to run to the end. It does not slow down and turn around when it gets closer to the location of a reinforcement that has already been taken.

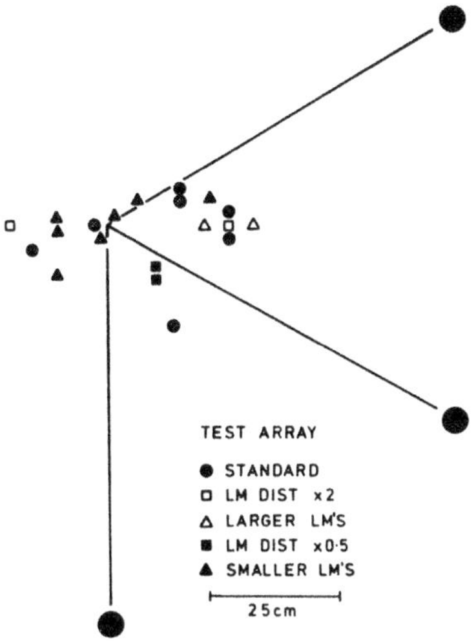

FIGURE 4. The small symbols are the centroids of the locations at which bees searched for food, which was normally located at the intersection of the three lines. The heavy dots represent the cylinders which provided the landmarks, in their standard positions. The different small symbols represent performance under different test conditions used by Cartwright and Collett (1983). The size or the distance of the landmarks was changed as shown in the key to symbols. The distance was doubled or halved; the size was trebled or halved. These changes were carried out independently. None of them affected the location of the bees' search, which was dominated by the angular separations of the landmarks.

Such a capacity for "ballistic" travel implies that the subject can anticipate the perceptual fields that would result from movement to various possible destinations. This would enable the animal to choose a direction and perhaps also a distance for travel to produce a particular stimulus array. The process is not well articulated and requires further analysis. However, it is not without a counterpart in time. Animals anticipate stimulus events in a reliable temporal sequence, and they will exert control over the events if given the opportunity (Honig, 1978, 1981; Weisman, 1984). Space and time are closely related for animals, as it always takes time to move through space. It is reasonable to suppose that they can learn to anticipate a particular location after moving in a certain direction for a certain period.

The anticipation of spatial fields in various possible locations is probably facilitated by the fact that spatial arrays change in lawful ways as the animal moves about. The visual angles of objects and their angular separation increase as they are approached. The rate of translation of images across the field is faster for nearer objects than for more distant ones. Particular positions in space are marked by the relative separations (or "bearings") of particular cues in the field, and these change in a predictable way for different locations. As far as I know, a detailed analysis of these changes is not available, but the problem can in principle be formulated empirically, so that the mystery may be taken out of the processes through an experimental analysis.

For such an analysis, we would do well to turn to Gibson's ecological approach to perception. Gibson (1950) suggested long ago that the stimulus array provides the cues to inform the organism of its location. There are orderly changes among such cues which comprise gradients in the array. For example, the retinal size of the elements of the array decreases and their density increases as a function of distance. Such cues support the perceived slant of a surface. He also called attention to the different speeds and directions of change of visual cues in the course of the subject's movement. Such changes are very different when the subject moves than when some part of the environment moves.

In Gibson's examples, the stimuli tend to be rather uniform -- trees, grass, even man-made objects like barrels. Therefore, the gradients of change with distance, slant or movement are orderly. Subjects should be able to discriminate order from randomness in such arrays. This can be tested with modern video and computer technology. For example, a pigeon should be able to discriminate an array of vertical lines whose interline distances change in an orderly fashion, as they would on a slanted surface, from an array in which the same distances are randomly ordered.

When an animal moves, the elements of an array move across the visual field at different rates as a function of their distance. Such a pattern can be simulated on a screen by computer-driven displays. Animal subjects ought to be able to discriminate an orderly gradient of displacement that would be produced by its own movement, from a set of elements moving in the same direction and with a similar distribution of speeds, but in a random or distorted pattern.

A further, important task is an analysis of the critical components of distal arrays. What is the optimum number of discrete distal cues in the array? We have seen that bees do quite well with three landmarks. Would further, redundant, cues assist the subject, or would they cause confusion? If there are redundant cues, can some be removed with no loss of accuracy? If both local cues and distal stimuli are available, will either block or overshadow the other?

Studies of this sort would admittedly manipulate and in some cases impoverish the stimulus array. One of the arguments in favor of working with distal cues is that a cognitive determination of position and movement is presumably based on a rich set of stimuli. But analytic work frequently requires an initial reduction of a situation to its essentials, so that fundamental mechanisms can be studied in isolation. This will most likely be the case for the analysis of distal environmental cues. We will need to determine which cues are necessary, which are sufficient, and which are redundant for the subject to recognize its location and to decide upon a route of travel.

REFERENCES

Balsam, T.D. and Tomie, A. (Eds.), 1985. "Context and Learning". Lawrence Erlbaum Associates, Hillsdale, N.J.

Beatty, W.W. and Shavalia, D.S., 1980. Spatial memory in rats: Time course of working and the effect of anesthetics. Behavioral and Neural Biology, 28:454-462.

Cartwright, B.A. and Collett, T.S., 1983. Landmark learning in bees. Journal of Comparative Physiology, 151:521-543.

Collett, T.S. and Cartwright, B.A., 1982. Eidetic images in insects: Their role in navigation. Trends in Neurosciences, 6:101-105.

Gibson, J.J., 1950. "The Perception of the Visual World". Houghton Mifflin, New York.

Gibson, J.J., 1979. "The Ecological Approach to Visual Perception". Houghton Mifflin, New York.

Hearst, E. and Jenkins, H.M., 1974. "Sign-tracking: The Stimulus-reinforcer Relation and Directed Action". Monograph of The Psychonomic Society, Austin, Texas.

Honig, W.K., 1978. Studies of working memory in the pigeon. In "Cognitive Processes in Animal Behavior", S.H. Hulse, H. Fowler and W.K. Honig (Eds.), Lawrence Erlbaum Associates, Hillsdale, N.J.

Honig, W.K., 1981. Working memory and the temporal man. In "Information Processing in Animals", S. Spear and R.R. Miller (Eds.), Lawrence Erlbaum Associates, Hillsdale, N.J.

Mazmanian, D. and Roberts, W., 1983. Spatial memory in rats under restricted viewing conditions. Learning and Motivation, 12:123-140.

Menzel, E.W., 1973. Chimpanzee spatial memory organization. Science, 12:943-945.

Menzel, E.W., 1978. Cognitive mapping in chimpanzees. In "Information Processing in Animals", S. Spear and R.R. Miller (Eds.), Lawrence Erlbaum Associates, Hillsdale, N.J.

Morris, R.G.M., 1981. Spatial localization does not require the presence of local cues. Learning and Motivation, 12:239-260.

Olton, D., 1982. Spatially organized behaviors of animals: Behavioral and neurological studies. In "Spatial Abilities. Development and Physiological Foundations". M. Potegal (Ed.), Academic Press, New York.

Olton, D.S. and Papas, B.C., 1979. Spatial memory and hippocampal function. Neuropsychologia, 17:669-682.

Olton, D.S. and Samuelson, R.J., 1976. Remembrances of places passed:

Spatial memory in rats. Journal of Experimental Psychology: Animal Behavior Processes, 2:97-116.

Roberts, W., 1979. Spatial memory in the rat on a hierarchical maze. Learning and Motivation, 10:117-140.

Roberts, W., 1981. Retroactive inhibition in rat spatial memory. Animal Learning and Behavior, 9:566-574.

Spetch, M. and Edwards, C.A., submitted. Spatial memory in pigeons (Columba Livia) in an "open-field" feeding environment.

Sutherland, R.S. and Dyck, R.H., 1984. Place navigation by rats in a swimming pool. Canadian Journal of Psychology, 38:322-347.

Sutherland, R.S. and Lynggard, R.C., 1982. Being there: A novel demonstration of latency spatial learning in the rat. Behavioral and Neural Biology, 36:103-107.

Suzuki, S., Augerinos, G. and Black, A.H., 1980. Stimulus control of spatial behavior on the eight arm maze in rats. Learning and Motivation, 11:1-18.

Weisman, R.W., 1984. Representations in pigeon working memory. In "Animal Cognition". H.L. Roitblat, T.G. Bever and H.S. Terrace (Eds.), Lawrence Erlbaum Associates, Hillsdale, N.J.

Acknowledgements

Preparation of this chapter was assisted by Operating Grant no. AO-102 from the Natural Sciences and Engineering Research Council of Canada. The Faculty of Arts and Science of Dalhousie University generously supported travel to the NATO meeting on Spatial Orientation in Animals and Man, where a prior version of this material was presented as a lecture. I am grateful to M. Spetch and C. Edwards for permission to illustrate unpublished material from their research.

A COMPARATIVE APPROACH TO COGNITIVE MAPPING

Jacques VAUCLAIR

Centre National de la Recherche Scientifique, Institut de Neurophysiologie et de Psychophysiologie, Département de Psychologie Animale,
31, chemin Joseph-Aiguier, F13402 MARSEILLE CEDEX 09

This paper raises several issues which appear to be of importance in the present discussion of orientating abilities and spatial coding.

The first issue concerns cognitive mapping - a consideration of this process appears important in order to understand spatial organization, and thus an extensive definition is provided. Furthermore, reference is made to the original concept of a "cognitive map", which serves to outline the nature of spatial coding : namely an organization of paths including their associated objects or events.

I then propose to examine the acquisition of spatial abilities (in humans and animals) ; such a study reveals the role played by various factors (notably active movement) in spatial behaviour.

Finally, two phenomena (compass navigation and in particular path integration) are presented as possible alternative mechanisms involved in orientating capacities.

A DEFINITION OF COGNITIVE MAPPING

The principal elements of the definition of cognitive mapping can be found in Pick and Rieser (1982), who propose the following extensive definition. Firstly, this process consists of a knowledge of the "set of spatial relations among all the objects in a space" (p. 118). Secondly, "cognitive mapping implies being able to operate on spatial knowledge in a way that is analogous to viewing the space from different station points" (p. 118). A third requirement is that "cognitive mapping is implicated when the updating is done without perception, when, for example, all the spatial relations cannot be seen because the space is too big or too complex... it is dark... one is blind, or the movement is imaginery" (p. 118), i.e. when the updating is done in an inferential way. Clearly, one very important benefit of such mental updating would be the capacity to plan alternative routes.

To this definition, I shall add a fourth component - that is, the possibility that in cognitive mapping, sequential information (gathered, for example, through exploratory movements) is read into a simultaneous system (Levine et al., 1982). As a result of this property, all path segments become equally available, and thus make possible the use of new segments to perform, for example, a detour or a short-cut.

This rather theoretical approach will of course apply only to the highest levels of achievement of spatial organization.

Since the term "cognitive map" was first coined by Tolman (1948), it is interesting to contrast Tolman's view with those of some of his followers. It is clear that for Tolman, a map is formed of two related pieces of information : first, locations and their connecting paths within a space, and secondly, and simultaneously, objects or events associated with the given space (as demonstrated for example by latent learning). This two-fold characteristic of maps has largely been overlooked by Tolman's followers, notably by O'Keefe & Nadel (1978), who define space in Kantian terms, as an "absolute space which exists in the absence of objects" (p. 86).

The space in which an organism moves, however, is not an empty container. Indeed, organisms move and orient in a space which is defined by the nature and relative positions of objects (see also Menzel, 1978). For example, both latent learning and homing experiments show that certain places (such as the food box, the nest or the burrow) have properties which can markedly influence the orientating or reorientating behaviours.

As a direct illustration of this notion, a study will be briefly described which investigated detour behaviour in golden hamsters. This experiment (Vauclair, 1980) attempted to replicate Tolman and Honzik's (1930) findings obtained with rats.

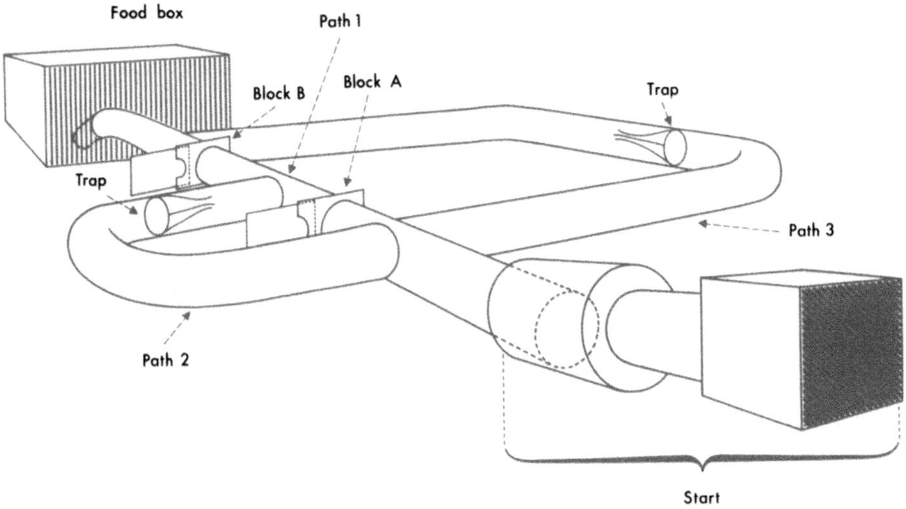

Figure 1 : Maze used to test detour's abilities of hamsters.

Tolman and Honzik's original maze was an elevated maze with three different paths leading to a goal box. After a short learning period, rats used path 1 more frequently than path 2, and path 2 more frequently than path 3. So, when the rats encountered a block at point A, they took path 2 to reach the goal. The rats are said to have behaved with insight, when route 3 was chosen over route 2 after a block was placed at point B.

My apparatus was a three-dimensional maze in which the starting box was replaced by the hamster's actual nest (cf. figure 1). On each trial, a dozen nuts were left in the food box in order for the hamster to fill its pouches and return to its nest. Two experimental conditions were used : in the first condition, the animals encountered the blocks during the outward trip ; in the second condition, the blocks were placed in position for the animals' return trip to the nest, that is after they had received the hazelnuts.

The results showed that on the first trial the hamsters did not behave with insight in either condition. A differential effect of the two conditions appeared, however, on the second trial. In Condition 2 subjects were able to make the necessary detour as from this second trial, while in condition 1, subjects needed up to five trials and numerous repetitions before selecting the correct path. The determining factor appears to be the fact that the nest functionned as the goal in condition 2.

In short, the above study demonstrates the role of the different locations and the pathways between them in spatial organization.

THE ACQUISITION OF SPATIAL ABILITIES

If we look now at the process of acquiring a spatial orientation system, we should be able to see some of the constraints which define the scope and the form of the spatial organization. Piaget (1954) found that the way in which a young infant searches for a hidden object in the object permanence task, is dependent on his prior interaction with the object, i.e. he searches where his own activity had previously uncovered it. Piaget described this behaviour as egocentric. The object permanence task is a good example of the importance of movement in coding spatial relations. For example, in a series of experiments with 9-month-old infants, Bremner (1982) showed that the typical error of Stage 4 subjects (i.e. after a displacement, the object is still searched for at the place it has previously been found) could be partially overcome when the infant, rather than the experimental array, was moved. Moreover, it has been demonstrated for the same task that 11-month-old infants searched more accurately after self-produced movements, than when passive movements were involved (Benson & Uzgiris, 1981). Finally, Acredolo (see Acredolo, volume 2) showed the same phenomenon in 1-year-olds, and found in addition that self-produced movement increased visual attention (as measured by visual tracking) towards relevant environmental stimuli.

Besides object permanence tasks, the facilitating effect of active versus passive exploration has also been demonstrated in tests of memory for spatial location. Results obtained with preschool (3-7 year olds) and school age children (9-10 year olds) indicate that the memory of the

3-7 year olds is better in the active exploration condition, although no differences were found between the two conditions for the older children (Feldman & Acredolo, 1979 ; see also, Herman, 1980). The importance of locomotion in spatial coding is also observed in adults ; for example, Pick & Rieser (1982) report that blindfolded adults are much more accurate at pointing to familiar objects from a novel station point after an actual movement to the new place, than after simply imagining that movement.

From this brief survey of human studies, we can stress two points :

1) In human infants, there is gradual progression from a self or ego-centric referent to an external referent, between roughly 6 and 16 months of age.

2) Active exploration has a positive effect on spatial organization and memory later in development, and in adulthood, even if spatial information can be efficiently encoded via knowledge of a projective or a Euclidian space.

In what respect is locomotion constraining the gathering and processing of spatial information ? Here again, we have some interesting data from the literature on humans. Experiments have shown that search strategies of very young children are dependent on the order in which objects have been hidden. For example, Cornell & Heth (1983) found that in 1-year-olds, the serial order of hiding was the most important determinant of search strategies, whereas for the 3-year-olds, this order was inconsequential. We can note that such a process could be equivalent to route learning, which requires an association (correlation) between movement patterns and specific environmental stimuli.

One can tentatively summarize the development of spatial orientation in humans as follows : the first form of spatial coding established is based on the sequence of paths associated with the encountered events. We will call this form "the coding of events", in line with Cornell and Heth (1983). The second form which can be seen by 3 years of age, is characte-rized by its independence of sequential events, and by a coding elaborated on spatial considerations, i.e. the proximity of object, and later on, Euclidian geometry.

SOME COMPARATIVE DATA

In studies with animals, the concept of cognitive mapping was strongly revived by the work of Menzel (Menzel, 1973). His study on memory of objects hidden in different locations, in chimpanzees, is one of the most impressive reports of animals' sophisticated spatial abilities (see also Boesch & Boesch, 1984). Furthermore, this experiment illustrates, quite obviously, some of the components of the map I have presented. The best way to report it is to quote Menzel : "A single member of a group of six chimpanzees was carried about the field, and allowed to watch as an experimenter hid up to eighteen pieces of food in natural cover. On each trial a different set of hiding places and a different route was used. Following this procedure... the experimenters turned the whole group loose... Not only did the informed animals find virtually all of the food (the uninformed group members scoring only by searching where the informed animal did, or begging from it directly), but they also took a route that bore no detectable relationship to that along which they had been carried and which was not vastly less efficient in terms of its overall travel distance than that which they might have followed if all of the foods were visible at the time of response". (Menzel, 1984, p. 518).

Two important features of the chimps' behaviour can be emphasized : firstly, the organization of the paths taken by the animal follows a least distance principle ; secondly, the structure of the animal's path is different from the route along which it had been carried. This last point is crucial for a discussion of the role of locomotor activity in spatial coding, since it suggests that an elaborate form of map is characterized by its freedom from any past locomotor experience. Does this consideration mean that, in some cases, spatial organization can be independent of any prior contact with the spatial layout ? I find this question hard to answer in any definitive manner. The only remark I can make is that the above experiment with chimpanzees is in no way a test of how a map is acquired ; in fact, given that the chimps were moving in a highly familiar environment, their behaviour reflected the rather sophisticated knowledge they had of their surroundings.

It is worth briefly mentioning another important factor in map acquisition - the time frame. Ellen (1980) in his studies with rats on the Maier 3-table task, showed that temporal events associated with locomotion must follow one another fairly closely in time, in order to give rise to simultaneous patterns in the rat's experience.

Recent studies on food-caching birds (tits and especially nutcrackers, Sherry, 1984) reveal that their capacities for remembering hundreds of caches over long periods of time (up to 6 months for nutcrackers) are extensive. Interestingly, the literature reporting the birds' abilities to store and recover food has drawn mostly on the process of "working memory". The behaviour of the birds, however, (in terms of the number of memorized locations and in terms of time intervals) falls outside the scope of that usually described as working memory - i.e. a short-term storing device (Olton, 1979). Despite the fact that few references (e.g. Balda & Turek, 1984) have so far been made in this domain to cognitive mapping, I would suggest that this might be useful.

One advantage of this approach would be to gain better insight into the food-caching and food retrieving processes. It might allow us, for example, to examine more closely the role of self-movement in the bird's memory. As an illustration of my suggestion, I wish to report some observations of Vander Wall (1982). According to this author, nutcrackers which did not hide food themselves, but were allowed to watch the other birds hiding seeds, found caches at levels above chance, but well below the levels of the active birds ; it appears then, that even if the memory seems based mainly on visual information, observing the cache site alone is not sufficient to establish a good memory of its location.

In short, memory is of course necessary to store the set of locations. However, since memory's properties (such as the primacy effect) only partially explain the retrieving behaviour, reference has to be made to an active mapping process in order to understand how information about locations is abstracted and organized.

CONCLUDING REMARKS

There is good reason, then, to interpret the learning of a particular location, irrespective of the route travelled to get there, as a demonstration of cognitive mapping (we saw that the possibility to operate on spatial relations was an essential component of this process , it seems however, that some categories of behaviours requiring travel between two points (such as the homing situation) could be dealt with by alternative hypotheses. At least two mechanisms might be called upon here. The first concerns the use of a compass, which could provide the organism with a powerful orientation system : see for example Baker (this volume) for studies on man and Wiltschko & Wiltschko (this volume) for experiments on birds. The second, and more exciting mechanism, refers to the unique use of vestibular cues to orient adequately. Investigations of short distances homing in hamsters have shown (Etienne et al., 1985, and Etienne, this volume) that these rodents can successfully return to their nest after passive transport, and without the use of visual, acoustical, olfactory, tactile or geomagnetic cues. Such a system is referred to as path integration, and could ressemble the inertial guidance described by Barlow (1964). This field of investigation was pioneered by Beritachvili-Beritov (1963) with studies on rats, dogs and humans ; capacities demonstrated by congenitally blind people to travel over short distances could rely on such processes (see for an illustration, Landau et al., 1981).

A more extensive and detailed understanding of path integration may help to resolve the controversies surrounding the construction of cognitive maps in the absence of locomotor behaviour (see Ellen, 1979). Above all, it might open new perspectives into what Potegal (1982) describes as a "fascinating conjecture", namely the transition from egocentric to purely spatial strategies, at both ontogenetic and phylogenetic levels.

REFERENCES

Balda, R.P. and Turek, R.J., 1984. The cache recovery system as an example of memory capabilities in Clark's nutcracker. In "Animal Cognition". H.L.Roitblat, T.G. Bever and H.S. Terrace (eds), Lawrence Erlbaum, Hillsdale, N.J.

Barlow, J.S., 1964. Inertial guidance system as a basis for animal navigation. Journal of Theoretical Biology, 6 : 76-117.

Benson, J.B. and Uzgiris, I.C., 1981. "The role of self-produced movement in spatial understanding". Paper presented at the Biennal Meeting of the Society for Research in Child Development, Boston, 2-5 April.

Beritachvili-Beritov, I.S., 1963. Les mécanismes nerveux de l'orientation spatiale chez l'homme. Neuropsychologia, 1 : 233-249.

Bremner, J.G., 1982. Object localization in infancy. In "Spatial Abilities". M. Potegal (Ed.), Academic Press, New-York.

Boesch, C. and Boesch, H., 1984. Mental map in wild chimpanzees : an analysis of hammer transports for nut cracking. Primates, 25 : 160-170.

Cornell, E.H. and Heth, C.D., 1983. Spatial cognition : gathering strategies used by preschool children. Journal of Experimental Child Psychology, 25 : 93-110.

Ellen, P., 1979. The hippocampus and operant behavior. In "Precis of O'Keefe and Nadel's The hippocampus as a cognitive Map". The Behavioral and Brain Sciences, 2 : 500-501.

Ellen, P., 1980. Cognitive map and the hippocampus. Physiological Psychology, 8 : 168-174.

Etienne, A.S., Teroni, E., Maurer, R., Portenier, V. and Saucy, F., 1985. Short distance homing in a small mammal : the role of exteroceptive cues and path integration. Experientia, 41 : 122-125.

Feldman, A. and Acredolo, L., 1979. The effect of active versus passive exploration on memory for spatial location in children. Child Development, 50 : 698-704.

Herman, J.F., 1980. Children's cognitive maps of large-scale spaces : effects of exploration, direction, and repeated experience. Journal of Experimental Child Psychology, 29 : 126-143.

Landau, B., Gleitman, H. and Spelke, E.,1981. Spatial knowledge and geometric representation in a child blind from birth. Science, 213 : 1275-1278.

Levine, M., Jankovic, I.N. and Palij, M., 1982. Principles of spatial problem solving. Journal of Experimental Psychology : General, 111 : 157-175.

Menzel, E.W., 1973. Chimpanzee spatial memory organization. Science, 182 : 943-945.

Menzel, E.W., 1978. Cognitive mapping in chimpanzees. In "Cognitive Processes in Animal Behavior". S.H. Hulse, H. Fowler and W.K. Honig (Eds), Lawrence Erlbaum, Hillsdale, N.J.

Menzel, E.W., 1984. Spatial cognition and memory in captive chimpanzees. In "The Biology of Learning". P. Marler and H.S. Terrace (Eds), Springer-Verlag, Berlin.

O'Keefe, J. and Nadel, L., 1978. "The Hippocampus as a cognitive Map". Clarendon Press, Oxford.

Olton, D.S., 1979. Mazes, maps and memory. American Psychologist, 34 : 588-596.

Piaget, J., 1954. "The Construction of Reality in the Child". Basic Books, New York (originally published in 1937).

Pick, H.L. and Rieser, J.J., 1982. Children's cognitive mapping. In "Spatial Abilities". M. Potegal (Ed.), Academic Press, New York.

Potegal, M., 1982. Vestibular and neostriatal contributions to spatial orientation. In "Spatial Abilities". M. Potegal (Ed.), Academic Press, New York.

Tolman, E.C., 1948. Cognitive maps in rats and men. Psychological Review, 55 : 189-208.

Tolman, E.C. and Honzik, C.H., 1930. "Insight" in rats. University of California Publications in Psychology, 4 : 215-232.

Sherry, D.F., 1984. What food-storing birds remember. Canadian Journal of Psychology, 38 : 304-321.

Vander Wall, S.B., 1982. An experimental analysis of cache recovery in Clark's nutcrackers. Animal Behaviour, 30 : 84-94.

Vauclair, J., 1980. Le rôle de la propriomotricité dans l'apprentissage d'un labyrinthe chez le hamster doré. L'Année Psychologique, 80 : 331-351.

DETOUR AND SHORTCUT ABILITIES IN SEVERAL SPECIES OF MAMMALS.

Nicole CHAPUIS

Department of Animal Psychology
Institut of Neurophysiology and Psychophysiology
CNRS-INP 9
31, chemin Joseph-Aiguier
13402 Marseille Cedex 9
FRANCE

This study is concerned with two properties of cognitive mapping. The first is plasticity, by which I mean the ability of an animal to reorganize its previous experience of a given situation. Thus, for example, when modifications are introduced into a familiar spatial task, some animals can find original solutions ; they do not become lost and they can still reach the goal. The second property is optimalization. It entails the choice and the planning of the best adapted solution : for example, taking the most direct of several possible ways to reach a goal.

In addition, I am particularly concerned with the question of how cognitive maps are built up, and what strategies of response various species of mammals employ here. My detour and shortcut experiments were conducted in "free" situations, i.e. in openfields in which no route was laid out, so animals could choose between various available ways to reach the goal. For the most part, they were conducted outdoors, in a natural landscape. The subjects were dogs, cats, hamsters and horses.

Their response strategies were examined in relation to previous information collected by animals in their own displacements : that is to say kinesthetic and visual (etc.) information originated from the landscape.

DETOUR EXPERIMENTS

The first set of experiments consisted of detour tasks, in which animals had to circumvent an obstacle, or to move away from the goal in order to reach it. Such tasks have long been, since Kohler (1925), to constitute tests of insight learning. Detour abilities have been demonstrated in several species of mammals and also in human infants by Piaget (1937) and by Lockman (this book, vol. 2). Several factors have been shown to affect this ability. Among them are whether or not the goal is visible from the choice-point, the nature of the cues whereby the goal can be located, and the length, the number and the complexity of paths leading to the goal.

Our experiments were done on cats (Poucet, Thinus-Blanc and Chapuis, 1983), dogs (Chapuis, Thinus-Blanc and Poucet, 1983), and on horses with Tournadre (note 2). In an openfield, one or several screens were placed between a starting-point and a baited feeding-bowl (figure 1).

Arrangement of the screen(s) was devised so as to leave two routes ; these routes differing in their respective lengths (short or long) and/or angular deviation (narrow or wide) related to the starting-point - goal axis. Two conditions were used : in the first one, the goal could be seen from the choice-point (transparent screen), whereas in the second, it could not (opaque screen). The aim was to assess the part played by these two factors - length and angular deviation - in a visually guided condition (goal visible) and in an "inferential" condition (goal hidden).

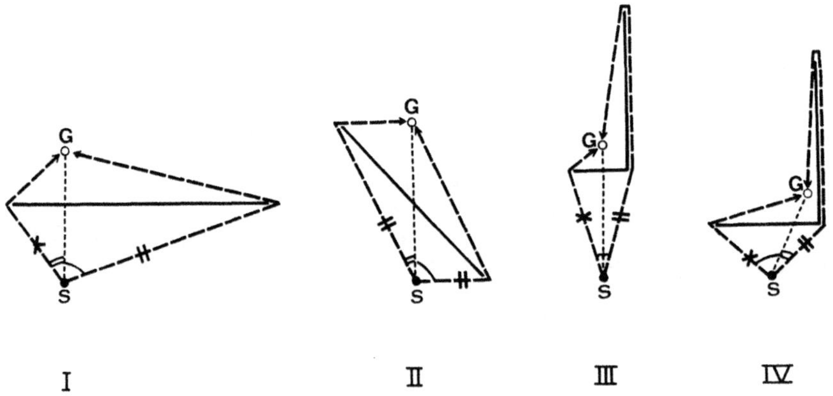

Figure 1. Overviews of the apparatus, with location of the starting-point (S), feeding-bowl (G), screen (solid) and routes (dashes) in the four situations. Adapted from Chapuis *et al.*, 1983 ; copyright The Experimental Psychology Society.

Four experimental situations were presented to the animals, as shown on figure 1, in the two conditions : goal visible and goal hidden.

SITUATION I

Angular deviations and lengths were both relevant and convergent, *i.e.* the shorter way was the less divergent and the longer way was the more divergent from the goal direction.

SITUATION II

The two ways differed by the angular deviations but were similar in length.

SITUATION III

The angular deviations were equal but the lengths were different.

SITUATION IV

Angular deviations and lengths were both relevant but opposed : the shortest way was the more divergent and the longer way was the less divergent.

Subjects could reach the feeding-bowl by either of the two ways and were allowed to eat food regardless of the way they had gone. Prior to the beginning of testing, the animals were trained to run from a starting-point straight to a feeding-bowl which was not blocked by a screen. Following this initial training period, the screen was introduced. Subjects were tested for eight days with twelve trials per day. After six daily trials, the location of the screen and the feeding bowl was changed symmetrically about the axis defined by the starting-point and the goal. Each way (as defined by the parameters in figure 1) appeared equally often to the right and the left of the symmetrical axis. At the beginning of the daily test, before the first trial, and after the change of the location of the screens (before the seventh trial), the animals were led from the starting point to the goal along each of the two routes in order to show them the place of the feeding-bowl behind the screen.

SPECIES STUDIED	PARAMETERS STUDIES	Angular deviation and length SITUATION I	Angular deviation SITUATION II	Length SITUATION III	Angular deviation versus angle SITUATION IV
	WAY CHOSEN	less divergent and shorter way	less divergent way	shorter way	shorter and more divergent way
CATS	visible goal condition	81.25 % ***	70 % **	65 % **	51.25 % no preference
	hidden goal condition	75 % ***	93.75 % *** (w/out distal cues : 62.5 %) *	83.75 % ***	85 % ***
HORSES	visible	81.10 % ***	72.2 % *+	76 % **	58 % no preference
	hidden	92 % ***	66 % *	64 % *	63 % no preference
DOGS	visible	95 % ***	71.7 % +	53.3 % no preference	50 % no preference
	hidden	96.7 % ***	45 % no preference	75 % +	68.3 % +

(Student's t test : $^{+}$ p < .10 ; * p < .05 ; $^{*+}$ p < .02 ; ** p < .01 ; *** p < .001)

TABLE I. Percentage of trials on which the animals chose various different ways of circumventing an obstacle.

The general procedure of the experiment was the same for the three species. But, of course, the test-situation itself was modified (for example in size) to fit each species. The experimental frame also was different since the eight cats were tested in a large indoor room while the six dogs and the nine horses were tested outdoors : dogs in a large meadow and horses in a paddock.

As Table I shows, in situation I cats selected the shorter and less divergent path, whether or not the goal could be seen. In situation II, where the angular deviation from the goal was the only factor, cats used the available parameter in the both conditions. Two tests were done with the opaque screen. In the first one, the visual cues of the room were available while in the second one they were not available because of black curtains on the walls. The trend to choose the less divergent way was greatly reduced by reference to the scores obtained previously. So the visual cues seemed to be used by cats to localize the place of the goal hidden behind the opaque screen. In situation III, the shorter path was preferred in the both conditions although a significant difference appeared between the results in the two conditions. Lastly, in the conflictual situation - situation IV - a preference appeared for the shorter way only when the goal was hidden. So, both angular deviation and length were relevant parameters for this experiment when they were alone or associated. When they were opposed, the shorter way could not be selected when the goal was perceived from the choice point. Thus the visible goal appeared to act as a "perceptual anchor" and to eliminate the use of distance information.

The results of horses were similar to cats' responses except in situation IV where they did not show any preference even when the goal was hidden.

In contrast, the behavior of dogs was a bit different since, on the whole, opposed behavioral strategies were used by the same subjects : the dogs preferred the shorter way when the goal was hidden but not when it was visible. They more often chose the less divergent way when the goal was visible than when it was not. Their performance in situation II, in the hidden goal condition, seems surprising since shortcut results (to be presented later) have shown that dogs were able to evaluate accurately the direction of a non-visible goal. The difference between dogs and cats (and horses) can most likely be attributed to differences in the number of environmental stimuli available to the animals. The cats were tested in an empty room with differentiated walls ; a large window was at South, a blind wall at West, a small window at North, and a door at East. The horses were tested in an area which has a fence and several buildings around it as stimuli. The field in which the dogs were tested was devoid of prominent visual distal landmarks. To assess accurately the correct position of the hidden goal was rather complex for dogs. It is likely to imply the use of an egocentric reference framework, the body axis, which was oriented towards the middle of the screen, at the starting-point, to know the position of the goal (left of right) with regard to this body axis. The lack of preference of dogs in situation III when the goal could be seen in the middle behind the screen support our previously expressed idea of a "perceptual anchor".

Thus, two classes of mechanisms appeared to be involved when the goal was hidden and when it was visible. When the location of the hidden goal must be remembered, animals were able to take into account and integrate the information collected during the prior exploratory runs in order to plan the shortest route. The direct visual perception of the goal, on the other hand, allowed the animals to use a guidance strategy that led the subject to reduce the distance between itself and the goal as quickly as possible, which, in this case, was the simplest strategy.

SHORTCUT EXPERIMENTS

The second set of studies investigated the animals' ability to take shortcuts. The first experiment (Chapuis and Varlet, note 1), was carried out with seven dogs in a 3 hectare meadow. The animals' task was to find meat hidden at two points (A and B), when released from a third (D) (figure 2). On each trial, the dog had been taken previously on a leash to the two baited points by a different path that led indirectly from A to B via D, i.e. along the DA-AD-DB-BD route. They were shown the food at points A and B without being allowed to eat it.

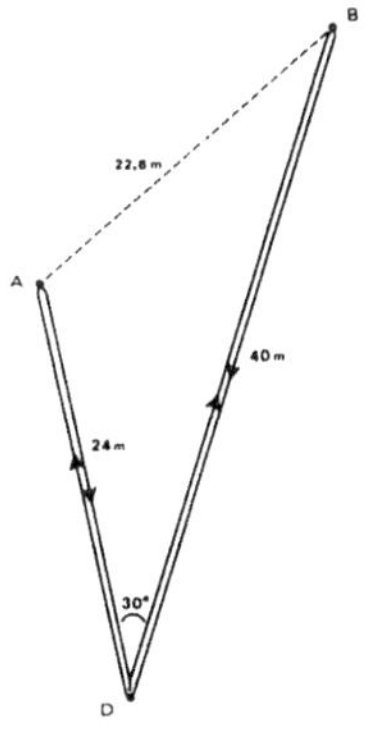

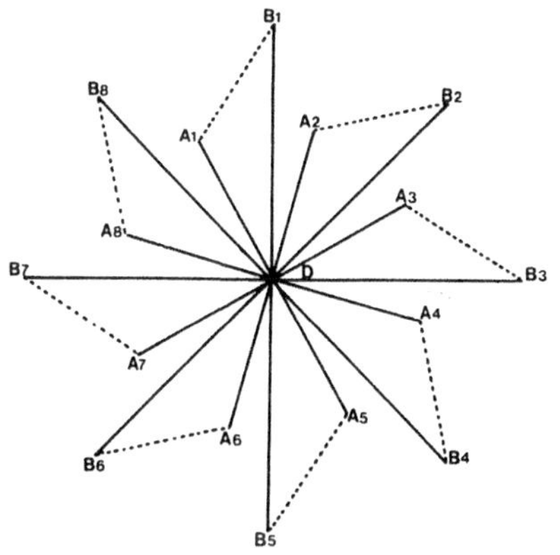

Figure 2. One standard experiment situation. Course of one trial. D corresponds to the starting-point, A and B to food-points. DA-AD-DB-BD vectors (solid) represent exploratory runs (dogs on the leash) and AB segment (dotted) the optimally oriented task run ("direct shortcut"). From Chapuis and Varlet, note 1 ; copyright The Experimental Psychology Society.

Figure 3. Experimental set-up showing the arrangement of the eight standard situations (DA_nB_n) on one field (ground HET). From Chapuis and Varlet, note 1 ; copyright The Experimental Psychology Society.

The experiment was carried out in two different fields, one nearly uniform - it was covered mainly with thyme - and the other heterogeneous, containing visual features such as bushes, puddles and trails. Within each field, each pair of baited points was changed eight times to different A and B places as shown in figure 3. In 96 % of the 224 trials, the dogs took a shortcut between the baited points rather than returning to the start.

A qualitative analysis of the characteristics of the shortcut trajectory induced me to distinguish between the "direct shortcut" (strategy I) which consisted of a straight line between A and B and the second and the third classes, which included trajectories making a smaller or a wider angle than the angles DAB or DBA ; so the path was either inside the triangle DAB, and corresponded to a strategy II or "inside shortcut", or the path was outside DAB, and corresponded to a strategy III or "outside shortcut". Strategy IV consisted of following the indirect route ADB (or BDA) which had been shown to the dog in the exploration phase. As can be seen in figure 4, the dogs used strategy I more frequently in the uniform than in the heterogeneous field.

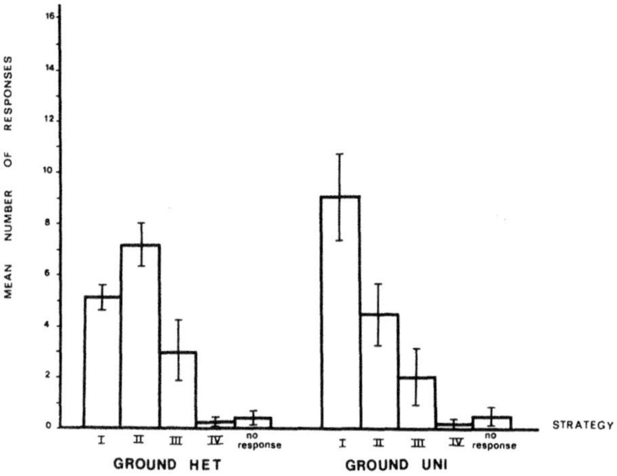

Figure 4. Frequency of different classes of responses (± S.E.M.) produced by the seven subjects tested 16 times on each type of field, heterogeneous (HET) and uniform (UNI). I : direct shortcut, II : inside shortcut, III : outside shortcut, IV : return. From Chapuis and Varlet, note 1 ; copyright The Experimental Psychology Society.

In the litter field, the animals had a greater tendency to use an "inside shortcut" (strategy II), heading for a point which intersected the path from the start to the second baited point rather than going directly to the baited point or returning to the start.

The shortcut responses presented in figure 5 are an example of the trajectories of one subject on the two grounds, heterogeneous and uniform. With this dog, shortcut responses were distributed only between strategies I and II.

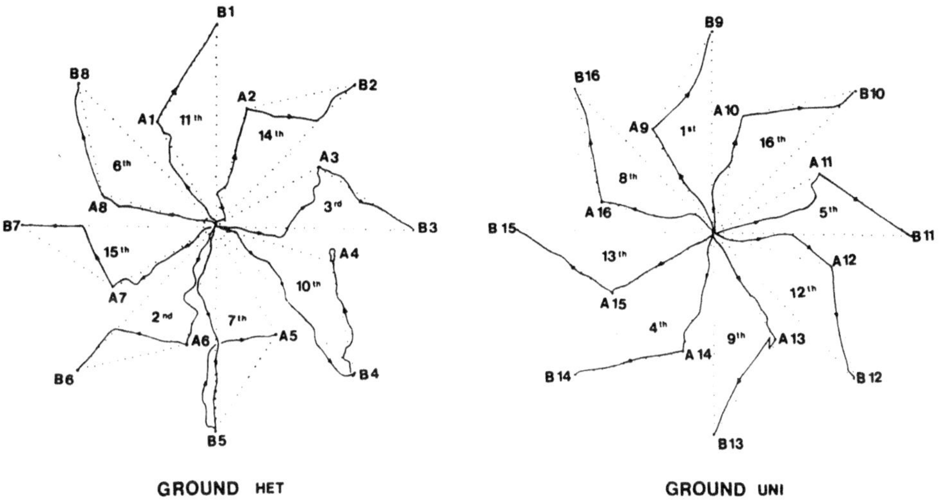

Figure 5. The map of the sixteen trials performed by one subject (number 3) which used the strategy of the "direct shortcut" in situations A_1B_1, A_3B_3, B_4A_4 and A_8B_8 on ground HET and in the eight situations on ground UNI. This subject used the strategy of the "inside shortcut" on the four other situations on ground HET, i.e. A_2B_2, B_5A_5, A_6B_6 and A_7B_7. The numbers noted in each DA_nB_n triangle correspond to the temporal order in which tests were presented to the subject. From Chapuis and Varlet, note 1 ; copyright The Experimental Psychology Society.

The trials in which the shortest path was taken suggest that dogs are able to evaluate, along a continuum, the direction of an invisible goal by integrating motor and/or sensory cues obtained during an early but indirect visit to the same goal.

On the other hand, I do not think that the use of the inside shortcut is due only to a misestimation of the direction of the second goal, since (if such were true) the distribution of exterior and interior shortcuts would be equivalent. The dogs' behavior could be interpreted as a compromise between returning to the start and heading for the second goal. From an adaptive point of view, it could correspond to the pursuit of safety. Depending on the meaning of the word safety, we can suggest two clearly related hypotheses. According to the first hypothesis, the probability of finding the line DB is greater than the probability of finding point B if the dog is not certain of the direction of the goal B when it is at the goal A. According to the second view, in using this strategy, the dog is able to find known territory by the shortest possible new way, without moving away from the goal as would be the case if it had used the path ADB. Observations of human orientation incline one towards the latter two hypotheses. Studies by Pailhous (1970) have shown, for example, that when Paris taxi-drivers found themselves in an unknown area, they did not attempt to drive directly towards their goal, but took the shortest possible route to reach a well-known major route, which they used to help them towards their goal.

It was also observed that dogs' shortcut responses were more accurate in the uniform field than in the heterogeneous one since direct shortcuts were more often taken than inside shortcuts in the first case and vice-versa in the second case. To encode the location of the two goals A and B, the dog must attend either to distal cues, to information obtained from locomotion (kinesthetic and visuo-kinesthetic cues), or to an inter-relation between them. In the heterogeneous field, they could of course also attend to proximal cues during their previous run on the path DA-AD-DB-BD. Note that no cue was directly related to the goals. So, in the uniform field, while at goal A, dogs had to decide the most direct way to take on the basis of the kinesthetic information and/or distal cues acquired during the previous exploratory run. Conversely, when the field contained landmarks, they did not need to work out the precise direction of the second goal B, since they could also use immediate local landmarks on the path DB.

A complementary experiment was carried out with six naive dogs. Here, we varied the size of angles DAB and DBA during the experiment. This experiment was conducted in a heterogeneous field. The results are consistent with the previous results since dogs took the inside shortcut in 38 % of the responses, and the direct shortcut only in 21 % of the cases.

On the other hand, in shortcut experiments with hamsters conducted in a wheel-maze (Chapuis and Lavergne, 1980), a strategy very similar to the inside shortcut was observed. We called it "intercepting strategy". It occurred in the case where the task was too difficult to be solved by a direct shortcut, that is when the previous exploratory displacement was very complex and also when no environmental cues were available.

Behavioral constants can thus be found among various species of mammals ... even including humans. Even if a species has the ability to integrate information in order to take the most direct or the shortest route to a goal, the animals do not necessarily use that ability. They may prefer to choose the surest route (one with which they never loose sight of the goal, for instance) or to head for a route to the goal with which they are already familiar.

All these results show the plasticity of the behavior of the mammals studied and also point to an optimalization of the responses, since the subjects succeeded in detour and shortcut tests. The concept of optimalization needs however to be enlarged beyond its classical acceptation in orientation studies, which is the taking of the shortest way to reach the goal, implying the saving of time and kinesthetic energy. Factors contributing towards minimization of the risks, such as the risk of becoming lost or vulnerable to predators, should also be taken into account. The notion of "least risk" does not always correspond to the "least effort" principle discussed by Tolman (1932), since the trajectories of animals in a given task sometimes cannot be analysed only with physical criteria such as the shortest time or the shortest distance.

In short, we can say that in the present study the animals adapted their behavior depending on their perceptual and cognitive abilities, their motivations, the variety and complexity of the landscape and whether or not they were given the possibility of acquiring kinesthetic information as well as information provided by landmarks.

REFERENCES

Chapuis, N. and Lavergne, F. 1980. Analysis of space by animals. Visual and proprioceptive cues in complex spatial tasks by hamsters. Neuroscience Letters Supplement, 5 : 171.

Chapuis, N., Thinus-Blanc, C. and Poucet, B. 1983. Dissociation of mechanisms involved in dogs' oriented displacements. Quarterly Journal of experimental Psychology, 35B : 213-219.

Kohler, W. 1925. The mentality of apes. New York : Harcourt Brace Co.

Pailhous, J. 1970. La représentation de l'espace urbain. L'exemple du chauffeur de taxi. Paris : Presses Universitaires de France.

Piaget, J. 1937. La construction du réel chez l'enfant. Neuchâtel : Delachaux et Niestlé.

Poucet, B., Thinus-Blanc, C. and Chapuis, N. 1983. Route-planning in cats related to the visibility of the goal. Animal Behaviour, 31 : 594-599.

Tolman, E.C. 1932. Purposive behavior in animals and men. New York : Appleton-Century-Crofts.

Notes

1. Chapuis, N. and Varlet, C. Shorcuts by dogs in natural surroundings. Quarterly Journal of experimental Psychology (In press).
2. Tournadre, V. 1984. Analyse des paramètres impliqués dans les déplacements orientés chez le Cheval à travers des expériences de détour. Mémoire de Maîtrise de Psychologie, Université de Dijon.

Acknowledgments

These experiments were carried out on the grounds of the Centre Régional d'Elevage et de Production d'Animaux de Laboratoire in Rousset, France.
I would like to thank Professor Emil Menzel from the State University of New York, and Dr. Charlene Wages, from Georgia State University in Atlanta for their suggestions on this manuscript.

MOVEMENT THROUGH SPACE AND SPATIALLY ORGANIZED BEHAVIOR

Jeanne M. Stahl, S. Bernard Bagley,
James L. McKenzie, and Phyllis L. Hurst

Morris Brown College, Atlanta University Center

A renewed interest in the concept of cognitive maps (Tolman, 1948) has been stimulated by recent behavioral and neuropsychological research (Stahl & Ellen, 1973, 1974; O'Keefe & Nadel, 1978). Of particular interest is the question of how cognitive spatial maps are acquired (Olton, 1977). The literature suggests that movement through space is crucial for efficient performance by organisms on spatial tasks which require the establishment of an hypothesized "cognitive map," or "orienting schema" (Neisser, 1976). Our research with the Maier 3-table spatial problem demonstrated that exploration of the entire apparatus was necessary for rats to solve the problem (Stahl & Ellen, 1974). Exploration of only portions of the apparatus is not sufficient (Ellen, Soteres & Wages, 1984). A clarification of the role of movement and exploration in spatially organized behavior would advance significantly our understanding of cognitive processes. Thus, the experiments reported here were designed to determine whether active exploration or merely passive movement through space is necessary for solving the Maier 3-table spatial problem.

GENERAL METHODS AND RESULTS

For this and subsequent experiments, subjects were young adult male Long-Evans hooded rats bred in the college animal facility. The apparatus was the Maier 3-table task which was developed to test the ability of rats to combine information learned during two isolated experiences to reach a goal (Maier, 1929). The apparatus was similar to that shown by Ellen in this volume. For standard testing, animals were given fifteen minutes each day to explore the entire three tables and elevated runways as a group. During exploration, no food was available on any table. The second daily experience involved feeding the rats on one of the three tables. The animals were then tested individually by placing them on one of the two remaining tables and observing their response. The response was scored as correct only if the rat ran directly to the food table. The feeding and start tables differed each day so that an animal could not solve the problem using a consistent turn or place response across days of testing. Each day, the animal had to recall the spatial relationships of the unbaited tables, integrate that information with the recently-acquired knowledge that the particular table on which it was just fed was now baited, and reach the goal table directly from the start table.

The first study involved thirty adult rats tested in five experiments designed to determine under what conditions exploration is crucial to the solution of the 3-table problem. Briefly, previously naive rats which were fed and tested once daily on the task without the opportunity to explore the apparatus performed at chance levels (50% correct). Two additional learning tests were added, with corrections allowed to provide additional movement around the apparatus daily. The animals still failed the first daily test, which required the use of cognitive spatial skills, but passed the second and third tests daily, which were essentially learning trials. This data clearly showed that the feeding experience alone was not sufficient for solving this problem even after the rats had a considerable amount of experience running to and from the different tables during daily tests. Even 102 days of this type of partial exploration of the apparatus did not provide an experience or information equivalent to what a rat gains from exploring the entire apparatus within a 15-minute period of time.

Analysis of response data revealed that most subjects had developed a consistent turn response. Allowing a correction following an error, and thus inadvertently rewarding a simple turn response, could have encouraged a learning through association which precluded a more complex cognitive solution to the problem. To eliminate this possibility, a second group of naive rats was tested with no exploration but was also not allowed to correct an error. Thus, animals were not rewarded for using a turn strategy. Again, these rats did not solve the problem without exploration even though they were rewarded only on days on which they ran directly to the correct table. Thus, rewarding the animals for a constant turn could not have interfered with a correct solution in this case.

We repeated our experiments using two groups of experienced rats which had performed well on this test when given standard testing which included daily exploration. We withdrew the daily exploratory experience. After 54 days of testing, exploration was reinstated for 30 days. Results showed that both groups of rats which had performed well when initially given testing with exploration subsequently performed at only chance levels when the exploratory experience was eliminated. They could never pass the test again. The figure below shows that animals that never regained the ability to pass the first (cognitive) test daily performed quite well on the third test, thus demonstrating that subjects were motivated to run to a food reward and retained the ability to learn the path to the food within the daily test session. Looking for a clue as to why these animals had lost the ability they once had to solve this problem, we examined the types of

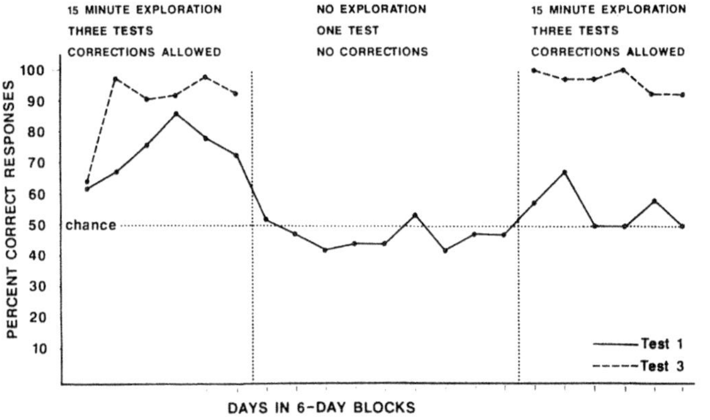

Performance of rats given standard testing (exploration, feeding, 3 tests with corrections) and then given a month without tests, then tested with no exploration, 1 test and no corrections, tested again using standard testing

responses and errors made and latencies and running speeds. Four seconds is about as fast as a rat can run the eight feet between the start table and the feeding table. In the present study, rats allowed to explore the apparatus prior to testing had combined latency/running speeds of 4 to 5 seconds. In contrast, the next figure shows that subsequent tests without exploration resulted in a marked increase of nearly 300 percent. These long latencies were due to a combination of increased latencies to leave the start table and hesitation at the intersection of the three runways (choice point). When exploration was reinstated prior to testing, the latency/running scores returned to baseline levels (4-5 seconds) even though the animals continued to perform at chance levels. An analysis of the type of errors and responses made showed that subjects developed a turn preference which persisted even when exploration was reinstated and the problem again became solvable. This suggested to us that once the animals had the experience of exploring the apparatus, they then knew the tables were connected and that food was available on one of them, but without exploration before feeding the rats did not know which table was without food since they had been fed on a different table each day. In this situation, they were essentially hunting for the food during the test. The rats developed the strategy during the no-exploration condition to optimize reinforcement since turning in one direction was rewarded fifty percent of the time. The strength of intermittent reinforcement maintained this turning response even when exploration was reinstated and a cognitive spatial solution again became available. This suggested that the simple body turn, which eventually led to reinforcement, became a habit which was highly resistent to extinction and one which later interfered with the use of a more efficient response. However, the fast latency/running speeds during the final 36 days of testing with exploration suggested another interpretation. The turn habit did not interfere with a cognitive solution to the problem. It merely controlled the response made to reach the food. A distinction is being made here between a cognitive and a performance deficit. The animals retained their ability to locate the food in space using a cognitive representation of the layout of the tables and the knowledge gained from the most recent feeding experience. However, they persisted in using the now well-developed turning response for a combination of reasons. (1)Due to the large number of daily tests without exploration, the turn response which the animals adopted had developed into

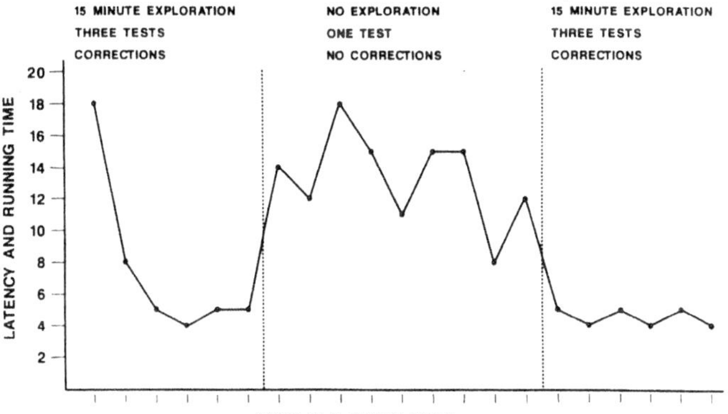

Latency and running time for rats given standard testing (exploration, feeding, 3 tests with corrections) and then tested with no exploration then tested again using standard testing

a persistent stereotyped response that was highly resistent to extinction. (2) This stereotyped response was maintained by reinforcement since either one or two turns in the same direction always led to a food reward. This was possible since we allowed an animal to correct a response following an error. (3) Even when the first turn led to the control table (no food), another turn in the same direction led to the food table. Two sequential turns took only about 8 seconds. This speed of access to the reward (4-8 sec) contrasted favorably with the animals' recent experience without exploration. During tests without exploration, rats took 8-18 seconds to reach the first table and at least another 4 seconds to correct an error on half of the tests. Thus, an already highly persistent body turn response was maintained when exploration was reinstated by reinforcement and the speed of access to reinforcement. This interpretation is consistent with other reports of the striking persistence of a well-learned response, even one which delays reinforcement. For example, hungry rats ran right over a pile of food in order to repeat a learned response to reach a goal and to be rewarded by a much smaller .15gm pellet of food (Stolz & Lott, 1964).

Although the above experiments demonstrated that exploration is necessary for a rat to solve the 3-table problem, it did little to clarify the function of this experience. The question remained: Is active locomotion necessary for solving this problem or is movement over the apparatus sufficient for solving this spatial problem? Thus, a second series of experiments involved a comparison of active vs passive transport across the 3-table apparatus and performance on the test. First, we wanted to determine whether rats can learn the spatial arrangement of the three tables and runways without locomotion but merely through passive movement across the runways and tables. Twenty naive rats were tested in several experiments. Six control animals given standard testing with exploration of the apparatus performed at between 85-100 percent correct. These animals provided model exploration patterns for animals in the passive transport groups who were individually transported by the experimenter in one of two styles of small carts. Their exploration exactly matched that of a control animal in terms of the order in which tables were visited.

Results demonstrated quite dramatically that passive transportation does not provide sufficient information for rats to solve the 3-table problem. All animals which were transported passively across the apparatus performed at the chance (50%) level even when their passive movement matched identically (in pattern and timing) the movement of animals which actively explored the apparatus and performed well on the test. The size of the transport cage, visual access to the surroundings, correction procedures, and time-out following incorrect responding did not enhance the performance of rats which had only moved passively across the apparatus. Furthermore, the active locomotion that they experienced during the test trial over thirty days of testing was not sufficient for them to solve the problem. This data indicated to us that active locomotion may be necessary initially for the formation of a cognitive representation of the layout of the test environment. It appears that passive movement through space is insufficient for the animal to form a cognitive representation of that space and that the animal must actively explore the environment before such a representation becomes important and available to the animal.

We hypothesized that once the animal has acquired a cognitive representation of the environment, as inferred from successful performance on the 3-table test, exploration may serve merely to inform the animal that all tables are unbaited and that food experienced previously on these tables is no longer there. This might be necessary to allow the most recent feeding experience to gain prominence and to prevent the rat from directing its behavior toward the previous day's food table. Thus, the purpose of the next experiment was to determine whether experienced rats which had actively explored the apparatus and had performed well on the 3-table test would maintain that performance when active locomotion was eliminated and replaced with passive transportation across the apparatus.

Six rats were allowed daily exploration of the 3-table apparatus and were tested for 30 days. After demonstrating a high level of performance on the test, they were placed into the passive transport condition. The data in the next figure hows that all subjects performed well when allowed active locomotion on the apparatus. However, their performance deteriorated when the passive transportation procedure was implemented. Only one rat performed at above chance during the twelve days of passive transport testing. Although we limited the passive transport testing to twelve days and did not allow a correction following an incorrect choice, most of the subjects were beginning to turn in a consistent direction by the end of the twelve days. When returned to standard testing with active locomotion allowed, they again performed at a high level and appeared to have regained their use of the spatial strategy.

For consistency of design, the animals in the above study had passive transportation matched to the same control animals which were used in the previous study. Since one rat did perform above chance in this condition, it occurred to us that the other animals might perform better if their transportation was matched to their own active locomotion patterns. This experiment is in progress at the time of this writing. The six animals used in the above passive transport experiment are being given additional passive transportation which, in this experiment, exactly matches their own active locomotion patterns recorded during previous tests which allowed the animals to explore. Four out of six of the animals now being tested are performing at well above chance levels of performance. This provides strong support for the notion that active exploration is necessary for the initial formation of the spatial map. Following its formation exploration merely serves to get the animals to the tables so they will have the experience of finding no food on any of the tables prior to the feeding experience. Once the animals have formulated a cognitive map using active

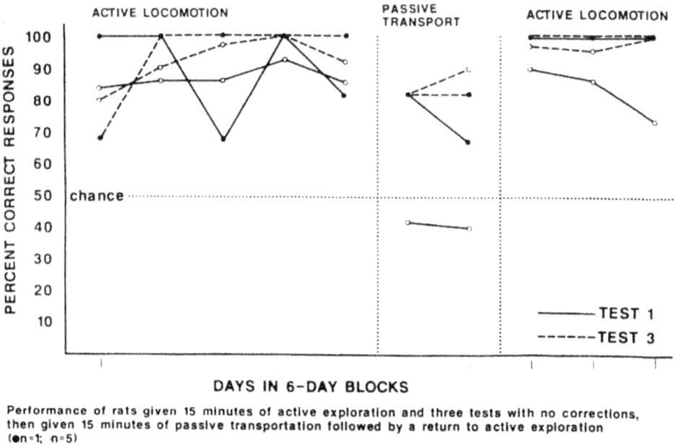

Performance of rats given 15 minutes of active exploration and three tests with no corrections, then given 15 minutes of passive transportation followed by a return to active exploration (●n=1; ○n=5)

exploration of the environment, passive transport followed by a feeding on one of the tables was sufficient for them to solve the problem.

SUMMARY AND CONCLUSIONS

In summary, the experiments presented in this paper lead us to the following conclusions. Movement through space has several functions which may change with the exposure of the rat to the test apparatus and the spatial problem. Initially, a previously naive rat learns, through active locomotion across the three elevated runways and tables, that the tables and runways are connected. At first the animal may merely learn of the connectedness of the tables but may not have established a cognitive representation of the layout of the specific tables and runways. This becomes necessary only when this information is required to solve a problem. A food-deprived rat next learns that all three tables are equal with respect to food--no table has food on it. It is not until the rat is fed on one of the tables and then moved to a table without food on it that a cognitive representation (map) of the layout of the specific tables and runways becomes essential to reaching the goal--finding food. This explains why some rats do not perform well on the test until the second or third day of testing. A goal (food) is necessary for providing a context in which it becomes important to remember the specific layout of the environment. We assume that some food-deprived rats will form a cognitive representation of the environment during the first exploration experience as a part of a food-hunting strategy. Other rats may do so as part of a strategy to satisfy their general exploratory drive or curiosity. This is suggested by the fact that some rats do perform well from the first day of testing. Usually, all normal rats can find the food directly once they have had the experience of exploring the three unbaited tables and runways and then being fed on one of the tables. We are suggesting that it is the daily feeding and testing experience which results in a strengthening of this cognitive representation of the environment and its storage in long-term (reference) memory for future use.

It would seem that, once this cognitive representation is well formulated and performance based on its use has been rewarded, the exploratory experience would no longer be necessary to solve the 3-table problem. Our preliminary data indicate that this is the case. We suggest that an experienced rat no longer requires active exploration to form or to maintain the cognitive representation of the test environment and to learn that the tables and runways are connected or that food is available within the environment. This was clearly suggested by the data showing that, when experienced rats had the exploratory experience eliminated, they still ran to a table within 8 to 18 seconds after the start door was opened. They would not have run this quickly if they did not know the table they were on was connected to a table with a reinforcer on it. The fact that these rats hesitated longer at the choice point indicated that they did not know which table had the food on it rather than whether food was available to them on one of the three connected tables. This data indicated that exploration functioned merely to expose the animals to the three unbaited tables to inform the rat that food was no longer available on any table prior to eating on a specific table. In other words, exposure to the unbaited runways and tables, not active exploration, is still necessary for an experienced rat to solve the problem since this experience informs the rat that no table is baited so that the memory of the food table for today, as opposed to yesterday, commands the rat's attention as the only table with food on it in an environment in which the other tables are no longer baited. This interpretation is consistent with a large body of data which demonstrates the resistence of spatial memory to forgetting and with Olton's (1977) suggestion that rats have the ability to reset working memory at the end of a trial on the eight arm radial maze. However, on the 3-table problem, it appears that the rat must change or disregard a detail in reference memory and depend on information in working memory. Without exposure to the unbaited tables, the animal does not have any way to know that the information in reference memory concerning the previous days feeding table is no longer useful. Exposure to the three unbaited tables informs the rat that all tables are now without food, making the next feeding experience more prominent and useful to the rat. We expect to test some of these ideas with experiments designed to provide experienced rats with the information necessary to solve the problem without active exploration of the apparatus and by manipulating the time between tests to see if the spacing of tests influences the usefulness of today's feeding experience in solving the 3-table problem.

References

Ellen, P., Parko, E. M., Wages, C., Doherty, D. and Hermann, T., 1982. Spatial problem solving by rats: Exploration and cognitive maps. Learning and Motivation, 13:81-94.

Maier, N. R. F., 1929. Reasoning in white rats. Comparative Psychology Monograph, 6:1-93.

Neisser, U., 1976. "Cognition and Reality". W. H. Freeman, San Francisco.

O'Keefe, J. and Nadel, L., 1978. "The hippocampus as a cognitive map". Clarendon Press, Oxford.

Olton, D. S., 1977. Spatial memory. Scientific American, 236:82-98.

Stahl, J. M. and Ellen, P., 1973. Septal lesions and reasoning performance in the rat. Journal of Comparative and Physiological Psychology, 84:629-638.

Stahl, J. M. and Ellen, P., 1974. Factors in the reasoning performance of the rat. Journal of Comparative and Physiological Psychology, 87:598-604.

Stoltz, S. B. and Lott, D. F., 1964. Establishment in rats of a persistent response producing a net loss of reinforcement. Journal of Comparative and Physiological Psychology, 57:147-149.

Tolman, E. C., 1948. Cognitive maps in rats and men. Psychological Review, 55:189-208.

This research was supported by a United States of America Minority Biomedical Research Support and National Institute of Mental Health grant (RR8006) to the first author.

STUDY OF COGNITIVE PROCESSES USED BY DOGS IN SPATIAL TASKS.

C. FABRIGOULE

C.N.R.S., I.N.P.9, 31, chemin Joseph-Aiguier, 13402 Marseille Cedex 09, France.

In this paper, I propose to describe one part of a research program concerning the processes through which animals gain knowledge of space. The behavioral adaptation of dogs to relatively unconstrained spatial learning tasks was studied in order to find out what animals' actions can reflect about the type of knowledge they are able to acquire concerning their surroundings. Before describing any dogs' results it is necessary to make some remarks about the principal characteristics of these experiments.

1. The domestication of dogs allowed us to use spatial learning tasks with no apparatus, and in which the movements of dogs are unconstrained. These tasks seemed more natural since the presence of apparatus must modify the way in which an animal perceives the spatial environment and limit considerably the movement it can make.

2. All the tasks had multiple solutions, that is to say the animal could obtain the maximum reward in a number of different ways. A particularly interesting characteristic is the fact that these tasks allow individual differences to appear, which would remain hidden in tasks with a unique solution.

3. The data presented here will not include group comparison but analysis of the evolution of individual subjects' behaviour.

4. The dogs used in a given experiment were from the same litter (generally consisting of six siblings). The experiments began when they were three to six months old. The animals were not deprived of food but fed only once a day.

EXPERIMENT I

The first experiment (Fabrigoule, 1974) was a progressive elimination of food experiment. We used the same procedure as Lachman & Brown (1957), Lachman (1971) and Buhot & Teule (1971), but without any alleys in the apparatus. The set-up (Figure 1) consisted of four identical food-bowls placed at equal distances on the arc of a circle with the starting-position at the center. At the beginning of each daily session all the bowls were baited with pieces of meat. The subject was then taken to the end of a short tunnel which led to the starting-point and was released. As soon as the subject had eaten the meat in one bowl it was taken back to the end of the tunnel. If he returned to a bowl that he had already emptied, he was allowed to go to another bowl until he found one which still held meat and so on until he had emptied the four bowls.

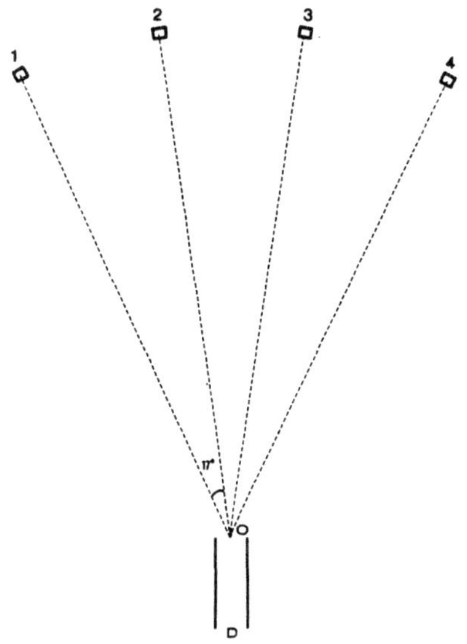

Figure 1 : Four-bowls elimination apparatus. 1, 2, 3, 4 : bowls ; D : release point ; O : choice point.

	% of return visits to peripheral bowls	% of return visits to central bowls	% of coming back to the departure
A	62,3	18,6	19,1
B	82,6	14,9	2,5
C	82,1	9,8	8,1
D	87,2	8,8	4,0
E	82,8	13,2	4,0
F	87,6	10,9	1,5

Table I : Percentages of return visits to peripheral and central bowls.
A, B, C, D, E, F : dogs.

The dogs all reached a criterion of three or less errors in nine sessions after 16 to 41 days of training. These animals reached the criterion relatively slowly because after the third day, they developed a tendency to retrieve the food first from the peripheral bowls and then to return to the peripheral bowls before visiting the central ones. Table I shows that the percentage of errors made on the peripheral bowls was much higher than that made on the central bowls. Given this tendency, we also looked to see whether the dogs alternated their visits to the two peripheral bowls. Table II shows that alternation was very strong in the second trial of the session, and the same tendency appeared in the third trial, whether or not the animal had alternated in the second trial. No systematic alternation was observed however, at the beginning of training, as can be seen in figure 2. At the start of training, repetitions were observed which decreased after the third day. The alternation tendency was thus the first result of learning. Buhot & Teule (1971) have observed a similar tendency at the beginning of training. We examined the sequences of visits made when the animals had reached the criterion and during overtraining. Figure 3 summarizes all the different successful sequences used after reaching the criterion. Here no stereotypy was found but there was some regularity in that individual dogs often used different orders from one experimental unit to another apart from the first two visits, which were usually made to the same two bowls.

	44 + 11	41 + 14	444 + 111	441 + 114	411 + 144	414 + 141
A	21	38	6	13	5	12
B	11	76	3	6	6	21
C	14	72	3	9	4	27
D	11	94	3	6	8	20
E	10	93	0	10	3	39
F	13	91	1	12	1	20

Table II : Comparison of repetition and alternation behaviour in the second and third trials.

EXPERIMENT II

The set up of the second experiment (Figure 4) consisted of six food-points arranged in a circle with a radius of 30 m about the starting point. D5 is the access route used by the experimenter to lead the dog to the starting point. For six days, each individual dog was led to the center and then from the center to each food point (1, 2, 3, 4, 5, 6), where he could eat the food previously put there, and back to the center. Thus, in the course of pretraining, dogs experienced only the six radii going from the center to the food points. On the seventh day and for the following 34 days the experimenter led the dog to the starting point, unleashed it and remained at the starting point till the dog had visited every food point. With this procedure we wanted to know if the dogs, after preliminary exploration of the radii alone, would be able to reorganize the information collected and replan their movements to find a shorter path leading to every food point.

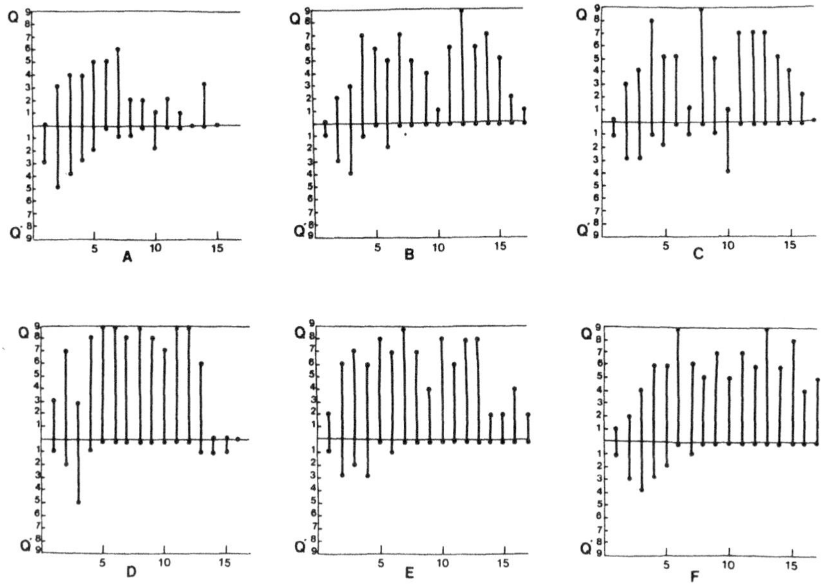

Figure 2 : Evolution of repetition and alternation in the second trial of an experimental unit. X : 16 blocks of 9 experimental units. Y : Q = alternation (14 and 41); Q' = repetition (11 and 44).
A, B, C, D, E, F : dogs.

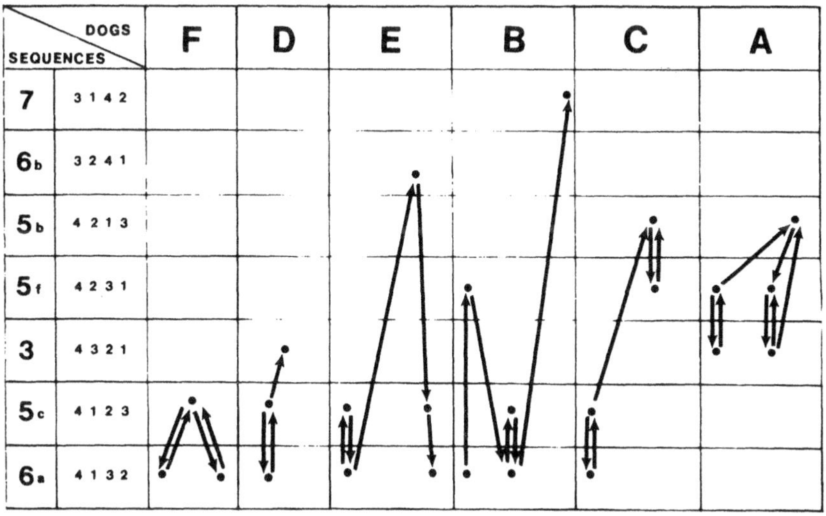

Figure 3 : Resume of all the different succesful sequences of visits used after reaching the learning criterion.

Out of the six dogs used, two succeeded on the first day of free-foraging in finding the six food points without revisiting any of them. Two dogs succeeded on the second day, one dog on the third day, and the remaining dog succeeded only on the fifth day. However, even on the first day, each dog found at least four food points without making any errors and when a difficulty arose, it concerned the last or the last two food places. Figure 5 shows the various sequences of visits used by individual dogs during 34 days. Small circles represent the use of circular sequences of visits. Most of the dogs used various sequences at the beginning without making any errors. Dog A remained very variable throughout this stage. Dogs D, E and F tended with overtraining to use the same sequence day after day. Dog B used the same sequence of visits throughout this stage. Despite the fact that this animal succeeded on the first day of free foraging, one wonders whether he had really acquired a good knowledge of the situation, or whether he was simply engaged in some kind of stereotyped behaviour. Dog C was successful on the first day of learning using a circular sequence of visits, and continued to use such sequences throughout this stage. These circular sequences had different points of departure and two opposite directions of rotation, so their use certainly reflects a good knowledge of the situation.

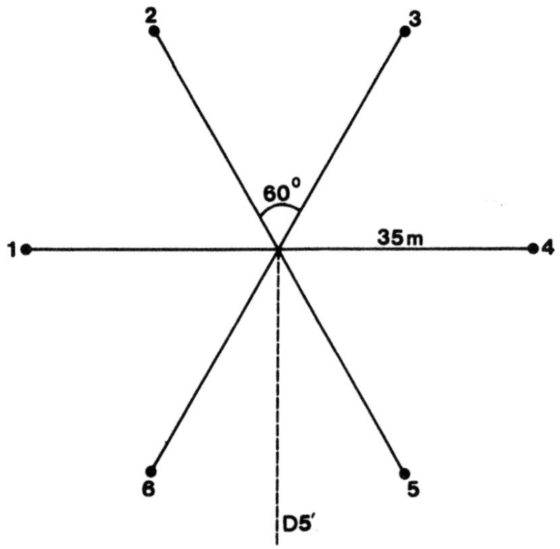

Figure 4 : Circle apparatus.

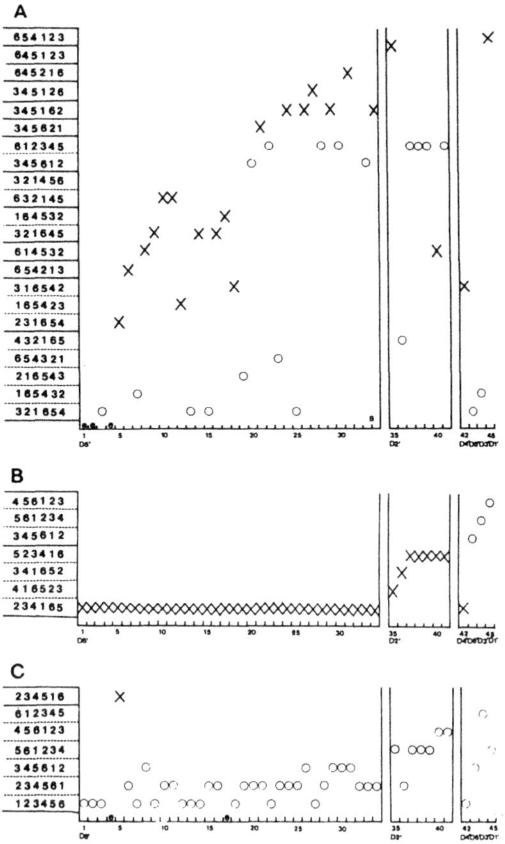

Figure 5a : Sequences of visits used by individual dogs during training (Dogs A, B, C).

From the 35th day of training until the 45th day, different paths were used to lead dogs to the starting point. All the dogs then showed a very good ability to replan their movements ; they all used different sequences of visits from those used in the previous stage without making any errors. The results of this experiment thus show that dogs are quite able to reorganize information collected when travelling along some routes in order to find their way between different food points. They are also quite able to cope with changes in the situation after a period of overtraining, even in cases where they seem to be using stereotyped behavior.

DISCUSSION

In the circle experiment, dogs were able to reorganize the information collected along the radii and find a shorter path leading to every food-point. These results are comparable to those reported by Menzel (1973) on Chimpanzees, which were capable of finding 18 food-points taking different, shorter routes than those they were originally shown by the experimenter. Dogs' performance levels are thus compatible with cognitive theories such as those developed by Tolman (1948), O'Keefe and Nadel (1978) and Menzel (1978).

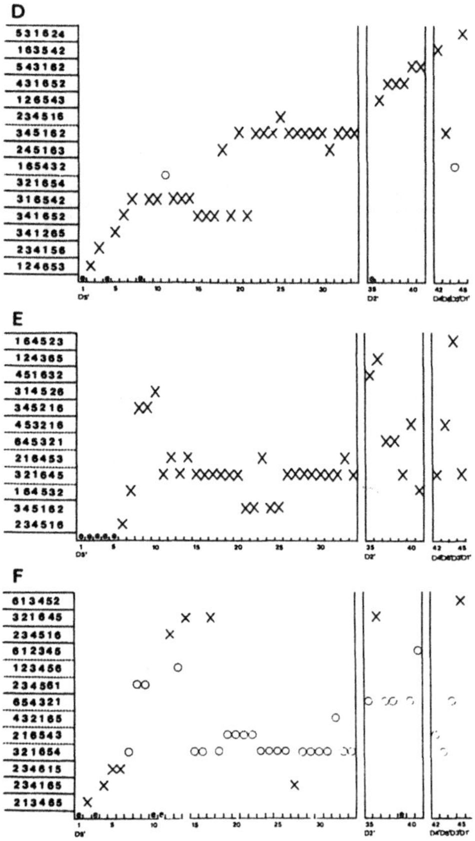

Figure 5b : Sequences of visits used by individual dogs during training (Dogs D, E, F).

In these theories, the assumption that animals are able to reorganize the information collected during their travels endows them with the ability to acquire a sort of representation of their surroundings which is not only a memory of the various movements already effected, but an overall representation of relationships between distinct elements. The gap between the movements effected and the spatial representation allows an animal a great deal of flexibility in the choice of paths. Animals' ability to take short-cuts (Tolman et al., 1946) and detours (Chapuis et al., 1983), to use the least-distance principle (Menzel, 1978) and to vary their routes (Olton, 1979; Fabrigoule and Maurel, 1982) have therefore been taken to reflect the use of cognitive processes in spatial situation.

Concerning the interpretation which can be made about the degree of variability of subjects' behaviour, the individual differences observed in our experiment call for some remarks. Subjects which succeed day after day (like dog A) in using different sequences of visits obviously need a representative memory which is not only a memory of the movements already effected. The opposite argument might be tempting, i.e. that those subjects which used stereotyped sequences of visits (like dog B) are "Hullian" subjects, having acquired only a series of S-R connections (Hull, 1943). Our results show that this assumption needs to be tested by changing the way the situation is presented to the subjects. What is very interesting in our data is the fact that, whatever the dogs did during training, they were all able to reorganize their travels when taken to the center by another route. In that situation the stereotypy demonstrated by dog B throughout the training phase, and by dogs D, E and F in the course of overtraining, does not indicate that non cognitive processes were involved; this seems to be rather a matter of individual preferences.

This shows again the importance of the distinction made in Tolman's cognitive theory between performance and competence : animals do not learn to do something as they are said to do in most S-R theories, but they acquire information about their surroundings. The way in which this knowledge is translated into actions is certainly dependent on the actual motivations of the subjects, but also on other factors, including some individual tendencies to be variable or more stereotyped.

From other data on dogs (Fabrigoule & Sagave, to be published), it seems that individual differences in the degree of variability are more likely to be observed when the task is an easy one like the circle task. Using the same procedure with an irregular pattern of six food-points, we found that the behaviour of dogs was more homogenous ; their revisiting patterns were all quite variable at first, becoming more regular with overtraining, with first visits to the same three or four food points in the same order. The task difficulty seems then to constrain the behaviour of dogs and to considerably reduce the expression of individual differences.

The data obtained in the progressive elimination of food experiment shows how interesting it can be to analyse the evolution of individual strategies right from the beginning of training to an overtraining stage. Contrary to the circle task which turned out to be very easy to master, this four-bowl elimination of food experiment was found to be difficult. The fact that the dogs were taken back to the starting-point in the course of the daily experimental session certainly added to the difficulty of the task. The central bowls may also have been difficult to discriminate between ; they were identical and their position in relation to the initial orientation of the animal was certainly not easy to distinguish. When faced with this difficult task, dogs adopted successive strategies reminiscent of the "hypothesis behavior" of Krech (1932). The successive strategies observed in our task did not differ completely from each other. What seems to occur is a progressive shift from one strategy to another, each strategy deriving partly from the previous one.

CONCLUSION

These data can certainly be more fully accounted for by a cognitive theory than by any S-R theory. The dogs' behaviour shows that they were able to reorganize information collected when travelling along some routes to find their way out of these routes. The variability of their behaviour shows that they certainly acquired a sort of representation of their surroundings which is not only a memory of the movements already effected. The successive strategies adopted in a difficult task like the progressive elimination of food task seem to reflect the ability of subjects to progressively reorganize their behaviour on the basis of the information collected.

In theories developed about spatial behaviour the utilization of cues, as well as the utilization of various systems are often conceived of as being mutually exclusive. But the type of strategies observed in our data, the occurrence of some regularities at certain stages in learning (in overtraining in the case of easy tasks, in the course of training when the task was difficult) suggests that the animals at some given moment used some external cues provided by the surroundings together with kinesthetic cues, and the respective importance of different types of cues seems to have evolved along with the reorganization of behaviour. Further investigations on the development of individual strategies in relatively unconstrained spatial tasks would certainly help us to understand how different types of cue utilization interact in the course of learning.

REFERENCES

Buhot, M.C. and Teule, M., 1971. Apprentissage d'élimination progressive chez le Hamster doré (Mesocricetus auratus). Cahiers de Psychologie, 14: 135-165.

Chapuis, N., Thinus-Blanc, C. and Poucet, B., 1983. Dissociation of mechanisms involved in dogs' oriented displacements. Quarterly Journal of Experimental Psychology, 35B: 213-219.

Fabrigoule, C., 1974. Recherches expérimentales sur l'apprentissage spatial chez le Chien. Apprentissage d'élimination progressive. Cahiers de Psychologie, 17: 91-110.

Fabrigoule, C. and Maurel, D., 1982. Radio-tracking study of foxes' movements related to their home range. A cognitive map hypothesis. Quarterly Journal of Experimental Psychology, 34B: 195-208.

Hull, C.L., 1943. "Principles of Behavior". Appleton Century, New-York.

Krech, I., 1932. The genesis of "hypothesis" in rats. University of California Publications in Psychology, 6(4):45-64.

Lachman, S.J., 1971. Behavior in complex learning situations involving three levels of difficulty. Journal of Psychology, 77: 119-126.

Lachman, S.J. and Brown, R., 1957. Behavior in a free choice multiple path elimination problem. Journal of Psychology, 43: 27-40.

Menzel, E. W., 1978. Cognitive mapping in Chimpanzees. in "Cognitive Processes in Animal Behavior". S.H. Hulse, H. Fowler and W.K. Honig (Eds.), Lawrence Erlbaum Associates, Hillsdale, N. J.

O'Keefe, J. and Nadel, L., 1978. "The hippocampus as a cognitive map". Clarendon-Oxford University Press, Oxford.

Olton, D.S., 1979. Mazes, maps and memory. American Psychologist, 34: 583-596.

Tolman, E.C., 1948. Cognitive maps in rats and men. Psychological review, 55: 189-208.

Tolman, E.C., Ritchie, B.F. and Kalish, D., 1946. Studies in spatial learning. I. Orientation and the short-cut. Journal of experimental Psychology, 36: 13-24.

ROLE OF THE SPATIAL STRUCTURE IN MULTIPLE CHOICE PROBLEM-SOLVING BY GOLDEN HAMSTERS.

Marie-Christine BUHOT and Hélène POUCET

Department of Animal Psychology
Institut of Neurophysiology and Psychophysiology
CNRS-INP 9
31, chemin Joseph-Aiguier
13402 Marseille cedex 9
France.

Among the studies which have been devoted to spatial organization in animals, many have used spatial problem solving, especially the multiple-choice food elimination problem, to elucidate diverse questions. With this paradigm, behaviour can be analysed not only in terms of performance, but also in terms of strategy, which could depend in varying degrees on the maze configuration, and on the availability of extra-maze cues.

Crannel (1942), then Lachman (1965, 1966, 1969a, 1969b, 1971) and Lachman and Brown (1957) introduced this multiple-choice elimination procedure. It consists of placing the rat in a maze including at least three diverging alleys from a common starting platform. A piece of food is placed at the end of each arm, and the animal has to learn to find and eat these rewards in the course of successive trials, without returning to a previously visited arm no longer containing any food. A win-shift strategy is thus required here, the optimal strategy being of course to choose each arm once without repeating any choice, i.e. to explore the maze exhaustively and economically. These pioneers raised the question as to whether the spontaneous alternation generally observed in a T-maze really reflected the tendency to take successively the most divergent ways (Lachman, 1969b). This behaviour was thought to result rather from stimulus satiation (Glanzer, 1953) and/or a search for novel stimuli (Dember, 1956).

Ammassari-Teule and one of the present authors used the same procedure with golden hamsters in a spindle-shaped maze (Buhot and Teule, 1971 ; Teule, Buhot and Durup, 1972 ; Ammassari-Teule, 1981 ; Ammassari-Teule and Durup, 1982). The maze consisted of three alleys diverging from a common starting platform, and finally reconverging towards a common goal-box. The performances differed depending on the location of the 3 rewards which were either 1) at the 3 maximally divergent points, 2) at the common goal box or 3) towards the end of the separate paths before they reconverged. The performances were higher when the reward locations were separated, i.e. one in each arm, than when one reward was provided at a single common point.

Using the same elimination paradigm, using the well-known eight-arm radial maze with a central choice platform, Olton (1977, 1978, 1979 ;

Olton and Samuelson, 1976) further investigated spatial memory in rats. In such a maze, the rats very rapidly learned to collect all eight rewards located at the end of the arms, with a very low number of reentrances into the same arm (Olton and Samuelson, 1976). The rats furthermore succeeded in extended radial mazes comprising, for instance, seventeen arms (Olton, Collison and Werz, 1977). The rats' high performances were attributed to the use of extra-maze visual cues, which they associated in space with the arms of the maze. Olton suggests that a short-term working memory is sufficient for a rat to perform this task accurately, that is the ability to memorize the immediately preceeding paths, which are integrated as a list of independent items (Olton, 1978). According to this hypothesis, the rat does not need to build an overall representation of the maze. This task can thus be successfully performed in such a maze by rats without requiring cognitive mapping or reference memory processes (Olton, Handelmann and Walker, 1981).

In order to study organizational processes in rat memory, Roberts (1979) used a hierarchical radial maze consisting of eight primary alleys radiating from a central platform, with three secondary alleys leading off from the end of each primary one. According to Roberts, the rats were able to construct the spatial characteristics of the maze and integrate them by cognitive mapping, which implies the existence of separate memory systems for the primary and secondary alleys.

Lastly, and more recently, Horner (1984), using a slightly different procedure, has studied the ability of rats to perform a food-searching task in three spatially quite different mazes, each comprising eight differently arranged arms baited at their extremities. These mazes were 1) a classical radial maze, 2) an anti-radial maze where the rewards were placed towards the end of the convergent arms, and 3) a parallel maze. Horner shows that the maze structure influenced the rats' performance in that it produced different initial patterns of exploration, which were either appropriate (like alternation) to the task or not.

Continuing along these lines, the present experiments were focused on the influence of the spatial characteristics of the maze on hamsters' performances and strategies, in a classical food-search learning task. The mazes used differed either in their metric or geometric characteristics, or in their topological organization. This study focused particularly on the analysis of the spatial and temporal components of the behaviour, as possible means of defining certain strategies.

These experiments were performed by six groups of golden hamsters, each of which was tested in a different maze. The six different mazes, which were all designed with one starting alley and four goals, were placed in a room providing extra-maze cues.

Three of the mazes, forming the "High Divergence series" (Figure 1), had goals which were far apart from each other and very different configurations. Among these three topologically different mazes, one (top of the Figure) had a single choice-point and four independent arms leading to the four goals ; another one (left-hand side of the figure) had two choice-points, the second of which led towards the two median goals. In the third maze (right-hand side of the figure), the first choice-point had two arms, each leading to a second two-arm choice-point.

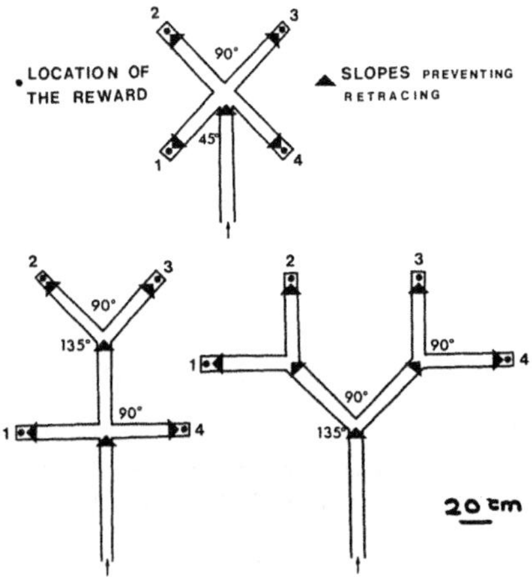

FIGURE 1. Top view of the mazes used in the "HIGH DIVERGENCE" (HD) series.

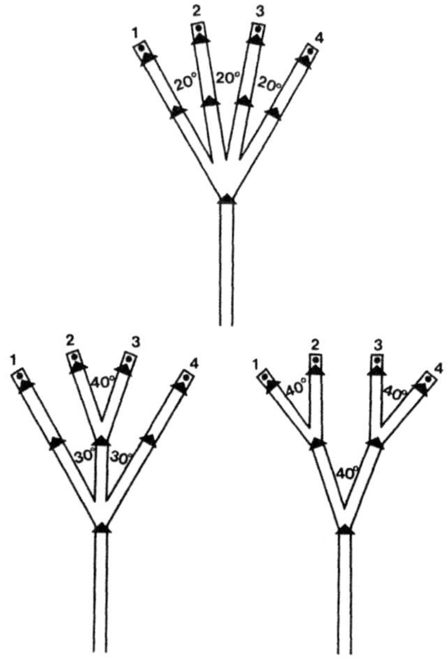

FIGURE 2. Top view of the mazes used in the "LOW DIVERGENCE" (LD) series.

The latter two mazes were thus hierarchically organized, so that the animal eventually had to take the first part of an alley twice to visit two different goals.

In the second three mazes, forming the "Low Divergence series" (Figure 2), the location of the four goals was identical. The only difference between them was the organization of the paths, each maze was topologically identical with one of the mazes of the "High Divergence series".

The animals performed one test a day, comprising at least four different rewarded runs. At the end of each run, the subject was removed and replaced at the starting point. The learning criterion was to achieve three consecutive daily tests with no error, i.e. without repeating any previously performed run.

The performances (Table I) were first analysed through the mean number of days taken to reach the learning criterion. They mainly showed strong differences between the two series of mazes, the Low Divergence series (bottom of Table I) being significantly less well performed than the High Divergence series (top of Table I). This suggests that the distance between the goals was a decisive factor.

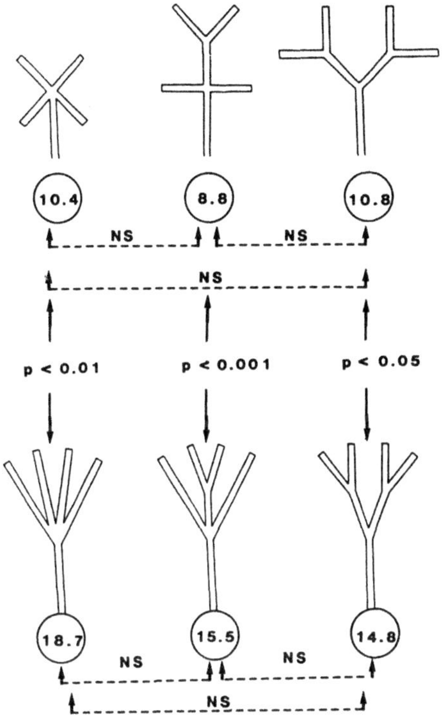

TABLE I. PERFORMANCES.
Mean number of DAYS until the learning criterion ; comparisons between the maze groups (Student's t).

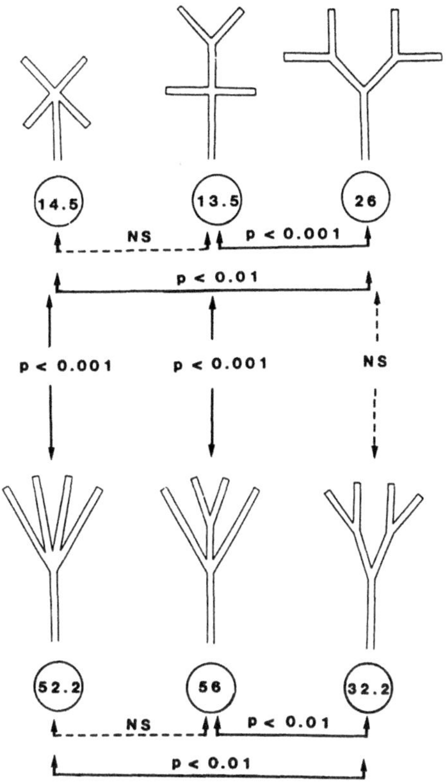

TABLE II. PERFORMANCES.
Mean number of ERRORS until the learning criterion ; comparisons between the maze groups (Student's t)

With regard to the mean number of errors made before reaching the learning criterion (Table II), the results showed both within and between series differences. Within each series, the first two mazes (the one starting with a four-choice point, left-hand side of the Table, and the other with a three-choice point, middle of the Table) did not differ significantly, whereas they both differed from the third in the series (right-hand side of the Figure) but in opposite respects : the third maze in the High Divergence series (top of the Table) yielded the worst performances of this series, whereas the third maze of the Low Divergence series (bottom of the Table), gave rise to significantly less error than the two first mazes in this series.

The performances achieved in these mazes showed different levels of difficulty for the animals, which can be accounted for by the differences in maze geometry and topology.

What do the results tell us about the animal's use of its experimental environment ?

Besides the differences in performance levels which are directly related to the difficulty of the task, it is interesting to consider the strategies, or simply the spatial components of the behaviour, produced in cases without error, i.e. the sequences comprising only the four rewarded runs. This analysis was conducted in order to study qualitatively and in greater detail the relationship between the spatial characteristics of the maze and the spatio-temporal properties of the behaviour.

These patterns can first be analysed through the degree of divergence, taking into account the distance between the successively visited goals. This has previously been used as a measure of the alternation tendency (Lachman,1969b).

The degree of divergence of a pattern corresponds to the sum of the intervals (empty spaces) between successively visited goals : two adjacent goals have one interval between them ; two goals separated by another one have two intervals between them. As an example, starting on one side of the maze, the "clockwise" pattern consists of taking successively the adjacent paths towards the opposite side. In this example applied to our mazes, the successive goal numbers may be either "1-2-3-4" or "4-3-2-1". If the sequence of visits to the various goals is "4-3-2-1", the degree of divergence is equal to 1 (passage from goal 4 to goal 3) + 1 (passage from goal 3 to goal 2) + 1 (passage from goal 2 to goal 1) = 3 which is the lowest possible degree of divergence. From one of our mazes to another, different locomotory patterns are required to provide the same degree of divergence.

The following example "2-4-1-3" is a pattern corresponding to the highest degree of divergence (= 7). The particularity of such a pattern is that it reveals a tendency to alternate distant goals, unlike the previous example. The following diagram shows more precisely the general procedure used to calculate this measure for the pattern "2-4-1-3" :

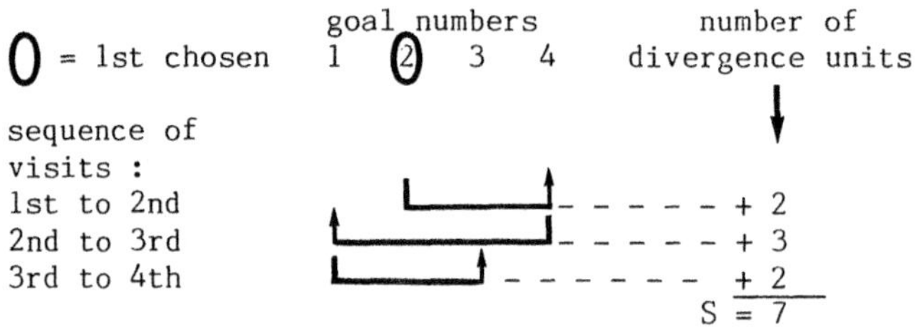

According to a classification of the twenty-four possible patterns in terms of degree of divergence, our results showed a strong tendency to produce patterns with a high degree of divergence with the three mazes in the second series (Low Divergence), whereas this tendency was weaker with the mazes in the first series (High Divergence). This difference will be discussed later.

The spatial properties of the successful patterns were further analysed with regard to goal positions in space and time. Each goal can be simply defined as being on the left or right : thus goals 1 and 2 are on the left, while 3 and 4 are on the right (see Figures 1 or 2) ; then they can be either the outer goals of the maze, like goals 1 and 4, or the inner ones, like goals 2 and 3.

On the basis of these simple definitions, the two above examples of patterns can be described as follows :
a "clockwise" pattern such as "4-3-2-1" forms a sequence by the regrouping between left and right ("r-r-l-l") and the inclusion of the two inner goals within the outer ones ("o-i-i-o") ;
"2-4-1-3" pattern is characterized by the alternation between left and right ("l-r-l-r") and the inclusion of the outer goals within the inner ones ("i-o-o-i").

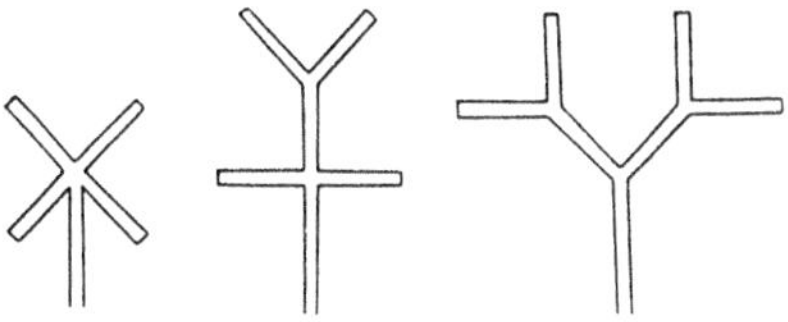

A	:	O O I I	O O I I	L R L R
B	L	---	---	---
	R	---	---	---
	O	93.3%	98.7%	---
	I	----	---	72.8%
C	:	4 1 3 2	1 4 2 3	2 4 1 3

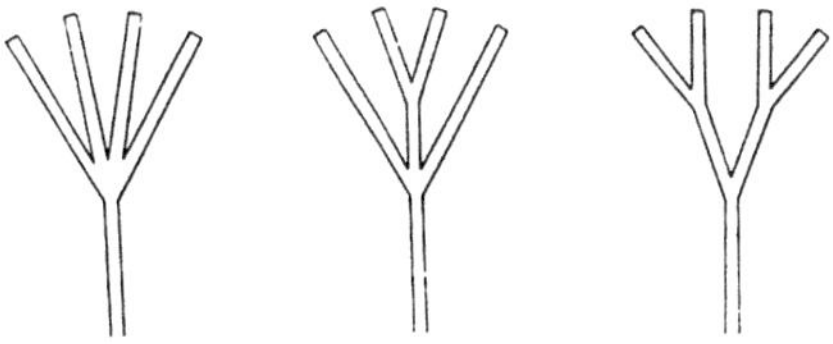

A	:	I O O I	L R L R O O I I	R L R L I O O I
B	L	---	---	---
	R	71.9%	---	---
	O	---	73.5%	---
	I	---	---	---
C	:	3 1 4 2	1 4 2 3	3 1 4 2

TABLE III. SPATIAL ANALYSIS OF THE SUCCESSFUL PATTERNS.
A : spatio-temporal tendency,
B : significant position (%) of the first choices,
C : most frequent patterns within each maze group.

Our data were analysed in this way, for each maze (Table III). The main spatio-temporal tendency (given in A) showed that the dimensions left/right and outer/inner were not always both taken into account. In the first series (top of the Table), the two left-side mazes were characterized by a regrouping between outer and inner goals, the two outer goals being taken first, whereas the third maze in the series showed a tendency to alternate between left and right. In the second series (bottom of the Table), the dimension inner/outer was taken into account in all three mazes, but left and right was also relevant in the two hierarchical mazes.

Depending on the position of the most frequently first chosen goal, which is given in B) as a percentage, the most frequent pattern (given in C) emerges. In more than ninety per cent of all cases, an outer goal was visited first in the two left-hand side mazes of the first (upper) series, while the predominant first choice in the third maze was an inner path. In the second series (bottom of the Table), a tendency to take first a right goal was revealed in the four-choice point maze (on the left), whereas the middle maze showed a preference to choose an outer goal first. No particular preference emerged in the last (right-side) maze.

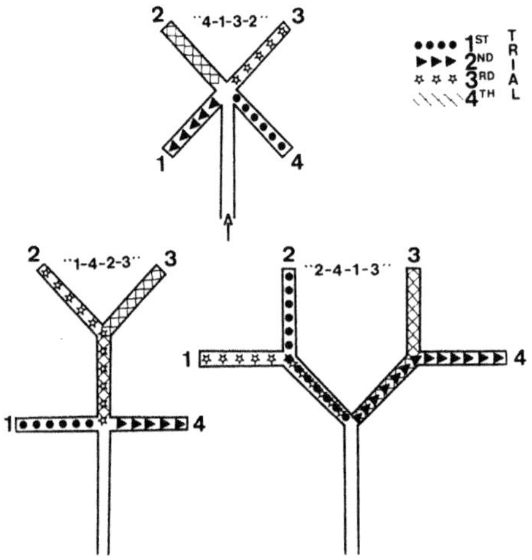

FIGURE 3. PATTERNS corresponding to each maze group. HD series.

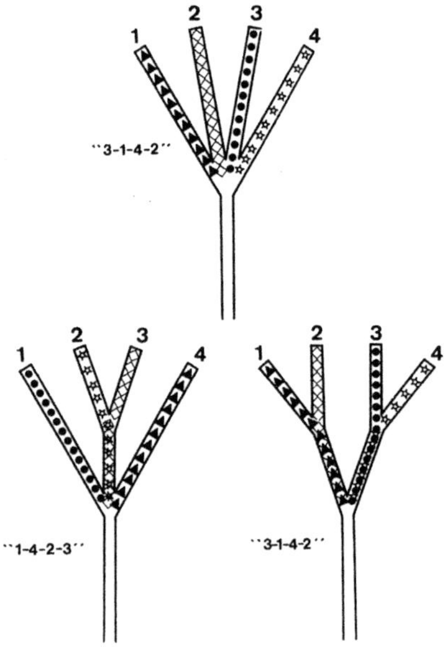

FIGURE 4. PATTERNS corresponding to each maze group. LD series.

The most frequent pattern recorded for each maze can now be built. It can be considered in relation to the spatial properties of the maze design. In fact, the main pattern was not the sole pattern produced. This most frequent pattern was rarely repeated successively since no stereotypy was observed in these experiments. In the first series (Figure 3), the following patterns emerged. In the cross-maze, the pattern "4-1-3-2" occurred most frequently (17.8%). The pattern representing (25.4% of all patterns) the left-hand side maze was symmetrical with regard to the outer/inner dimension and in certain respects was similar to the cross-maze pattern. In both these mazes the two most often first taken paths are called "outer". In fact, they had particular locations in space since they were either behind or beside the first choice point. On the other hand, in the second maze, the outer alleys were shorter. In the third (right-hand side) maze, where the first alley to be chosen was an inner one, an alternation was observed between left and right, which were well-separated in the maze : the pattern "2-4-1-3" occurred in 10.7% of the cases.

In the second series involving far less divergent alleys (Figure 4), the most frequent (29%) pattern generated by the first (upper) maze showed the highest degree of divergence : the choice of an inner alley first, followed by the two outer ones, corresponds to a strong tendency to avoid the clock-wise strategy, i.e. not to take adjacent alleys in succession. This suggests that, by its successive choices, the subject separated in time points which were spatially close, thus possibly making them easier to differentiate between. This interpretation is an attractive one, and therefore requires further investigation ! In the maze with three alleys leading off from the first choice point (left-hand side), the pattern most frequently (20,4%) produced was identical to the one of its counterpart in the first series, that is : choice of the outer alleys first, and alternation between left and right. It should be mentioned that the length of all the alleys in this series was identical, but that the inner paths may have been taken successively, possibly because they were joined by a common primary alley. In the last maze (right-hand side), the same pattern as in the (upper) classical maze was produced in 29% of the cases, this is consistent with an alternation between left and right, which in this maze consisted of two sub-spaces joined by a common primary alley.

To summarize these data, Figure 5 schematically illustrates the use of space in each maze as revealed through the spatial analysis of the successful patterns. This figure suggests the degree of dependence between the spatial behaviour of the animals and the spatial characteristics of the maze in which they have been placed. The operations of regrouping, inclusion or alternation are represented by arrows between the maze's sub-spaces which are shown as outer and inner (dotted areas) or left and right (empty areas).

In the High Divergence series (top of the Figure), the left-hand side and middle mazes thus presented a tendency in the subjects to consider the maze as being split in the middle along a perpendicular line with respect to the starting alley : this was indicated by the regrouping of the two outer alleys which were eliminated first. The third maze in the series also presented a splitting of the maze into two parts by alternating its left and right sides.

FIGURE 5. Schematic representation of the use of maze structure by the hamsters.

: OUTER/INNER, : LEFT/RIGHT,
: regrouping, : alternation,
: inclusion.

Performance of the second series was mainly characterized by a tendency to produce highly divergent patterns in all the mazes. This meant that both our dimensions were relevant, since neither of them alone was actually representative of the maze. An alternation between left and right was common in the middle and the right-hand side mazes, where the maze was split along its median axis ; the outer/inner dimension was also relevant.

What emerges from these data taken as a whole is the influence of the maze structure on the choice task. In the series where the goals were highly divergent, the subjects adopted spatial strategies which were highly dependent on the topological organization of their experimental space ; whereas, in the series where the goals (and the alleys) were closely gathered ahead of the starting alley, and thus may have been less easy to differentiate between, the subjects were less accurate and they adopted a more general rule, based on the maximum of divergence, mainly using alternation between distant goals. The animals acted as if they used time to bring together distant goals which were more discriminable and hence easier to memorize, and to separate in time goals which were spatially close to each other. Time, here, seems to have mediated the relationship between the animal and space. What is still not clear is the relationship between spatial strategy and performance. Was the level of performance directly related to the choice of a particular strategy, which itself probably depended on the spatial structure of the maze? The question is still open.

The main theoretical question finally concerns whether hamsters required cognitive mapping to succeed in these mazes, or whether only the memory of the last alley taken sufficed to avoid reentrances. First, the achievement of the learning was not accompagnied by any stereotypy. Secondly, the mazes were simple with regard to the number of goals (which were invisible from the choice point) ; but they were complex with regard to the organization of the alleys. Thirdly, the performances and the organization into sub-spaces differed from one maze to another. These three categories of arguments plead in favour of the animal's ability to attend to the structure of the maze and to refer to the available extra-maze cues, to perform the task successfully.

REFERENCES

Ammassari-Teule, M., 1981. Apprentissage d'élimination chez les petits Mammifères : pluralité, plasticité et préparation des systèmes de réponse. Thèse d'Etat. Document polycopié, 221p.

Ammassari-Teule, M. and Durup, H., 1982. Spatial learning in golden hamsters : relationship between food-searching strategies and difficulty of the task. Behavioural Processes, 7: 353-365.

Buhot, M.-C. and Teule, M., 1971. Apprentissage d'élimination progressive chez le Hamster doré (Mesocricetus auratus). Cahiers de Psychologie, 1: 135-165.

Crannell, C.W., 1942. The choice point behavior of rats in a multiple path elimination problem. Journal of Psychology, 13: 201-222.

Dember, W.N., 1956. Response by the rat to environmental change. Journal of comparative and physiological Psychology, 49: 93-95.

Glanzer, M., 1953. The role of stimulus satiation in spontaneous alternation. Journal of experimental Psychology, 45: 387-393.

Horner, J., 1984. The effect of maze structure upon the performance of a multiple-goal task. Animal Learning and Behavior, 12: 55-61.

Lachman, S.J., 1965. Behavior in a multiple-choice elimination problem involving five paths. Journal of Psychology, 61: 183-202.

Lachman, S.J., 1966. Stereotypy and variability of behavior in a complex learning situation. Psychological Report, 18: 223-230.

Lachman, S.J., 1969a. Behavior in a complex learning situation involving five stimulus-differentiated paths. Psychonomis Science, 17: 36-37.

Lachman, S.J., 1969b. Behavior in a three-path multiple choice elimination problem under conditions of overtraining. Journal of Psychology, 73: 101-109.

Lachman, S.J., 1971. Behavior in complex learning situation involving three levels of difficulty. Journal of Psychology, 77: 119-126.

Lachman, S.J. and Brown, C.R., 1957. Behavior in a free choice multiple path elimination problem. Journal of Psychology, 43: 27-40.

Olton, D.S., 1977. Spatial memory. Scientific American, 236: 82-98.

Olton, D.S., 1978. Characteristics of spatial memory. In "Cognitive processes in animal behavior". S.H. Hulse, H. Fowler, W.K. Honig (Eds.), Lawrence Erlbaum Ass.,J. Wiley and sons, Hillsdale, 341-373.

Olton, D.S., 1979. Mazes, maps, and memory. American Psychologist, 34: 583-596.

Olton, D.S., Collison, C. and Werz, M.A., 1977. Spatial memory and radial arm maze performance of rats. Learning and Motivation, 8: 289-314.

Olton, D.S., Handelmann, G.E. and Walker, J.A., 1981. Spatial memory and food searching strategies. In " Foraging behavior : ecological, ethological and psychological approaches". A.C. Kamil and T.D. Sargent (Eds), Garland STPM Press, New York, 333-354.

Olton, D.S. and Samuelson, R.J., 1976. Remembrance of places passed : spatial memory in rats. Journal of experimental Psychology : Animal Behavior Processes, 2: 97-116.

Roberts, W.A., 1979. Spatial memory in the rat on a hierarchical maze. Learning and Motivation, 10: 117-140.

Teule, M., Buhot, M.-C. and Durup, H. 1972. Apprentissage d'élimination progressive chez le Hamster doré (Mesocricetus auratus). Psychologie Française, 17: 175-182.

MEMORY PROPERTIES OF SPATIAL BEHAVIOURS IN CATS AND HAMSTERS

Bruno POUCET and Georges SCOTTO

Department of Animal Psychology
Institute of Neurophysiology and Psychophysiology
C.N.R.S. - I.N.P. 9, 31 chemin Joseph-Aiguier
F-13402 Marseille cedex 9, France

Recent theories about how animals and men use space have dealt with various mechanisms likely to underlie spatial behaviours (Shemyakin, 1962 Hulse, Fowler and Honig, 1978 ; O'Keefe and Nadel, 1978 ; Morris, 1981). Development of the theory according to which animals may use several different orientation mechanisms was the result of a controversy which lasted for forty years (cf. Thinus-Blanc, 1985). On the one side, investigators considered the organism as being merely reactive to specific stimuli guiding the subjects' responses (Hull, 1952). Another view was that animals built a cognitive representation of the environment, allowing them to generate place hypotheses about that environment, regardless of their own locations (Tolman, 1948). Recently, O'Keefe and Nadel (1978) have developed a theory according to which animals may use two types of mechanisms, i.e., "routes" and "maps". Routes belong to a "taxon system", which is thought to be governed by the laws of associative learning and to consist of producing cue-dependent egocentric responses, "guidances" and "orientations". Maps ("locale system") are thought to be cognitive representations of the environment which do not rely on any particular places or cues.

Among the properties assigned by O'Keefe and Nadel (1978) to these different orientation systems, one that needs to be specially mentioned here is the involvement of memory. As a matter of fact, the properties of routes and maps are linked to the amount of information to be stored and processed. While the information necessary for the use of guidances and orientations is restricted, by definition, to particular cues that are associated with either the goal direction or the response to be made, the map of an environment is built up of a set of place representations connected together according to rules which represent the distances and directions amongst them. Although a small number of stimuli are sufficient to help an animal in identify a place on this map, information is stored in such a way that animals can find their way from any part of the familiar environment and are able to locate any place in that environment, on the basis of the spatial configuration of the cues. Therefore, maps are characterized by a high information content, and function with a set of complex operations which reorganize the spatial information : the animal can thus find its way to the goal even when the latter is not signalled by a landmark. Because of the redundancy of information involved in the mapping system, the identification of a place in the environment does not depend on any particular cue or group of cues (O'Keefe and Nadel, 1979), and can be expected to be relatively insensitive to loss or disturbance of spatial information.

It is a different matter for an animal following a route, i.e., approaching or avoiding cues, or performing specific responses at specific places. Here, the animal's performance is clearly quite dependent upon the presence of the particular cues which determine its choices. Under these conditions, whole or partial removal of spatial information which is relevant to the localization of a goal would have a deleterious effect on orientation : the animal would not be able to orient itself correctly if the cues upon which it is used to relying in choosing its directions are no longer present. Hence the use of a map makes an animal's behaviour much more flexible than does following a route. Furthermore, the respective properties of maps and routes will mean that the difficulty experienced by animals in dealing with the memory-related components of experimental tasks will depend on whether they are using a mapping system or a route system.

Delayed reaction tasks (cf. Hunter, 1913) involve the memory component which is precisely the central point of the present study. Earlier psychologists presented this problem as one involving a response to a discriminative stimulus not physically present at the time of the response, and therefore requiring for its solution a capacity for symbolic or representative processes (Fletcher, 1965). Thus, if representative processes are implicated differently in map and route systems, one should expect animals to perform delayed reaction tasks with different levels of success depending on which of these two orientation systems they use : animals using the route system are likely to be unsuccessful while animals using a map should succeed. The present paper focuses on the results of two different studies with cats and hamsters, respectively, which showed that memory was in fact differentially involved in map and route orientation behaviours.

SPATIAL BEHAVIOUR OF CATS IN CUE-CONTROLLED ENVIRONMENTS

The first experiment was carried out with a view to find appropriate behavioural indices with which to characterize the level of spatial processing in cats (Poucet, 1985).

Two groups of six cats underwent spatial learning in a cross-maze set up in a cue-controlled environment surrounded by black curtains. The cats had to reach a baited goal F from any of three starting points S_1, S_2 and S_3 (Fig. 1). With the first group of cats (group I), the only available cue was a salient light-bulb placed distally from the goal. Although there was only one cue, this situation was called a "mapping situation" since the cats had to deduce the location of a non-signalled goal by taking into account the angular relation between the cue and the goal. This angular relation (135°) remained constant throughout the experiment (left side of Fig. 1). Another group of cats (group II) was exposed to a "guidance situation", in which the light-bulb cue was just above the goal throughout the experiment : thus, in order to reach the goal, the cats just had to move towards the cue.

Each daily session consisted of 12 trials, i.e., four from each starting point. At the beginning of each trial, the animal was placed at a given starting point. The subject was then free to choose its path to the goal, where it was rewarded with food.

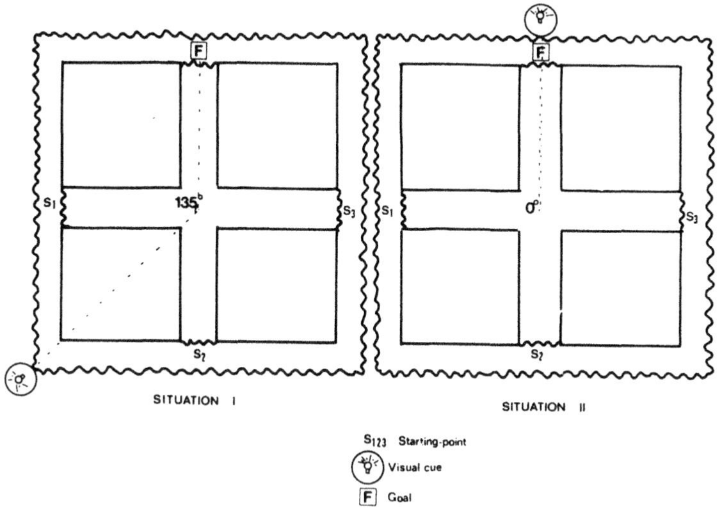

Fig. 1. Overview of the two experimental situations used in the cat experiment, showing the location of the visual cue by reference to the goal location. Left : "Mapping situation" (Group I). Right : "Guidance situation" (Group II). From Poucet, 1985. Copyright The Experimental Psychology Society.

If the animal made an error, it was not allowed to correct its choice, but was immediately removed from the apparatus and placed at a new starting point for the next trial. The order of starting points was randomized so that the cats could not rely on body turns or some other uncontrolled strategy. In addition, olfactory cues were neutralized. Lastly, the entire set-up was rotated every day in relation to the magnetic north and the experimental room.

In each situation, the animal was thus required to take into account the location of the only available visual cue, the light-bulb. During learning, a careful recording of head-movements, hesitation times and body orientations was performed (cf. Poucet, 1985 for details). These behavioural data showed the existence of two information gathering strategies within each group (Fig. 2). The first strategy consisted of collecting the relevant information right from the starting point, and was adopted by five cats in the mapping group (subjects B, C, D, E, F) and four cats in the guidance group (subjects G, H, J, M). The second strategy consisted of going directly to the central choice point and choosing a path from there, after several body reorientations. One cat in the mapping group (subject A) and two cats in the guidance group (subjects K, L) proceeded in this way.

After reaching the learning criterion, the cats were subjected to delayed reaction tasks, during which a time lapse was introduced between the moment when the relevant cue was still available and the moment when the subjects were allowed to make for the goal, after removal of the cue. Basically, the procedure consisted of a 30-second observation period during which the animals were restrained in a transparent cage at the sta-

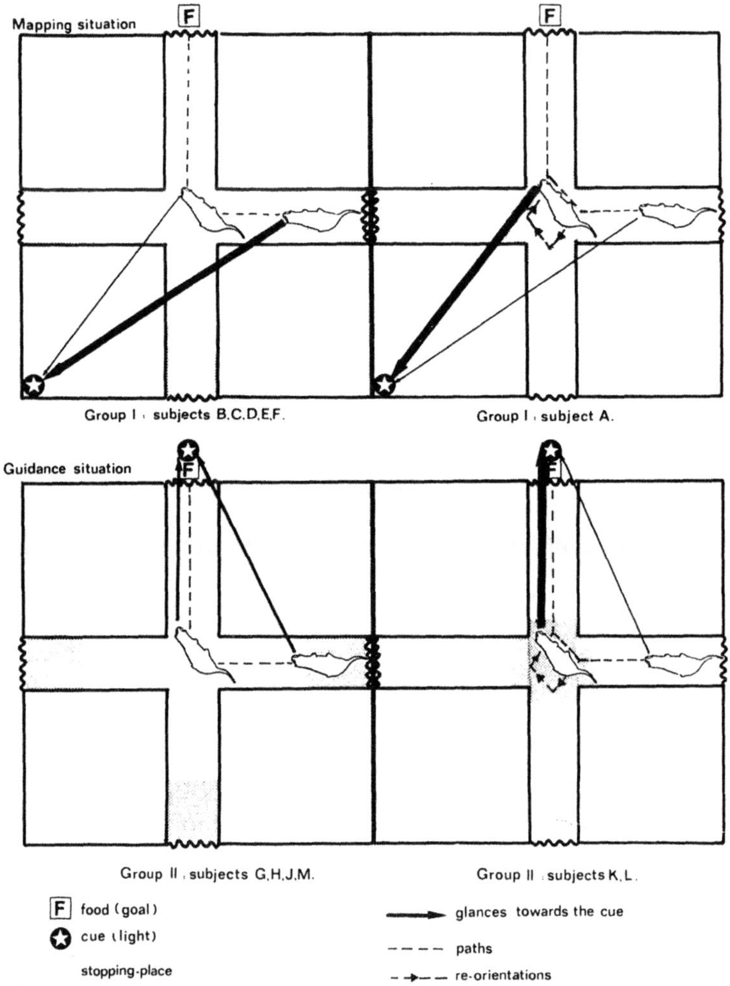

Fig. 2. Schematic representation of the different strategies of information gathering in each situation. From Poucet, 1985. Copyright The Experimental Psychology Society.

rting point. The cage was then covered with an opaque screen and the cue was removed. After an interval ranging from 8 to 60 seconds, the cat was released and allowed to make for the goal in the cue-less set-up.

The results (Fig. 3) show clearly that the performances of the subjects exposed to the guidance situation (group II) were affected much more strongly by the delay than those of the subjects in the mapping group (group I). All the cats in the mapping group succeeded with intervals of 8 and 15 seconds whereas only two cats in the guidance group reached the criterion with these intervals. Three cats in the mapping group succeeded even with a 1-min interval, while none of the subjects in the guidance group reached the criterion in these trials.

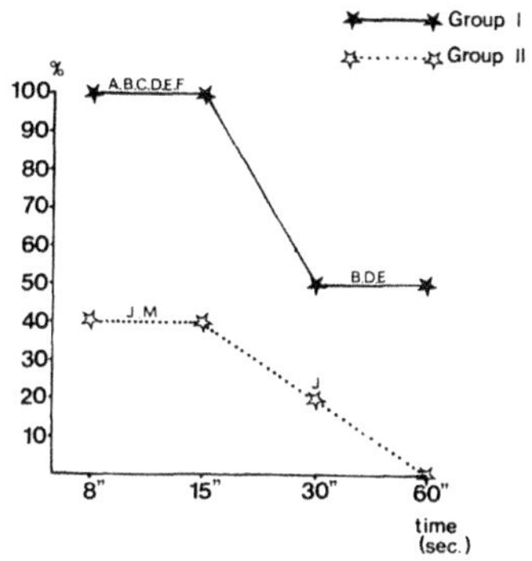

Fig. 3. Percentage of cats scoring above chance during delayed reaction tasks, for each delay. From Poucet, 1985. Copyright The Experimental Psychology Society.

Although these results were surprisingly clear-cut, they were not altogether unexpected. O'Keefe and Conway (1980) have reported that rats subjected to a delayed reaction task in a distributed-cue situation comparable to the present mapping situation could perform correctly even after an interval of 30 minutes, while rats exposed to a clustered-cue situation comparable to our guidance situation were markedly affected by even short delays. The present results strongly suggest that the guidance strategy relies to a large extent on non-representative processes since successful performance depends on the presence of the relevant cue. During the learning phase, the guidance cats had only to spatially associate the cue location with the goal location and could then rely on this relation in order to get to the food. All these subjects needed to do was to remember that there was food under the cue, and to take a route leading to the cue. No spatial processing was required to get to the goal, with the exception of the perceptual processes involved in locating the cue and moving towards it ; therefore no planning of the correct displacement was really necessary.

In the mapping situation, the goal arm was not signalled by any cue, and did not differ from the other arms in the maze, except that it occupied a particular location in relation to a visual cue. The difference between the guidance and mapping situations in the present experiment could be said to parallel the difference between the "visible" and "hidden" platform conditions with Morris' water maze (Morris, 1981). In the mapping situation, as well as under the hidden platform condition with the Morris water maze, the only means by which animals can locate the goal is to take into account the spatial relation between the cue and the goal in the former case or the multiple relations offered by the configuration of the extra-apparatus cues in the latter case. Thus, in both procedures the localization of the goal requires the animal to deduce the location of a non-signalled place --the goal-- in relation to the location(s) of signalled place(s) --the cue(s)--. In the present experiment, we showed that subjects could make this operation by using one of two different strategies. One subject (subject A) was observed to gather information at the central choice point, indicating that it was trying to look for a constant relation between the cue and the goal from a single point. This subject, when facing the relevant cue from the choice point, may have learned to make a turn with definite amplitude and direction. In contrast, the other subjects were observed to collect information at the onset of the trial, from the starting points. Our hypothesis is that these subjects were using a "map" of the environment, which allowed them to locate the goal in relation to their own location and to "plan" the correct displacement on that map. This planning might explain why these animals performed the delayed reaction tasks better than the other group, since by the time the relevant cue was removed, the correct path had already been determined. In conclusion, these results suggest that strategies based on the mapping system are relatively independent of immediately perceived cues.

MEMORY PROPERTIES OF SPATIAL STRATEGIES IN HAMSTERS

The previous experiment brought to light a strategy that did not look like either a guidance strategy or a mapping strategy. This third strategy, which was described as "looking for the relation between the goal and the cue from a single place, i.e., from the central choice point", and called "reorientation strategy", seems to involve the use of an egocentric frame of reference in which the constant relations between the cue, the choice point and the goal are stored. It may be hypothesized that by adopting such a reorientation strategy, the subjects can considerably reduce the task difficulty from the point of view of the processing of spatial information, since they use the same procedure at every trial, namely, always choosing at the same place.

Unfortunately, the cat experiment did not allow any further assessment of this reorientation strategy, since only one cat (subject A) in the mapping group and two cats in the guidance group (subjects K and L) used it. It should be noted, moreover, that these subjects were not among those which performed the delayed reaction tasks most successfully.

The following experiment with hamsters was designed to compare the properties of egocentric orientation (from a single place) with those of allocentric orientation (from several different places). For each group, the spatial relation between the cue and the goal was comparable to the one used in the mapping situation of the cat experiment, but the arrangement of screens above the maze required animals from different groups to collect information from either the choice point or the starting points. Basically, the apparatus (Fig. 4) was of the same type as before, except that it was reduced to fit hamsters (length of arms : 0.5m ; width : 0.07m). Screens were put over the arms of the maze so that, with group C (N=6), relevant information could be collected only at the central choice point. With group S (N=6), screens were placed above the central part of the arms and above the choice point so that the subjects could collect information only at the starting points. Group A (N=6) was trained in a maze with no screen at all.

By and large, the same procedure was used as previously. The hamsters had to learn to go from any starting point to the goal, using the spatial relation between a single cue and the goal (Fig. 4A). When they reached the goal, they were rewarded with food and allowed to explore

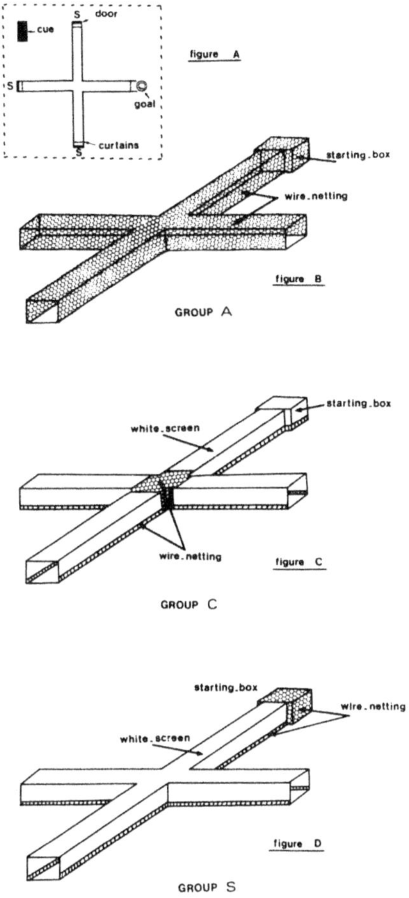

Fig. 4. Schematic representation of the experimental situations used in the experiments with hamsters.
A. Overview of the experimental area, showing the relative locations of the cue, the goal and the starting points.
B, C, D. Devices respectively used for groups A, C and S. The starting-box was removable so that it could be placed at another point ; in order to simplify the drawings, the surroundings (curtains, cue) are not represented (see Fig. 4A).

the extra-apparatus environment. The same precautions were taken as in the previous experiment to ensure that it was not possible for the animals to rely on olfactory cues, body turn strategies or other uncontrolled landmarks. The only available cue was a salient striped pattern placed at 135° to the goal as viewed from the choice point.

After the learning period, during which the animals had to learn to reach a baited goal from any of three starting points (9 trials/day, 3 trials from each starting point), delayed reaction tasks were conducted. Intervals ranging from 30 seconds to 15 minutes were introduced between the moment when the relevant cue was still available and the moment when the subjects were allowed to choose.

With groups A and S, the procedure was the same as that described in the cat study. The animals were placed at a starting point in a transparent cage and allowed to collect information for 30 seconds. The cage was then covered with an opaque screen for a predetermined period during which the cue was removed. At the end of this period, the animal was released and left to make its way to the goal in the cue-less set-up.

With group C, the procedure was slightly different since these subjects could pick up information only at the central choice point. In this case, the animal was placed straight away at the central choice point, in a special transparent restraining device. After a 30-second observation period, an opaque screen was put over this central device and the cue was removed. After a given time interval, the central device was completely removed and the subject had to choose the goal arm from the choice point with no cues available.

The mean number of days required by the subjects to reach the learning criterion (80% of correct responses) was 9.33, 12.83 and 15.64 for groups A, C and S, respectively. An analysis of variance revealed a significant difference between groups A and S, but no difference between A and C or between C and S.

The results of the delayed reaction tasks are presented in Fig. 5 which shows the number of subjects in each group which were successful with each length of interval. With intervals longer than two minutes, the performances of group C were affected. Working towards intervals of 8 minutes, no subject from group C was able to choose the correct goal arm, whereas the performances of group A and S were quite good, even with intervals of 12 and 15 minutes. No significant difference between groups A

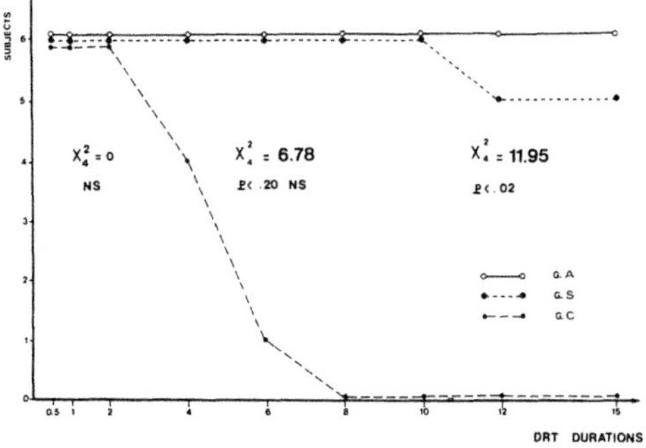

Fig. 5. Number of hamsters scoring above chance during the delayed reaction tasks, for each delay (in min.).

and S was observed (the results of the chi square statistical analyses are shown in Fig.5).

In all three groups in the present study, the only relevant visual cue was spatially dissociated from the goal location, and no landmark could be directly associated with the goal location. Thus, while the inter-group difference in the cat study was linked to the difference in location of the cue in relation to the goal, the difference between the groups in the hamster study was caused by the fact that the animals were required to collect the relevant information from different places, depending on how the screens were arranged above the arms of the cross-maze. Animals in group C were induced to adopt a reorientation strategy, that is, to give a specific response determined by the subject's position at the choice point in relation to the cue location. Thus, successful orientation system depended closely upon their understanding of a single spatial relationship viewed from a specific place. In a sense, orientation was thus founded upon the use of an egocentric framework. By contrast, group S subjects had to determine the goal location very early in each trial, right from the starting point. The main assumption was that under this condition, orientation would be based on an allocentric frame of reference wherein the goal is objectively located in relation to the cue whatever the animal's own location. Moreover, planning of the displacement was imposed by the arrangement of the screens above the maze, which made it impossible to check on the cue location at a latter stage. The processes involved in this type of orientation might therefore be closely related to those involved in cognitive mapping.

The results of delayed reaction tasks show that hamsters assumed to use a mapping strategy (group S) or free to use either strategy (group A) achieved considerably better results than hamsters required to use a reorientation strategy (group C). This difference does not appear to be linked to the difficulty of the initial learning task since there was no obvious relation between the number of days-to-criterion and the scores on delays. An alternative explanation is that animals master delays better in test trials if they have learned to cope with delays between seeing the cue and making the choice during training. Group S necessarily experienced such a delay in training. By contrast, group C always had the cue visible just before making their choice. However, this argument does not account for group A's ability to cope with delay. Thus, unless these results are merely due to uncontrolled procedural biases (we are currently conducting control experiments to check on this possibility), this investigation can be said to bring to light some general properties of spatial behaviours.

CONCLUSION

Our experiments with both cats and hamsters demonstrated that animals that were induced to adopt a mapping strategy were markedly less affected by delays than animals using either a guidance strategy or a reorientation strategy.

These results are in agreement with those obtained by O'Keefe and Conway (1980) with rats in comparable situations. The taxon behaviours are strongly dependent on the temporal sequence of landmarks which determine the responses : therefore, they rely more on the perceptual phenomena and the memory system would seem to encode "responses" rather "places".

On the other hand, maps (locale system), which are spatial representations, seem to indicate places of importance in the environment. Responses might depend on spatial relations between these places. The results of the present delayed reaction experiments appear to support the dissociation between taxon and locale strategies. If we assume that, under some condition (for example, when information is gathered from the starting points), the separation between the cue and the goal requires the animal to store the relations between the cue, the goal and the starting points in its memory, then, once the relevant cue has been located, the determination of the correct path can proceed on the basis of this representation (even when the cue has been removed as in delayed problems).

On the contrary, taxon strategies (guidances or orientations) do not require the animals to anticipate the movements to be made in order to reach the goal. The memory deficit shown by animals using taxon strategies in delayed problems thus suggests their orientation relied on cue-dependent responses rather than on places : when the cue was removed, the animal had no means of determining the appropriate response.

Before concluding, we would like to point out a methodological difference between the present procedure and that used by O'Keefe and Conway (1980), that deserves further consideration.

The mapping condition of O'Keefe and Conway's experiment involved a number of cues distributed around the cross-maze, whereas in the present experiments the mapping situations involved the use of a single cue spatially dissociated from the goal location. Hence, it could be argued that the present experiments might encourage a cue rather than a map strategy, since the environment in which the maze was placed was impoverished with the exception of the one salient cue.

As we stressed in a previous publication, however, in comparing the learning rates of subjects using different orientation strategies, a necessary prerequisite is to equate the number of loci defined by the cues (Poucet, 1985). This can be achieved by using either one or several cue-locus pairs, but it is not unlikely that the use of several cues in a situation designed _a priori_ to induce a guidance strategy (for example, three different cues disposed above the goal and above two starting points, respectively) would actually bias the animals towards adopting a mapping strategy, or some other uncontrolled strategy (such as a "conditional motor response", defined as a learned motor response associated with each starting point). At least the problems surrounding the use of several cues can be circumscribed by using a single cue. Although it may at first sight appear unwise to speak about mapping in experiments involving a single cue which is spatially dissociated from the goal, such experiments can indeed serve to assess animals' ability to deal with complex spatial relations. Therefore, if animals construct maps with such spatial relations, the study of the way these constitutive elements of maps are encoded makes it possible to draw more general conclusions about maps themselves. On another hand, as such experiments allow behavioural analyses of information gathering, they can be viewed as one of the possible means of precisely characterizing the orientation strategies. Poucet (1985) has already pointed out that it is crucial to define the various strategies actually used by different animals before comparing their respective properties. Obviously, any attempt for such comparisons would lead to misinterpretations if this condition is not fulfilled (for example, lack of behavioural analyses would have led to claim that subject A in the cat experiment was using a mapping strategy while it was actually using a reorientation strategy).

Finally, the present experiments raise several questions concerning the nature of the information actually remembered and used after a certain time by the animals which succeeded in the delayed reaction problems. Did that information actually concern the cue location or rather the goal location, or even the response to be made ? Answering these questions may be a necessary step towards understanding the processes involved in orientation.

REFERENCES

Fletcher, H.J., 1965. The delayed-response problem. in "Behavior of non-human primates". Schrier, Harlow and Stollnitz (Eds.), Academic Press, New york-London.

Hull, C.L., 1952. "A behavior system". Yale University Press, New Haven.

Hulse, S.E., Fowler, H. and Honig, W.K., 1978. "Cognitive processes in animal behavior". Erlbaum, Hillsdale, N.J.

Hunter, W.S., 1913. The delayed reaction in animals and children. Behavior Monographs, 2 (serial n°6).

Morris, R.G.M., 1981. Spatial localization does not require the presence of local cues. Learning and Motivation, 12 : 239-260.

O'Keefe, J. and Conway, D.H., 1980. On the trail of the hippocampal engram. Physiological Psychology, 8 : 229-238.

O'Keefe, J. and Nadel, L., 1978. "Hippocampus as a cognitive map". Clarendon Press, Oxford.

O'Keefe, J. and Nadel, L., 1979. Précis of O'Keefe & Nadel's The hippocampus as a cognitive map. The Behavioral and Brain Sciences, 2 : 487-533.

Poucet, B., 1985. Spatial behaviour of cats in cue-controlled environments. The Quarterly Journal of Experimental Psychology, 37B : 155-179.

Shemyakin, F.N., 1962. General problems of orientation in space and space representations. in "Psychological science in the USSR". B.G. Ananyev (Ed.), U.S. Office of Technical Reports, Arlington, Va.

Thinus-Blanc, C., 1985. A propos des "cartes cognitives" chez l'animal : examen critique de l'hypothèse de Tolman. Cahiers de Psychologie Cognitive, 4 : 537-558.

Tolman, E.C., 1948. Cognitive maps in rats and men. Psychological Review, 55 : 189-208.

THE ROLE OF INTRAMAZE STIMULI IN SPATIAL PROBLEM SOLVING

Charlene Wages

Georgia State University[1]
Atlanta, Georgia 30303

O'Keefe and Nadel (1978) proposed a neuropsychological mechanism by which environmental stimuli could be encoded into an "information structure" (p.78) such that the distance and direction between those stimuli were specified. They referred to this mechanism and its ensuing cognitive representation as a cognitive map (O'Keefe and Nadel, 1978, see p. 2). According to this hypothesis, organisms establish cognitive representations that locate stimuli in space relative to each other and to the total environment. This hypothesis stressed the importance of topographical relationships among environmental stimuli in defining places (Suzuki, Augerinos, and Black, 1980) and ultimately guiding spatial behavior (i.e., behavior which requires that an animal move through space to achieve a goal). The role of distal stimuli in spatial behavior has been shown in a number of studies (Hebb, 1949· Stahl and Ellen, 1974; Sutherland and Dyck, 1984). Presumably distal stimuli allow animals to use mapping strategies as opposed to egocentric or other non-mapping mechanisms in solving spatial problems because perceptually, distal stimuli do not change their relative positions as the organism moves. O'Keefe and Nadel (1978, p. 73) note, however, that this property of distal stimuli precludes their use in distinguishing among places in the environment and have cautioned investigators not to disregard the role of proximal stimuli in spatial behavior. Indeed, O'Keefe argued that "spatially separated intramaze cues can also serve as place cues" (1979, p340).

The purpose of the present experiment was to determine whether animals could solve a spatial problem when the only salient environmental stimuli available in the problem space were proximal, intramaze stimuli. Probes were used to determine the strategy used to solve the problem.

METHODS

Spatial Task

The task used was the Maier 3-table problem. This is a three stage problem in which the organism must combine information in the third stage with information that was acquired during the first two stages in order to perform correctly. During Stage 1, the animals were allowed to locomote as a group over the entire apparatus. During the exploration phase no food was on the apparatus. In the traditional test procedure all tables, obstruction screens and runways of the apparatus are black· thus, to learn the spatial relationships between the tables, an animal has to learn the spatial relationships among the environmental stimuli associated with each table. During Stage 2, a dish of food was placed on one of the tables which was designated as the food table for that day. The animals were placed as a group on that table and allowed to eat for 30-60 sec. A second dish of food, which served as an olfactory control, was placed on a second table and covered with wire to prevent the animals from eating the food should they have gone to that table on the test trial. During Stage 3, or the test phase, the animals were tested individually by placing them on the remaining table of the apparatus which was the start table for that day. In order to get further access to the food an animal had to return directly to the table on which it was fed. If it went to the table on which it was not fed, an error was scored and it was immediately removed.

Each table served as the food table an equal number of times during a testing sequence. Right or left turns led to the correct table equally often. Because of these restrictions an animal which adopted a turn strategy or a strategy of going to a given table whenever it was available would have scored at chance levels.

Subjects and Apparatus

Twenty-two Long-Evans hooded rats were assigned to either an INTRAMAZE or an EXTRAMAZE stimulus condition. These groups were further divided into groups which were tested on either a large apparatus with runways of 154 cm or a small apparatus with runways of 11 cm. In each condition the apparatus was surrounded by black curtains which were hung from the ceiling of the room so that they formed a rectangular enclosure for the the apparatus. For the large apparatus this enclosure measured 3.15 x 3.95 m. For the small apparatus this enclosure measured 2.31x2.16 m. In each room, the apparatus consisted of three runways radiating from a common center point. Each runway terminated at the entrance to an individual table. The angle between any two adjacent runways was 120 degrees.

The INTRAMAZE and EXTRAMAZE stimulus conditions were identical for the groups regardless of the size condition in which they were tested. For the INTRAMAZE stimulus group, each table and runway was marked with a distinctive set of wooden stimulus inserts. These inserts covered not only the surfaces of the tables and runways but also the front and back of the obstruction screens which fronted the tables. One table was covered with a solid gray insert; another was covered with a black and gray striped pattern; the third table was covered with a solid flat black insert. These inserts were on the apparatus during all phases. For the EXTRAMAZE group three distinctive stimuli were hung on the curtains surrounding the apparatus. Two rectangular poster boards (58x88 cm) were used. One was solid white while the other was a white and black striped slanted line pattern. The third stimulus was a 6-point star pattern cut from the same size white poster board.

RESULTS AND DISCUSSION

Standard testing

An analysis of variance with stimulus condition and size of apparatus as the between groups factors indicated that at the end of the 28 day testing period there were no differences between the groups either in terms of the type of salient stimulus present in the test area or the size of the apparatus on which the animals were tested.

The behavior of those animals which scored 68% or more over the 28 days of testing (a value which would have occured by chance only 4 percent of the time) was analyzed further. The bars with large stripes in Figure 1A represent the average performance of the nine successful animals on Days 25-28 of testing. Each of the INTRAMAZE animals were correct an average of 75% of the time over this period, while the animals in the EXTRAMAZE group were correct 78% of the time during this period. The range of individual scores is indicated by the black, vertical bar.

At this time the important question was whether or not the successful performance shown by both groups was based on representations that encoded the respective salient stimuli. If the strategies were based on such a representation, then the successful performance should have deteriorate when the stimuli were removed. If, however, the successful performance were based on representations that encoded uncontrolled, extraneous stimuli, then removal of the salient stimuli should have had no effect on performance.

Probe 1

The purpose of Probe 1 was to determine that the animals were, in fact, encoding the salient stimuli. In order to preclude the possibility that, following the removal of the salient stimuli, the animals would begin to develop a spatial representation incorporating stimuli which heretofore had not been involved, the animals were not allowed to explore the apparatus during the probe phase (see Herrmann, Bahr, Bremner and Ellen, 1982 for the role of exploration).

During the first probe period the salient stimuli remained for 3 of the INTRAMAZE animals and were removed for 2 of them. In the EXTRAMAZE condition, the stimuli remained for 1 animal and were removed for 3 of the animals. The performance of the individual animals during the probe is represented in Figure 1A by bars containing either no lines or small lines.

The responses of the individual animals to the probe were not uniform. One animal in each stimulus condition (IM4 and EM2) was unaffected by the removal of the stimuli. For these two animals, it is clear that the representation guiding the behavior was not dependent upon the salient stimuli which were provided. It is thus impossible to know if the successful problem solving of IM4 was in fact based upon a representation incorporating the intramaze stimuli.

The behavior of three animals (IM5, EM3 and EM4) deteriorated to below criterion levels when the salient stimuli were removed. For these animals, it is clear that it was the salient stimuli which were provided that were critical for the successful problem solving.

Three animals for which the stimuli were not removed (IM1, IM2,and EM1) continued to perform successfully even though they did not receive exploration prior to the test trial. The remaining animal (IM3) failed to perform at criterion level during this probe. The failure of this animal raised the possibility that the deterioration in performance shown by IM5, EM3 and EM4 was due to the lack of exploration rather than the removal of the salient stimuli. It was thus necessary to establish that it was the disruption of the test environment (i.e., removal of stimuli) rather than removal of the exploration which led to the failure of the animals for which the stimuli were removed. To accomplish this, a second phase of the experiment was conducted.

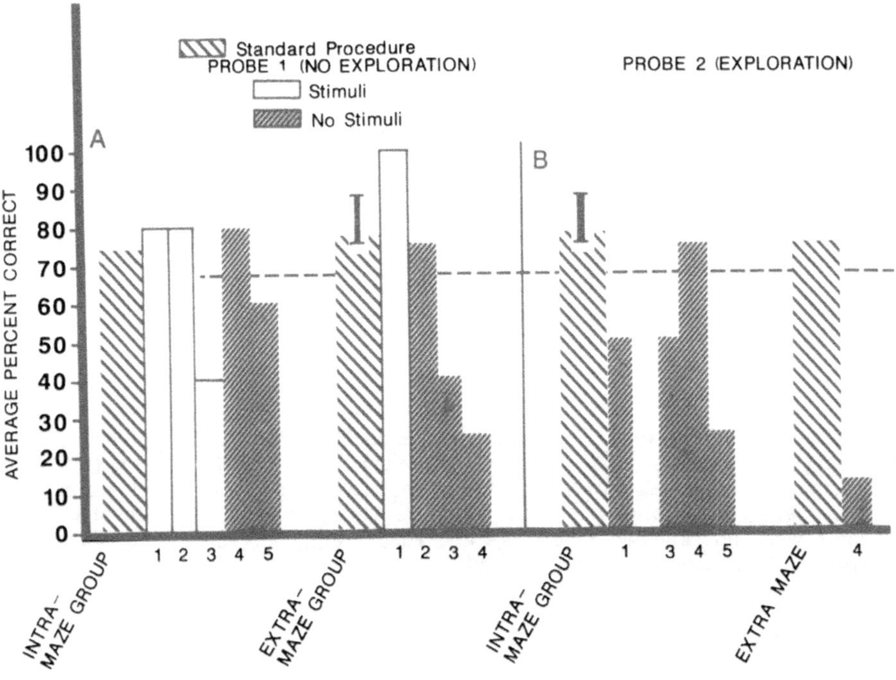

Figure 1. Performance during standard testing and during probe trials; A. last 4 days of standard testing, and Probe 1 B. second phase of standard testing and Probe 2.

Standard testing

The standard testing procedure was reinstated for all animals for 8 days. The appropriate set of intra- or extramaze stimuli was present for these animals during this eight day period. The performance during the eight day standard test period is indicated by the bars with large stripes in Figure 1B. Five of the nine animals scored above 68% for that period. For 4 of animals which were in the INTRAMAZE group the scores ranged from 75-88%. The single successful EXTRAMAZE animal was correct 75% of the time. These animals were then tested in a second probe phase, while the remaining animals were dropped from further experimentation.

Probe 2

During the second probe the animals were given the daily exploration phase, but the salient stimuli were removed. As was the case with Probe 1, IM4 was unaffected by removal of the stimuli thereby providing further support for the view that this animal's successful performance on the task was based upon a representation containing some extraneous stimuli other than those which were provided.

The performance of each of the other animals fell to below criterion levels, thereby indicating that it was the removal of the salient stimuli and not the lack of exploration during the first probe that was responsible for the deterioration of performance during that probe. Thus the success of IM1 and IM2 during the first 28 days of testing and during the first probe can clearly be said to depend upon representations involving intramaze stimuli.

The present experiment represents the first attempt at examining the strategies mediating performance on the traditional 3-table task. The results obtained reinforce the view that a performance score in a spatial task is not sufficient to identify the nature of the strategy mediating the performance (Olton, 1979). Rather, probe trials must be conducted to identify the mechanism involved.

The diversity of results obtained in these animals during the probes illustrates the kind of strategies involved in performance on the 3-table task. The findings are consistent with the view of Ellen, Parko, Wages, Doherty, and Herrmann, (1982) which suggested that succesful performance on the 3-table task could be achieved through the use of either a mapping or egocentric strategy. During the first probe, animals IM1 and IM2 did not need daily exploration of the apparatus in order to be successful. Ellen, Soteres, and Wages (1984) showed that when rats are unequivocally using a cognitive map strategy, merely feeding the animals on one of the tables prior to a test trial is sufficient to reactivate the representation of the test environment. The ability of animals IM1 and IM2 to solve the problem as long as the enviromental stimuli were present, even in the absence of exploration, suggests that a cognitive representation in which distance and direction between stimuli were specified existed for these animals.

In contrast, the failure of IM3 to perform successfully even in the presence of the salient intramaze stimuli (first probe) suggested that this animal did not have a cognitive representation which could be redintegrated by the daily feeding experience. Certainly to the extent that the animal could not perform successfully even when the stimuli were still in the environment, one could not argue that the animal had a cognitive representation of the distance and direction between those stimuli. The fact that both exploration and stimuli were neccessary for IM3 to perform successfully suggested that this was an animal which did not have a cognitive representation of the invariant features of the test environment; rather each day it had to learn the spatial relationship which depended upon the position of the stimuli with respect to itself as it moved through the environment. This type of strategy is illustrative of an egocenentrically-based non-mapping strategy.

REFERENCES

Ellen, P., Soteres, B., and Wages, C. 1984. Problem solving in the rat: piecemeal acquisition of cognitive maps. Animal Learning and Behavior, 12: 232-237.

Ellen, P., Parko, M., Wages, C., Doherty, D., and Herrmann, T., 1982. Spatial problem solving by rats· exploration and cognitive maps. Learning and Motivation, 13: 81-94.

Hebb, D. O., 1949. "The Organization of Behavior." John Wiley & Sons, Inc., New York.

Herrmann, T., Bahr, E., Bremner, B., and Ellen, P., 1982. Problem solving in the rat: stay vs. shift solutions on the three-table task. Animal Learning and Behavior, 10: 39-45.

O'Keefe, J. and Nadel, L. 1978. "The Hippocampus as a Cognitive Map." Clarendon Press, Oxford.

O'Keefe, J., 1979. Hippocampus function: does the working memory hypothesis work? Should we retire the cognitive map theory? The Behavioral and Brain Sciences, 2:339-343.

Olton, D., 1979. Mazes, maps, and memory, American Psychologist, 34: 588-596.

Stahl, J. M. and Ellen, P., 1974. Factors in the reasoning performance of the rat. Journal of Comparative and Physiological Psychology, 87: 598-604.

Sutherland, R. J. and Dyck, R. H., 1984. Place navigation by rats in a swimming pool. Canadian Journal of Psychology, 382: 332-344.

Suzuki, S., Augerinos, G., and Black, A., 1980. Stimulus control of spatial behavior on the eight-arm maze in rats. Learning and Motivation, 11:1-18.

Footnote

1. Current address: Francis Marion College, Department of Psychology, Florence, SC 29501.

RATS USE THE GEOMETRY OF SURFACES FOR NAVIGATION

Ken Cheng

University of Sussex

Radial maze studies give evidence that the rat can remember a number of locations in space by their relations to surrounding landmarks. Typically in these experiments, the rat is placed on the centre platform of a maze with a number (usually eight) of arms radiating outwards from the centre. It is allowed to go down the arms to collect food pellets placed beforehand at the ends of the arms. The rat, after only short training, seldom revisits an arm it has already been to. In successfully accomplishing this task, the evidence shows that the animals use primarily the relations of the target locations to the surrounding landmarks, and not the smell of food pellets, smell left on the arms of the maze, or stereotypic response patterns (Maki, Brokofsky, & Berg, 1979; Olton, 1978, 1982; Olton & Collison, 1979; Olton, Collison, & Werz, 1977; Olton & Samuelson, 1976; Roberts, 1984; Suzuki, Augerinos, & Black, 1980; Zoladek & Roberts, 1978).

The experiments reported here, given in fuller detail elsewhere (Cheng, submitted), address the question of what aspects of a goal's spatial relationships to surrounding landmarks the rat uses. Results show that the rat uses purely geometric information, the geometric relations between the target and the shape of the environment. A model is then presented in which the rat relies primarily on a geometric module containing only geometric information, and not non-geometric properties of surfaces and locales, such as reflectance characteristics (black vs. white), texture (rough vs. smooth), or smell (anise vs. peppermint), which are also called featural information here.

Systematic errors and reactions to spatial transformations indicate that the rat uses the geometric relations between the target and the shape of the environment in locating a place. Consider first the experimental logic behind the former. Within the rectangular environment in which the rats were tested (Figure 1A), they confused geometrically equivalent locations, those that stood in the same geometric relations to the shape of the environment. Figure 1 illustrates this geometric ambiguity. The surfaces making up the rectangular box differ in non-geometric properties: One long wall is white while the other is black. Distinctive panels differing in appearance, texture, and smell make up the corners. Suppose that the filled dot in Figure 1A is the location sought. Imagine a navigator using a map specifying only the shape of the environment, like Figure 1B, with the filled dot representing the target. This shows the metric configuration of surfaces as surfaces, but not what the surfaces themselves look like, the information depicted in Figure 1C. In this rotationally symmetric environment, two locations fit the specifications represented by Figure 1B: the correct location, and a rotational equivalent, located at 180° rotation from the target through the centre of the box (the open circle). The map in Figure 1B can be matched to the actual environment in two equally good ways, one 'correct' congruence, and one 'erroneous' one that arises when the record is rotated 180° with respect to the environment, with the top wall in Figure 1B lining up with the bottom wall in Figure 1A. The latter rotational error is a match solely on geometric grounds. If the target were specified by its relations to the arrangement of featural information, for example, as in Figure 1C, only one match between map and environment can be found. Here, when the map is rotated 180° with respect to the environment, the non-geometric information on the map would not match the non-geometric information in the environment, even though the surfaces on the map line up with the surfaces of the environment. The use of a shape record then, should lead to systematic rotational errors. Conversely, the existence of systematic rotational errors demands some explanation in terms of using purely geometric relations between the target and the shape of the environment. Ideally, using only a shape record at all times should lead to 50% correct searches and 50% rotational errors.

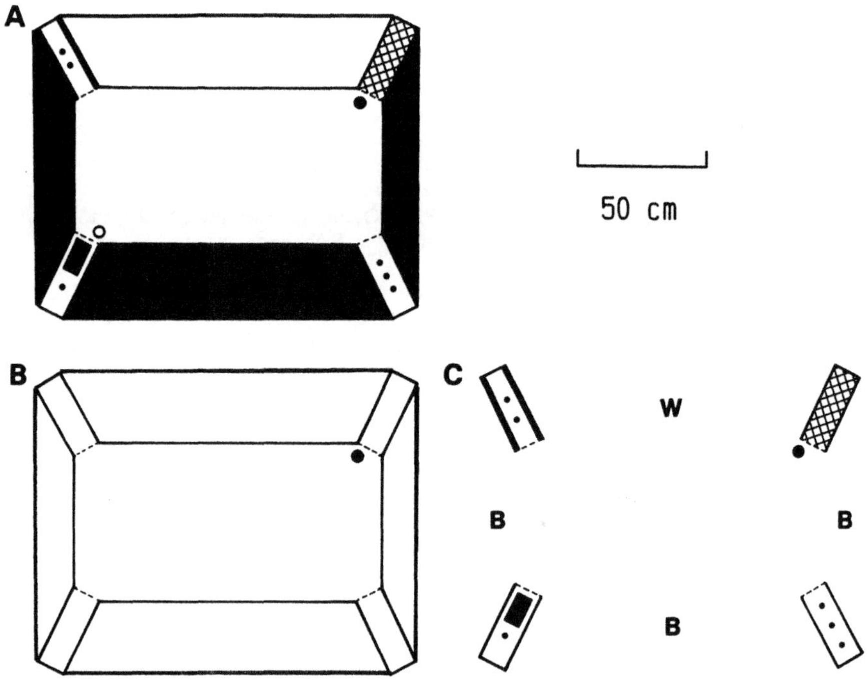

Figure 1. Geometric and featural information in the rectangular box used in the experiments. (A) A plan view of the environment. Three walls are black while one is white. Panels (11 cm X 31 cm) differing in visual, tactile, and olfactory features are in the corners. The filled circle represents the location of hidden food. (B) The shape of the environment, which contains geometric information. Specifying the target's location only by its geometric relations to the shape results in an ambiguous specification of the location. The open circle in the lower left in (A) stands in the same geometric relations to the shape of the environment as the target location, and cannot be distinguished from the target location on geometric grounds alone. (C) The arrangement of featural information, with the letters W and B representing white and black respectively. Specifying the target's relations to this information provides an unambiguous specification of the place.

EXPERIMENTS

In a working memory task in this rectangular environment, rats' performance approached this theoretical ideal. The location to be remembered on a trial was chosen at random for each trial from one of 80 locations. An animal was shown the location of food, was removed, and put back in an exact replica of the environment 75 s later, with the food buried. It must be stressed that the entire apparatus was rotated between trials and during the delay periods, in order to prevent the animals from using any spatial framework external to the experimental box. All rats reached asymptote on this working memory task quickly. When the three rats were run in the box shown in Figure 1A, each already had 192 trials in various versions of the rectangular box. All produced similar stable results from the start. The rats on average, out of 60 trials each, made 44% correct digs, 25% rotational errors, and 31% unsystematic errors. These figures do not differ across target locations. Each rat made rotational errors systematically and none made significantly more correct digs than rotational errors. To cast this quantitatively, on 69% of the trials, they dug at a location with the correct geometric relations to the shape of the environment (correct dig or rotational error). Of these geometrically correct digs, 64% stood in the correct relations to the arrangement of features as well. This amounts to 28% above chance. One interpretation then, is that the rats used the purely geometric relations between target and shape of environment on most trials, but rarely used featural information in addition.

Can rats use featural information at all in searching for a place? Reference memory tasks in the same environment show that they can. In these tasks, identical looking bottles of food were placed at the corners. The rat's task was to knock over the one at the correct corner, the only one from which food came out. For each rat, the same corner was always correct. The entire experimental apparatus was again rotated from trial to trial to prevent the use of spatial frames of reference external to the box. Four rats were tested both in the box shown in Figure 1A and in a box with all four walls black but containing the featural panels in the corners. The animals on the whole learned the task quickly (averaging 58% correct on the first 30 trials). Asymptotic performance was similar across rats and environments. Combining all rats in both environments, 74% of the first choices were correct, 21% were at the corner diagonally opposite the correct corner, and only 5% at the other two corners, adjacent to the correct corner. When the food stayed in one corner of the box from trial to trial, all rats learned to choose the correct corner in preference to its geometrically equivalent diagonally opposite corner, even when the white wall was rendered black like its opposite wall. Hence, rats can use the non-geometric information contained in the panels to distinguish between geometrically equivalent locations.

The rats still on the whole, however, made systematic rotational errors: 81% of all errors were diagonal errors (chance being 33%). To cast this quantitatively, on 95% of the trials at asymptote, the rats chose a geometrically correct corner, the correct corner or its diagonal opposite. This is 90% above chance. Of the geometrically correct choices, 77% were at the corner with the correct features, amounting to 55% above chance. This is statistically lower than the 90% figure at the .001 level. Hence, the rats distinguished geometrically distinct corners (adjacent pairs), those that stood in different geometric relations to the shape of the box, more often than geometrically equivalent corners (diagonally opposite pairs) requiring for their distinction the use of featural information.

When the rats had achieved success in the box with all walls black, they were further tested with the distinctive panels from the food corner and its diagonally opposite corner removed. This forced the animals to use featural information at some distance from the goal location. Each rat suffered a similar immediate decline lasting the 120 test trials. They continued to choose a geometrically correct corner, but failed to distinguish the two. On the last 50 trials, they chose the correct corner 47% and its diagonal opposite 53%. When the featural information near the geometrically correct locations were removed, rats failed to use the remaining featural information distant from the geometrically correct corners to disambiguate geometrically equivalent locations, although they still used the geometric relations between the target and the shape of the environment.

The animals were then retrained with all four panels present. Three relearned quickly and one continued to make diagonal errors as often as correct choices. For the three successful rats, geometric transformations altering spatial properties were carried out to investigate how featural information is used. Changing properties that are crucial to the animal ought to result in declines in performance. Fuller accounts of the logic behind geometric transformations are given elsewhere (Cheng & Gallistel, 1984; Cheng, submitted). Here, space permits but a brief summary. The idea was to test whether a change in the geometric relations between the target panel and the shape of the environment leads to declines. Two transformations that minimally changed the geometric properties in the arrangement of panels were used. In one, the affine transformation, each panel moved one corner over, all in the same clockwise or anticlockwise direction, with the food moving likewise and staying at the same panel as before. In a second, the reflection, adjacent panels along either the long walls or the short walls exchanged places. The food again remained at the same panel. Both transformations carry the target panel to a corner with different geometric relations to the shape of the environment. If the target corner had a long wall to the right of the short wall as one faces it before the transformation, it had a long wall to the left of the short wall afterwards. Under these transformations, as in the earlier work (Cheng & Gallistel, 1984), the performance of every rat dropped significantly, here on average from 91% correct in retraining to 47% correct.

Both the affine transformation and the reflection change the overall arrangement of panels as well as the geometric relations between the target panel and the shape of the environment. The rats, however, were not reacting to this change in the arrangement of featural information. They were exposed to another transformation that combined the affine transformation with the reflection, thus changing all the properties in the arrangement of panels changed by either the affine transformation or the reflection. Carrying out both a reflection and an affine transformation results in one diagonal pair of panels exchanging places. With this diagonal transposition, two panels that were adjacent along a short wall before the transformation were adjacent along a long wall afterwards, and the one formerly to the right was to the left afterwards. The food, however, stayed at the same panel and, more importantly, also stayed in geometrically the same kind of corner as before. No rat's performance dropped significantly under the transposition, where they averaged 80% correct. Hence, the overall arrangement of featural information can be altered without adversely affecting performance, but a change in the geometric relations between the target and the shape of the environment results in decrements. In heading to a target panel, the rats were sensitive to, and hence must have encoded, the geometric relations between it and the shape of the environment.

To summarize the results, in a working memory task, where the target location varied from trial to trial, the rats were geometrically correct on most trials, but made almost as many rotational errors as correct digs. In a reference memory task, where the food stayed in one location from trial to trial, rotational errors were much reduced, although still systematic. But the animals did use featural information, since they chose the correct corner more often than its geometrically equivalent diagonal opposite. This success depended on having the panels at the geometrically correct corners present. Removal of these, or movement of these under an affine transformation or reflection to a geometrically different corner, led to decrements in performance. The overall arrangement of features on the other hand, can be altered, as in the transposition, without adversely affecting performance.

A GEOMETRIC MODULE

The systematic errors that the rats make, the rotational errors, indicate that they do use the environmental shape in specifying a location in space. These errors are at locations that are correct in all their geometric relations to the shape, but wrong in their relations to the arrangement of featural information. The rats must have used the metric relations between the goal and the shape of the environment to do this. To explain the systematic rotational errors, I propose a unit in the rat's spatial representation in whose normal functioning such errors would arise. This unit of the mind, which I will call the metric frame, encodes only the geometric properties in the arrangement of surfaces as surfaces. It encodes the shape of the environment, including all geometric properties in that shape, but contains no featural information at all. Features are kept in other records. Their whereabouts are specified by address labels indicating where on the metric frame they are to be found. To find a desired feature, such as hidden food, the rat usually specifies that target's address on the metric frame, and then matches that address to the shape of the environment. When two matches between metric frame and perceived world are possible, ambiguity arises, leading to rotational errors. It must be stressed, however, that in all normal environments, the shape match would be unique, and an address on the metric frame unambiguously specifies a location in space.

The use of the metric frame alone suffices to account for the rat's successes in spatial tasks in almost all environments. But the rats here under some circumstances used featural information to distinguish between geometrically equivalent locations. The data tentatively suggest the following way of using features to help specify a place. A target place is still specified as an address on the metric frame. But in addition, some requisite features near the target are retrieved from the featural records and 'glued' onto the metric frame. The target must now match specifications on its geometric relations to the environmental shape plus specifications on some nearby features. Nearby features seem important since when the panels at the geometrically correct corners were removed, the rats failed to use the remaining features distant from the geometrically correct corners to distinguish between geometrically equivalent locations. The matching process does not use features independent of geometric address or the overall arrangement of features. A change in the geometric relations between a target panel and the shape of the environment (as in the affine transformation or reflection) led to decrements in performance, while overall arrangement of features can be changed (as in the diagonal transposition) without adversely affecting performance.

If something like the theoretical story I have sketched above is true, the rat's mind, in using landmarks to organize spatially guided behaviour, exhibits modularity (Fodor, 1983), a notion further explained by Neisser in these volumes. Different theoretical units record and organize different kinds of information. The metric frame records only the shape of the environment, and featural records deal only with certain features in the environment. The metric frame serves as a coordinating centre in organizing spatially guided behaviour. The locations and spatial relations among non-geometric features are checked through its mediation. Normally, only the address of the goal needs to be specified on the metric frame. To specify another feature near the goal means another retrieval step. In my view, this extra step, rather than differences in salience between geometric and featural information, is the reason why rats here sometimes used features less well in spatial orientation. The fact that the rats used featural information well under some circumstances, plus the particular fashion in which transformations on featural arrangement affect performance suggest some story along this line, even though details are unclear.

This is not to say that the only way the rat uses features is through the mediation of the metric frame. The records of features may surely be used sometimes without having to check up on their addresses. Discrimination experiments have shown that rats can under some circumstances head to a feature irrespective of its location. Rats placed in a pool of water can learn rapidly to head to a visible platform irrespective of its location (Morris, 1981), although they can also find a hidden platform that stays in one location. This makes teleological sense: sometimes, some things are desired irrespective of their location. The conditions under which the animal uses features in this mode rather than through the mediation of the metric frame are uncertain.

CONCLUSION

The experiments here show that the rat, in relocating a place by the use of landmarks, uses the geometric configuration of surfaces. In fact, I suggest that it relies primarily on a unit of the mind encoding only the geometric shape of the environment. Other records deal with non-geometric features, specifying their whereabouts by address labels indicating where they are on the geometric module or metric frame. If something like this is true, the rat's mind exhibits modularity in the organization of spatial knowledge. Different theoretical units encode and deal with different specific kinds of information.

ACKNOWLEDGEMENTS

The research reported here was done in partial fulfillment of the requirements for the degree of Doctor of Philosophy at the Department of Psychology, University of Pennsylvania. It was supported in part by a Biomedical Research Support Grant, RR 07083-18 Sub. 10. I am indebted to C.R. Gallistel for advice on this work.

REFERENCES

Cheng, K. A purely geometric module in the rat's spatial representation. Paper submitted to Cognition.

Cheng, K. & Gallistel, C.R., 1984. Testing the geometric power of an animal's spatial representation. In "Animal Cognition". H.L. Roitblat, T.G. Bever, & H.S. Terrace (Eds.), Lawrence Erlbaum Associates, Hillsdale.

Fodor, J.A., 1983. "The modularity of mind". MIT Press, Cambridge.

Maki, W.S., Brokofsky, S., and Berg, B., 1979. Spatial memory in rats: Resistance to retroactive interference. Animal Learning and Behavior, 7:25-30.

Morris, R.G.M., 1981. Spatial localization does not require the presence of local cues. Learning and Motivation, 12:239-260.

Olton, D.S., 1978. Characteristics of spatial memory. In "Cognitive Processes in Animal Behavior". S.H. Hulse, H. Fowler, & W.K. Honig (Eds.), Lawrence Erlbaum Associates, Hillsdale.

Olton, D.S., 1982. Spatially organized behaviors of animals: Behavioral and neurological studies. In "Spatial Abilities: Development and Physiological Foundations". M. Potegal (Ed.), Academic Press, New York.

Olton, D.S. & Collison, C., 1979. Intramaze cues and "odor trials" fail to direct choice behavior on an elevated maze. Animal Learning and Behavior, 7:221-223.

Olton, D.S., Collison, C., and Werz, M.A., 1977. Spatial memory and radial maze performance of rats. Learning and Motivation, 8:289-314.

Olton, D.S. & Samuelson, R.J., 1976. Remembrances of places passed: Spatial memory in rats. Journal of Experimental Psychology: Animal Behavior Processes, 2:97-116.

Roberts, W.A., 1984. Some issues in animal spatial memory. In "Animal Cognition". H.L. Roitblat, T.G. Bever, & H.S. Terrace (Eds.), Lawrence Erlbaum Associates, Hillsdale.

Suzuki, S., Augerinos, G., and Black, A.H., 1980. Stimulus control of spatial behavior on the eight-arm maze in rats. Learning and Motivation, 11:1-18.

Zoladek, L., & Roberts, W.A., 1978. The sensory basis of spatial memory in the rat. Animal Learning and Behavior, 6:77-81.

DISSOCIATION BETWEEN COMPONENTS OF SPATIAL MEMORY IN THE RAT DURING ONTOGENY.

Françoise Schenk, Institut de Physiologie, Faculté de Médecine, Bugnon 7, CH-1005 Lausanne, Switzerland -
FNRS grant no 3.335.082.

Moving between two fixed points in an environment may require various sensory informations (see Baker, Bingman, Etienne, Mitchell, Potegal, this issue). Moreover, this behaviour can be accomplished by different strategies, from the simple association learning to the more elaborate place strategies that require a representational system. O'Keefe and Nadel (1978) have proposed that the hippocampus acts as a cognitive mapping system encoding the absolute spatial properties of an environment and the location of the subject in this framework. This brain structure is a supramodal association area connecting the neocortex with somatomotor and visceromotor control systems (Swanson, 1983). In the absence of a functional hippocampus (e.g. following lesions), the theory predicts that the subject will tend to solve any spatial problem by taxon strategies, i.e. simple S-R associations.

As predicted, damages to the hippocampus induce long lasting deficits of place navigation in a circular pool (Morris et al., 1982; Sutherland et al. 1983). Simple association learning does not really compensate for this deficit, since the only improvement during training results from systematic swimming at a certain distance from the walls. However, more recent work disconnecting the hippocampus from its major afferent cortical input through the entorhinal area has indicated the existence of a spatial subsystem generating navigation routes through a training position (Schenk and Morris, 1985). The rats with lesion of the entorhinal cortex were not able to swim directly to the invisible platform from various starting positions in spite of extensive training, nor did they show a normal focalized searching pattern in the absence of the escape platform. However, they swum very often through the training position thus reaching a high score of passages through the "annulus" indicating the usual location of the escape platform.

This dissociation had not been observed after any other brain lesion; it was not predicted by the cognitive map theory either. As the emergence of cognitive skills, both developmentally and phylogenetically, seems to progressively replace more associative types of learning (Oakley, 1983), we decided to study the emergence of spatial abilities during development, in order to find out whether a similar dissociation might be evident at any age.

Rats, like all altricial Murid Rodents, have an immature central nervous system at birth; they are blind and deaf and incapable of motor coordination until 12-16 days postnatally (Campbell et al 1974). The first spontaneous explorations out of an artificial burrow are observed during the third week, and spatial orientation is likely to develop from this time in order to ensure spontaneous return to the nest. From 20-25 days of age, rats show evidence of spatial ability in an 8-arm maze (Rauch and Raskin, 1984). However, data from spatial discrimination learning (Bronstein and Spear 1972), position habit reversal (Harley and Moodey, 1973; Blue and Ross, 1982) and spontaneous alternation (Douglas et al. 1973) suggest that adultlike spatial behaviour develops in steps over the first 5 weeks of life.

At the same time, the hippocampal formation shows considerable developmental changes (Altman et al., 1973; Cotman et al., 1973; Meibach et al., 1981; Pokorny and Yamamoto, 1981). The improvement of spatial abilities during this time might be related to these neurophysiological changes.

For our experiments, we chose two tasks in which rats have to learn the exact location of a goal, i.e. the place navigation task (Morris, 1981) and the hole board task (adapted from Barnes, 1979). The path taken is in no way constrained and the subjects are free to adopt any of various strategies; they will only differ in their efficiency to reduce the latency to the reinforcement. The results of these different yet similar tasks (Fig. la and b) will be described below.

A. THE PLACE NAVIGATION TASK

As shown on Figure la, rats are trained to escape from water onto a small platform hidden at a fixed position in the pool. Since the light grey walls do not provide any visual cues, this position can only be localised by using distal cues outside the pool.

Escape efficiency is quantified in terms of latency, start orientation and distance swum to the platform (fig. la). A spatial bias to the training position is measured during probe trials, i.e. when the subjects are allowed to swim for

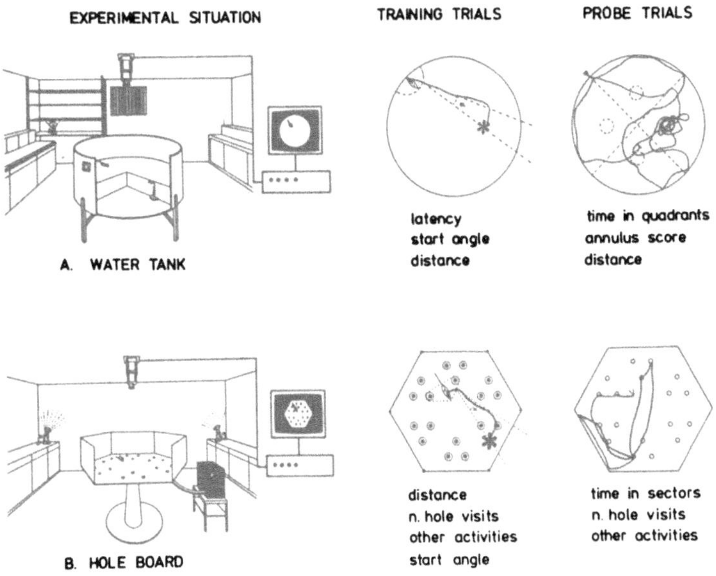

Figure 1 : Schematic drawings of the experimental set ups and charts of representative trials.

30-60 seconds in the absence of the platform. The number of passages through the exact position of the absent training platform is computed and compared with the number of passages through the other three untrained positions (accuracy score).

Four groups of 14 rats were trained for 5 daily sessions (4, 4, 8, 8 and 4 trials) from the age of 21, 28, 35 or 42 days in a small pool (diam 90 cm). Half of the subjects in each group were required to find the invisible platform by using distal room cues only ("place only" condition). The other half was helped by the presence of a small visible cue on top of the platform ("cue+place" condition).

These results have already been presented elsewhere (Schenk, 1985) and will be briefly summarized. Escape efficiency on the 5th day of training is shown on Figure 2, clearly indicating that young rats until 32 days of age were impaired on the "place only" task, while performing efficiently on the cued task. The accuracy of the orientation toward the platform when leaving the starting area increased with age, as indicated by a decrease of vector variance.

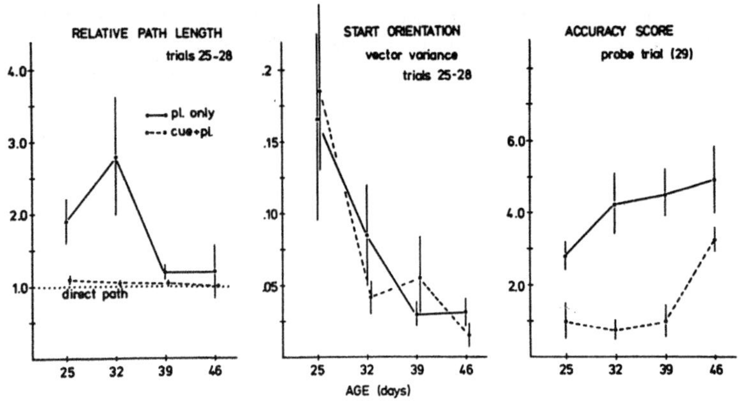

Figure 2 : Mean escape path length and mean of individual vector variance in the place navigation task as a function of age in session V. Mean accuracy scores during the immediately following probe trial.

The behaviour observed during the probe trial, immediately following the 28th training trial, was also age-dependent. The youngest rats trained in the "place only" task already showed a slight bias toward the training position (fig. 2, accuracy score). Whereas in the "cue+place" condition, a bias toward the training position was only evident in the group of the oldest rats.

Globally, these results suggest that there is a clear age limit, above 32 days, for the development of direct approaches from any starting position. On the other hand, knowledge about the training position, as indicated by the behaviour during the probe trials, seems to develop more smoothly and gradually, depending on training conditions.

B. HOLE BOARD TASK

This apparatus (fig. lb) has been adapted from Barnes (1979) with the addition of walls (40 cm high) to keep wild jumping woodmice in the experimental area (cross section 153 cm). The detachable metal fittings of the holes are interchangeable and washable. Their lower part is closed by a tightly fitted plug of foam rubber, covered with a second, loosely fitted disk of foam rubber. The fitting in the training position is connected to a flexible tube and closed by the second disk only. This disk is easily removed by rats of any age and the flexible tube allows escape to the subject's home cage.

The test subject is deposited 50 cm away from the wall, facing the center of the board (fig. lb). Six different starting positions are designed randomly for each session of 6 consecutive trials. Escape efficiency is inferred from path length and from the number of hole visits. Start orientation is calculated when the rat leaves the starting area, arbitrarily defined as a triangular surface including the two nearest holes (fig. lb, start angle). Probe trials are conducted with no hole connected to the home cage. A great diversity of behaviours are recorded during this test, as well as the time spent in each sector and the number of visits to each hole.

Similar experiments with adult rats (Schenk, in prep) have shown that they are highly motivated to disappear into the hole connected to their home cage. They showed a high frequency of nearly straight runs from any starting position to the training hole when trained in the "Place external" condition.

Four groups of 12 rats were trained for 5 daily sessions (3, 6, 6, 6, and 6 trials) from the ages of 18, 22, 26 and 30 days. Half of the subjects of each group were trained to find a hole related to cues outside the hole board, but disconnected from olfactory cues inside the enclosure since the board was rotated between trials (Place External). The other half was trained to locate a hole remaining in a fixed position on the board, but at different spatial locations relative to outside room cues (Place Internal); in this case, the training position was only signalled by cues from the board itself, mainly olfactory ones. An additional group of 6 rats, 42 days of age at the start of training, was tested in the "Place External" condition.

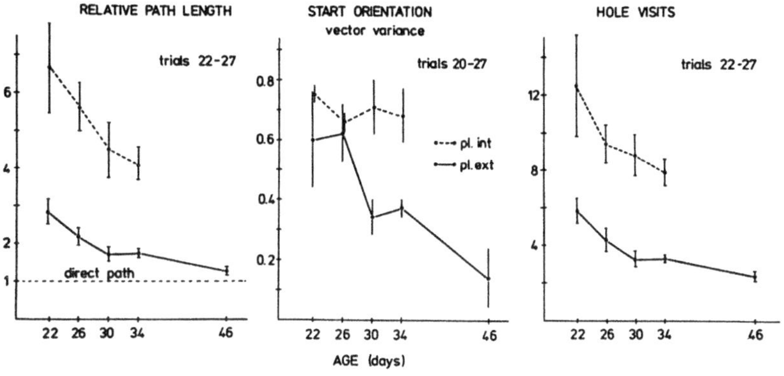

Figure 3 : Mean escape performance in the hole board task as a function of age in session V.

Only a brief summary of the results obtained from developing rats is given in Figure 3, as they will be presented elsewhere together with data from various cued procedures. Rats under 30 days of age were clearly less efficient than older subjects. The former showed little start orientation and followed long circuitous routes. However, it has to be pointed out that, eventhough start orientation was equally poor in both training conditions under 30 days of age, path length was considerably longer when the rats were trained in the "Place Internal" condition, and was accompanied by a higher number of hole visits. The distribution of these visits was biased toward the spatial position where the connected hole had been found on the first trial of a daily session. Surprisingly, this effect was significant only in the youngest rats, as shown by Friedman analyses of variance of the second and third trial of sessions 2-4. Thus, it appears that young rats persevere in visiting holes in the sector related to distal room cues where they have previously found an open hole.

A probe trial in which the training hole was connected only 80 seconds after the subject had been placed on the table was conducted on day 6, with a retention interval of 24h. All groups trained in the "Place External" condition spent most of their time in the training sector. Figure 4 shows that this tendency was very weak in the youngest rats during the first 24-second block of the test (two-way analysis of variance of the 4 youngest groups, age, $F = 1.10$, df 3/20, n.s.; time blocks, $F = 8.042$, df 2/40, $p<.01$; age x time interaction, $F = 3.38$, df 4/40, $p<.01$). The three youngest groups were hyperactive, they crossed a higher number of sections than the 35-day group did (two-way ANOVA, age, $F = 3.91$, df 3/20, $p<.05$; time, $F = 7.58$ df 2/40, $p<.01$; age x time interaction, $F = 2.14$, df 4/40, n.s.)

Detailed analysis of the time intervals spent away from this area indicated significant differences between the three youngest groups in spite of equal activity levels; the two youngest groups had a higher proportion of long intervals away from the training sector ($X2 = 7.0$, 2 df, $p<.05$). The behaviour of the oldest rats, 35 and 47 days of age, was very similar to the one observed in adults, i.e. consisting of an early phase of intense focalisation on the training position followed by a progressive spreading of the movements on the board.As observed in adult rats, no significant effect of the training sector was detected in any group trained in the "Place internal" condition (not shown here).

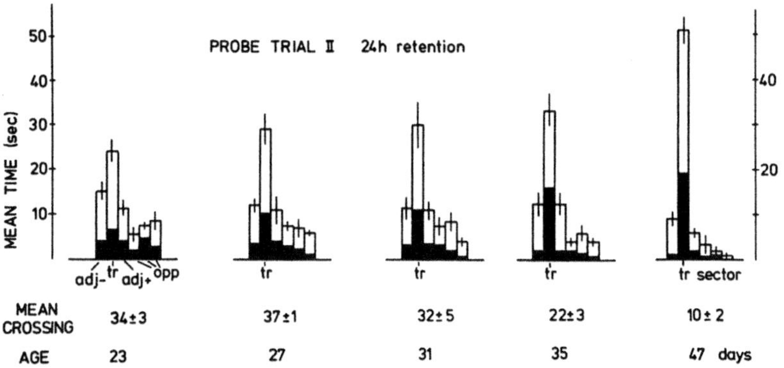

Figure 4 : Mean time spent in each sector of the hole board during the 72-second probe trial in the "Place external" groups. Values during the first 24-sec time block are shown in black.

DISCUSSION

The results from the hole board task confirm those obtained with the place navigation task. In addition, they provide complementary information on spatial learning abilities at an earlier age. It appears clearly that three-week old rats learn to recognize "places of interest" such as the spatial position of the hole leading to the nest even in the absence of proximal cues. However, they take circuitous routes, visiting many holes on the way, before reaching this place.

In contrast with the progressive improvement of place recognition, there is a clear age limit, around 30 days, for the development of starting orientation, with further amelioration over the next week. This age relation is similar in both tasks, thus not due to a particular stress, a motor deficit or an immobilization response (Adams and Jones, 1984). Moreover, when a visible cue is placed on top of the platform, the youngest rats follow a direct path much more often; similar evidence is also found in an ongoing experiment on the hole board.

In conclusion, young rats under 30-35 days of age do not show complete spatial abilities. If they rely on distal cues outside the test area exclusively, they do not follow direct routes to reach a goal in spite of obvious motivation. In addition, they do not show an organized searching behaviour; instead, they express a more diffuse knowledge of a previously reinforced spatial position, what is evident from an increased tendency to remain in this area whenever they happen to be there.

In other words, it seems that cognitive spatial processing in the rat is not possible before the end of the 5th week of life. During the same period, there is a strong upsurge of "rough and tumble" play (Baenninger, 1967), and it has been suggested that this phase might be necessary for the development of normal spatial abilities such as learning about the position of a moving target (Einon, pers. communication).

Contrary to our expectation, no impairment of the place navigation task induced by lesion of the hippocampal formation was completely similar to the deficit of the young rats; only lesions of the parietal cortex seemed to produce a rather specific impairment of starting orientation (Kolb et al., 1983). However, one should expect that functional properties of the hippocampus such as place unit fields in the pyramidal CA3 and CA1, shown to be related with spatial learning abilities (Barnes, 1979; Barnes et al. 1983), undergo critical changes during the 5th week of life. The functional development of the subiculum might also be a critical step, since neurons in this region show strong directional properties (Ranck, 1984).

REFERENCES

Adams, J. and Jones, S.M., 1984. Age differences in water maze performance and swimming behavior in the Rat. Physiology and Behavior, 33, 851-855.

Altman, J., Bruner, R.L., and Bayer, S.A., 1973. The Hippocampus and behavioral maturation. Behavioral Biology, 8, 557-596.

Baeninger, L.P. 1967. Comparison of behavioral development in socially isolated and grouped rats. Animal Behaviour, 15, 312-323.

Barnes, C. A., 1979. Memory deficits associated with senescence: a neurophysiological and behavioral study in the rat. Journal of Comparative and Physiological Psychology 93, 74-104.

Barnes, C.A., Mc Naughton B.L., and O'Keefe, J., 1983. Loss of place specificity in hippocampal complex-spike cells of senescent rats. Neurobiology of Aging, 4, 113-119.

Blue, J.H., and Ross, S., 1982. Spatial discrimination learning by young rats. Bulletin of the Psychonomic Society, 19, 35-36.

Bronstein, P.M., and Spear, N.E., 1972. Acquisition of spatial discrimination in rats as a function of age. Journal of Comparative and Physiological Psychology, 78, 208-212.

Campbell, B.A., Misanin, J.R., White, B.C., and Lytle, L.D., 1974. Species differences in ontogeny of memory : indirect support for neuronal maturation as a determinant of forgetting. Journal of Comparative and Physiological Psychology, 87, 193-202.

Cotman, C., Taylor, D., and Lynch, G., 1973. Ultrastructural changes in synapses in the dentate gyrus of the rat during development. Brain Research, 63, 205-213.

Douglas, R.J., Peterson, J., and Douglas, D.P., 1973. The ontogeny of a hippocampus dependent response in two rodent species. Behavioral Biology, 8, 27-37.

Harley, C.W., and Moody, F.L., 1973. An age related position reversal impairment in the rat. Physiological Psychology, 1, 385-388.

Kolb, B., Sutherland, R.J., and Wishaw, I.Q., 1983. A comparison of the contributions of the frontal and parietal association cortex to spatial localization in rats. Behavioral Neurosciences. 97, 13-27.

Meibach, R.C., Ross, D.A., Cox, R.D., and Glick, S.D., 1981. The ontogeny of hippocampal metabolism. Brain Research, 204, 431-435.

Morris, R.G.M., 1981. Spatial localization does not depend on the presence of local cues. Learning and Motivation, 12, 239-260.

Morris, R.G.M., Garrud, P., Rawlins J.N.P., and O'keefe, J, 1982. Place navigation impaired in rats with hippocampal lesions. Nature (London), 297, 681-683.

Morris, R.G.M. 1984. Development of a water-maze procedure for studying spatial learning in the rat. Journal of Neuroscience Methods, 11, 47-60.

Oakley, D.A. 1983. Learning capacity outside neocortex in animals and man : implications for therapy after brain injury. In "Animal models of human behaviour : conceptual, evolutionary and neurobiological perspectives". G.C.I. Davey (ed.) 247-266. Chichester : Wiley.

O'Keefe, J., and Nadel, L., 1978. "The hippocampus as a cognitive map". Oxford University Press.

Pokorny, J., and Yamamoto, T., 1981. Postnatal ontogenesis of hippocampal CA1 area in rats : I Development of dendritic arborization in pyramidal neurons. II Development of ultrastructure in stratum lacunosum and moleculare. Brain Research Bulletin, 7, 113-130.

Ranck, J.B. Jr. 1984. Head direction cells in the deep cell layer of dorsal presubiculum in freely moving rats. Society of Neurosciences Abstracts, 10, 176.12.

Rauch, S.L., and Raskin, L.A., 1984. Cholinergic mediation of spatial memory in the preweanling rat : application of the radial arm maze paradigm. Behavioral Neurosciences, 98, 35-43.

Schenk, F., 1985. Development of place navigation in rats from weaning to puberty. Behavioral and Neural Biology, 43, 69-85.

Schenk, F., and Morris, R.G.M., 1985. Dissociation between components of spatial memory in rats after recovery from the effects of retrohippocampal lesions. Experimental Brain Research, 58, 11-28.

Sutherland, R.J., Wishaw, I.Q., and Kolb, B., 1983. A behavioural analysis of spatial localization following electrolytic, kainate- or colchicine-induced damage to the hippocampal formation in the rat. Behavioral Brain Research, 7, 133-153.

Swanson, L.W., 1983. The Hippocampus and the concept of the limbic system. In W. Seifert (ed.) "Neurobiology of the Hippocampus", 3-19.

SPATIAL AND NONSPATIAL STRATEGIES IN THE ORIENTATION OF THE RAT.

Dorothy Einon, University College, London.
Chantal Pacteau, Universite Louis Pasteur, Strasbourg.

There are many cues which the rat might use to locate itself in space. It might orient towards a specific visual stimulus, locate a goal as next to a specific cue, or in relationship to two or more cues. Or since its distant vision is good but acuity poor (Dean, 1978) it may use the broad geometry of the environment and locate a place in relation to a major boundary, (see Cheng this volume).

The rat is nocturnal. It travels along well defined paths within its immediate environment, using natural boundaries for more distant travel. It would be surprising if the cues it uses to navigate were exclusively visual: auditory, olfactory, vestibular, and perhaps magnetic information almost certainly play a role.

In associative tasks rats do not respond equally to all potential cues. Inded they may show little evidence of learning about a reliable stimulus in the presence of a more salient cue (see Dickinson, 1980). Selective attention (Reynolds, 1961) is, however, unlikely to be a feature of spatial localization. In the radial arm maze the rat visits each arm of the maze without repetition. It is a task rats learn sooner if multiple cues are available (Gilbert and Innes, 1983), but where performance can be maintained even though information from any one sense modality is lacking (Suzuki et al, 1978; Olton and Collinson, 1979; Dean, 1979).

Perhaps we should not be surprised by this: after all place cells in the hippocampus continue to fire after the removal of visual cues associated with that place (O'Keefe and Conway, 1978). In other words once a task is well learned there may be little need for over-determination.

In a radial arm maze, rats learn remarkably quickly. Traditional conditioning experiments take much longer. That animals need more extensive training to find one cued arm in the radial maze, (Einon, 1980; Olton & Pappas, 1979) than to pick up food from all eight arms, suggests these are task rather than apparatus induced differences.

In our first experiment we examine this in more detail. It is well established that rats choose spatial rather than intramaze cues in the radial arm elimination task, but what do they use to locate a single position? Rats were trained to move from each of the eight arms of a radial maze to a food source located in another arm. Between trials the whole maze was rotated and specific pairs of arms interchanged. For one group (the spatial group) the target was always a specific place in the room. The spatial position of the starting point, the turning angle required, and the actual arm in which the reward was located were all varied from trial to trial. For a second group the target arm was always physically the same arm. Its position in space, its relationship to adjacent arms, the animal's starting position and the angle through which it had to turn all varied from trial to trial. For the third group the target arm was always located at 90 degrees to the left (or right) of the starting arm. All other potential cues (specific arm, spatial position, and the relationship to other arms) were completely irrelevant. We predicted that spatial location would be the easiest task.

Details of this experiment are presented elsewhere (Einon, 1980). It suffices to say that the spatial task was very difficult indeed for the rats to learn. The median number of trials to reach a criterion of 7 out of 8 trials correct in two successive 8 trial blocks for each of the three tasks was: spatial 392; intra-maze cues 108; response 64 trials. The group differences are highly significant.

Why should spatial learning be so difficult? In fact traditional visual discrimination tasks in a T maze involved quite extensive training, and produced analagous results: learning supported primarily by visual cues is more difficult than response learning. It should perhaps be noted that rats were able to learn the 8 arm elimination task in this apparatus within 10 trials.

Were they using visual spatial cues? The experiments which have shown that such cues are primarily involved in the elimination task have all tested animals after considerable training, but Gilbert and Innes (1983) have shown that intra maze cues may be as important as extra maze cues during the learning of this task. Overshadowing and blocking experiments (Dickenson 1980) might lead us to suppose that the stimuli used late in training would be those used earlier, but it would perhaps be unwise to generalise between conditioning experiments and spatial navigation.

The distinction between training and performance is important. The radial arm maze is frequently used to assess spatial memory following physiological and environmental manipulations. In nearly all cases it is learning rather than performance which is assessed in such studies. Until we have established that visual spatial cues are of primary importance during the learning of the elimination task the assumption that poor learning reflects disturbance of spatial memory is clearly premature. Removing salient cues during training certainly disrupts learning more than performance (Dale, 1980) as indeed does olfactory bulbectomy (Hall and Macrides, 1983). It is thus possible that although scent marking and response patterning play little role in maintaining performance in the radial arm maze, they may have a large influence during learning.

In our second experiment we examine this. Olfactory cueing could in principle provide obvious cues: the animal simply has to avoid those arms which it has marked or visited. Well trained rats do not rely on such cues, but do they use them in the initial stages of training?

Rats (N=42) were trained in four radial arm mazes, designed to give different levels of olfactory cueing. If such cues are important, then mazes which maximize the opportunity for scent marking should be easier to learn. Furthermore, since there is some indication that reducing the cues available to the rat induces greater reliance on response patterns, (Dale 1980), we predict that when olfactory cues are absent, simple response strategies, such as visiting adjacent arms will be more prevalent.

The mazes are described elsewhere (Einon, 1989; Einon, Sindon & Pacteau, in prep). All four were the same size and construction, but two had 14cm high walls and two were open-sided. No maze was covered. One open and one walled maze was placed 5cm above a large pool of circulating water. The other two were placed inside the pool of water, so that water washed over the floor to a depth of .5cm. Thus two groups of rats walked through the maze, two paddled 'ankle deep'. Food pellets were placed around the pool to mask the smell of those used for reinforcement.

Moving water almost certainly reduces olfactory cues, while the presence of walls probably increases them. Certainly our rats spend a considerable amount of time rearing (and possibly marking) the entrances to the walled arms. Rats in the walled dry maze (n=11) have the maximum olfactory cues, (floor and arm entrances) while for the rats trained in the wet open maze (n=10) olfactory cues should be minimal. The groups trained in the wet walled maze (N=11) and the dry open maze (N=10) are intermediate. In the walled maze, water tended to remain within a particular arm, so olfactory cues were potentially available. It was also possible to mark arm entrances since the water was shallow.

Animals were trained to a criterion of no more than 2 errors in three days, or a minimum of 6 days. Following this they were given one disruption trial. After four choices, the rat was removed, from the maze for two minutes then returned for the final four choices. If olfactory cues assist learning we predict that trials to criterion and numbers of errors should be in the order: Dry Wall Wet Wall = Dry Open Wet Open.

Analysis of variance performed on the error scores, and the trials to criterion both showed significanmt group differences. Rats learned faster in closed than open mazes, and in dry rather than wet ones. There was no interaction. The mean trials to criterion are given in Table 1.

TABLE 1

	OPEN WET	OPEN DRY	WALL WET	WALL DRY
Total animals	10	10	11	11
Trials to criterion	8.4	7.6	6.1	5.5
Number of rats with patterns:				
Day 5	4	6	5	6
Criterion day	8	9	8	5
Error with response pattern (a)	9.3	8.3	9.0	8.8
Errors without response pattern (b)	14.8	10.0	12.2	9.5
Diff in error rates (b-a)	5.5	1.7	3.2	0.7
Rats showing decrement following disruption	10	5	3	0
Errors following disruption	2.8	0.7	1.4	0.0

Although significant the differences are small. In the absence of olfactory cues the wet open maze animals learn, suggesting that spatial cues are used in learning. However more detailed analysis revealed that error rates grossly overestimated learning. Most animals in the wet open condition visited adjacent arms, (Dale, 1979 reports a smilar finding following blinding): their solution thus depends on response patterns not spatial cues. In Table 1 the number of animals taking up the adjacent arm strategy on day 5 and the day criterion is reached are shown for each of these groups. Also reported are the error rates for the rats using and not using the adjacent arm strategy. Note that in the walled dry condition, it makes little difference whether animals use a response strategy, but in the wet open condition there is little evidence of learning in the absence of such strategies.

The results of the disruption trial (Table 1) reinforce this: there is little disruption in the dry walled maze, and considerable disruption both in the number of errors made, and the number of animals showing a decrement in performance, in the open one. In other words, animals who are unable to use scent trails rely almost exclusively on response patterns to solve the task, implying that at this early stage in learning, spatial cues are of little importance. Obviously, the rat comes to rely upon these cues, unless our rats are very different from those reported elsewhere in the literature.

In a final experiment we looked at a later stage in training using a task in which response patterning was impossible. We took two of the present groups of rats (those trained in the wet open, and dry open conditions) and gave them a new task in which only four of the eight arms were baited. Adjacent arms are not consistently reinforced, and olfactory information, if present, should be less reliable. In choosing four from eight baited arms, the rat must avoid not only the arms it has previously visited, but also those that are never baited. If olfactory information is essential in learning, the task should be difficult in the dry maze: impossible in the wet. In fact both groups learned. Those in the wet maze took longer but although the difference was significant, it is extremely small: Wet 38.5, Dry 36 trials.

Clearly our animals can use spatial cues. In fact the results of the last part of this experiment are quite consistent with earlier findings in the radial arm maze: after considerable training rats do appear to rely more heavily upon visual cues.

In conclusion we would like to make three points. First it would be unwise to assume that because an animal uses one set of cues in one task, it will use the same cues in another task which appears superficially similar. The extensive literature on 'preparedness', has of course made this point more forcibly than we can, but it may explain in part at least why in our first task, and in Cheng's task, rats appear to ignore obvious visual cues. Perhaps in finding one place, a single visual cue is not as reliable as geometric, or olfactory information. Second, just as the pigeon uses many sources of information to home successfully, (see Wiltscho this volume) so the rat almost certainly uses many different sources of information to find its way about. That at any one stage of training, or in any one apparatus or task one source of information seems to dominate, does not necessarily rule out the possibility that other cues are being employed. The finding that rats have been shown to choose spatial cues following mastery of a radial arm maze, yet appear only minimally impaired by blinding reinforces this point. Finally we should like to point out that rats learn best if they have available different sources of information. Any lesion, drug treatment, rearing environment, or genetic manipulation which cuts down the potential information to the rat, may make learning more difficult. Failures on the radial arm maze, as in other mazes, may reflect such a reduction in information. They can not be assumed to reflect underlying deficits in either spatial or memorial abilities.

References

Dale R H I 1979. The role of vision in the rat's radial maze performance Unpublished doctoral disertation, University of Western Ontario.

Dean, P, 1978. Visual acuity in hooded rats; effects of superior collicular or posterior neocortical lesions. Experimental Brain Research, 18, 433-45.

Dickinson, A, 1980. Contemporary Animal Learning Theory. Cambridge University Press, Cambridge.

Einon D F, 1980. Spatial memory and response strategies: Age, sex and rearing differences in performance. Quarterly Journal of Experimental Psychology, 32, 473-489.

Hall, R D and Macrides, F, 1983. Olfactory bulbectomy impairs the rats radial-maze behaviour. Physiology & Behaviour, 30, 797-803.

Gilbert, M E & Innes, N K, 1984. The influence of cue type and configuration upon radial arm maze performance in the rat. Animal Learning and Behaviour. 11, 373-380.

O'Keefe, J & Conway D H 1978. On the trail of the hippocampal engram. Physiological Psychology 8 229-238.

Olton, D S & Collinson C. 1979. Intramaze cues and 'odour trails' fail to direct choice behaviour on an elevated maze. Animal Learning and Behaviour 7, 221-223.

Olton D S & Papas B C 1979. Spatial Memory and hippocampal function. Neuropsychologia 17, 669-682.

Reynolds F 1961. Attention in the pigeon. Journal of Experimental Analysis of Behaviour, 4 203-208.

Suzuki S, Augerinos G & Black A H, 1978. Stimulus control of spatial behaviour on the eight arm maze in rats. Learning and Motivation, 11 1-18.

SECTION II. NEUROETHOLOGY AND ETHOLOGY

NEUROETHOLOGY: TOWARD A FUNCTIONAL ANALYSIS OF STIMULUS-RESPONSE MEDIATING AND MODULATING NEURAL CIRCUITRIES

J.-P. Ewert (Neuroethology Department, FB 19, University of Kassel, D-3500 Kassel, Federal Republic of Germany)

Neuroethology is concerned with the analysis of the neural substrates and mechanisms that underlie behavior, and hence involves studies on: signal recognition and localization, sensorimotor interfacing, modulatory functions and storage, and motor pattern generation. Neuroethology strives to explore general rules of neurobiological organization through a comparative approach, thus allowing insight into the evolution of functional principles that determine behavior of a species (Bullock 1983).

Preliminary results toward such an approach in vertebrates, have been reported in studies on tectal functions that are controlled by various prosencephalic influences. It has been suggested that certain pathways (Fig.1) exert gating effects on the excitability of tecto-bulbar/spinal output-neurons in deciding as to whether or not a behaviorally meaningful stimulus can be taken as a target for an orienting movement (Ewert 1980; Chevalier et al. 1984). In anurans, the anatomic substrate appears to be provided by a "striato-pretecto-tectal" and a small "striato-tegmento-tectal" pathway (Fig.1) (Wilczynski and Northcutt 1983; Neary and Northcutt 1983). In reptiles and birds, the former (striato/pallido-pretecto-tectal) is prominent while another has been observed in the form of a small "striato-nigro-tectal" pathway (Reiner et al. 1980). According to Wilczynski and Northcutt (1983), such a "dual-root system" is the ancestral tetrapod condition (see also Reiner et al. 1984). In mammals, the comparable pathway mediated by the substantia nigra is a prominent circuit in this context (besides the main basalganglio-ventrolateralthalamo-motocortical pathway), while the one via pretectum may have been lost (Wilczynski and Northcutt 1983).

There are other prosencephalo-tectal pathways ("pallio-thalamo-tectal") which in amphibians have their origin in the medial pallium, a phylogenetic precursor of the mammalian hippocampus (Herrick 1925, 1933), and which appear to be involved in the modulation of stimulus-response mediating circuitries (For definitions see Krasne et al. 1979).

During the last two decades, working on common toads we have collected data toward a functional description of prosencephalo-tectal circuitries that are responsible for deciding whether a visual moving stimulus can be taken as a releaser of prey-catching behavior (Ewert 1984). The toad's prey-catching sequence - orienting toward prey, stalking, binocular fixation and snapping - is mediated by innate releasing mechanisms (Ewert 1985) characterized by recognition and localization properties. The activation of each of the ballistic action patterns of the sequence requires stimulus recognition based on configurational visual features of the sign stimulus; the choice of action, i.e., the catching strategy involves in addition perceptual operations that locate the stimulus in space. The release of each action thus depends on the ongoing visual signals; activation of an action pattern is not a prerequisite for another action to be started.

In the course of the neuroethological analysis of these processes, the following aspects have emerged:

(i) The x-y coordinates of visual space are mapped in the CNS in a multiple manner, involving a sort of "visual matrix", that is a representation of the visual world in which the analysis of sign stimuli obeys certain rules

- stimulus features are analyzed by functional units (assemblies) of cells inherent in thalamic/pretectal-tectal networks

- certain interactions between feature-analyzing cells specify perceptual operations regarding signal discrimination and localization

- the results of information processing within a cell assembly are expressed by the firing of output-cells (specified neurons) that exhibit properties required for stimulus recognition and localization, respectively, and transmit these messages to corresponding bulbar/spinal motor pattern generating circuitries

(ii) Innate releasing mechanisms are stimulus-response mediating command systems

- a command (releasing) system is composed of neurons specified for feature selection and localization (command elements) in a certain combination

- the message of a command element is transmitted by several neurons belonging to the same class ("population code")

- activation of a motor pattern generator requires simultaneous appropriate inputs from all the command elements of a command system

- modulations of command systems regarding behavioral "state", habituation, and learning are concerned with internal loop-operations that involve different prosencephalic (diencephalic, telencephalic) structures

In the following, I shall provide experimental evidence in support of these themes with particular emphasis on structure/function relationships in retino-pretectal/tectal circuitries. Such an approach toward a neurobiological analysis of goal oriented action patterns in a lower vertebrate, including its modulation by state- and storage-related mechanisms, may be of interest in a "Study Institute on Cognitive Processes and Spatial Orientation in Animal and Man". More specifically, I shall refer to the question of whether an innate "prey schema" that has emerged during evolutionary history is represented by a particular cell, or a cell assembly inherent in a network. Another important question is the way learning can be accounted for, i.e., whether what psychologists refer to as memory is merely a cell or a functional unit of interacting cells having particular feature-coding properties.

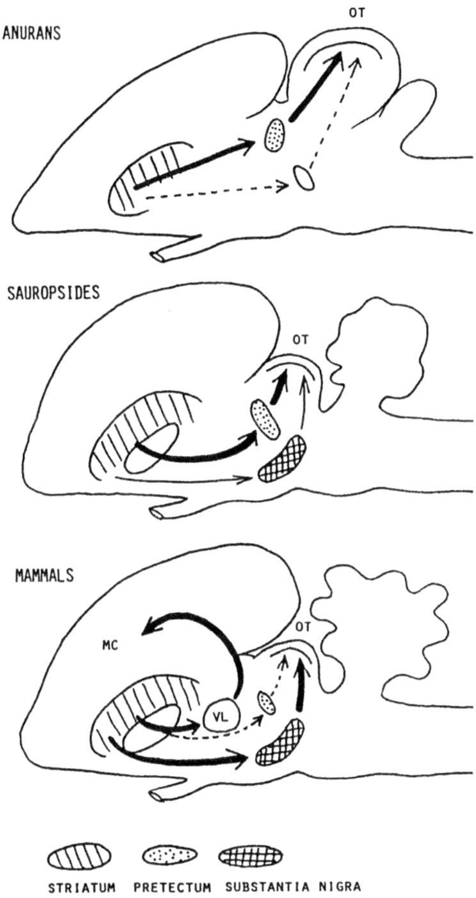

Fig.1. Disynaptic pathways by which basal ganglionic telencephalic structures may influence tecto-bulbar/spinal output neurons. Anuran amphibians: "striato-pretecto-tectal" and "striato-tegmento-tectal" pathways. Sauropsides: "striato/pallido-pretecto-tectal" and "striato-nigro-tectal" pathways. Mammals: "striato-nigro-tectal" pathway. For explanations see text (Modified after Reiner et al. 1984).

SIGNAL DESCRIPTION

Definition of Recognition

In comparative ethology, stimuli are referred to as sign (or key) stimuli when they elicit fixed patterns of behavioral responses, provided the motivation of the animal is appropriate. The stimulus/behavior relationships are mediated by so-called releasing mechanisms (RM) having recognition properties which can be innate (IRM), and these may be extended or modified by individual experience (Lorenz 1943, Tinbergen 1951, cf. also Barlow 1985). Releasing mechanisms perform sensory analytical operations (related to stimulus recognition and localization) and have motor pattern triggering functions.

Given that a juvenile prey-naive toad (the tadpole is vegetarian), after metamorphosis immediately with transition to terrestrial life, responds to certain features of moving visual stimuli with motor patterns related to prey capture (Traud 1983, cf. also Ewert 1985), we may conclude that this happens upon recognition, i.e., a decision-made by comparison between the incoming signal with sort of an innate "prey-image" that has emerged during evolutionary history. Recognition, here, is defined as the classification of stimulus distributions from the visual environment into innate classes of functional significance (Ewert and v.Seelen 1974).

Configurational Stimulus Features

A precondition for the neurobiological analysis of visual prey recognition is a quantitative description of the prey signal. Experiments using dummies have shown in common toads that stimulus area extensions in relation to the direction of movement are critical parameters (Ewert 1968, 1969, see review 1984). This can be demonstrated by a simple experiment: a small 3mm x 30mm stripe moving worm-like in direction of its long axis signals prey, whereas the same stripe looses its prey-signal, or is even treated as threat, if the long axis is oriented perpendicular to the direction of movement - that is, antiworm-like. Using square or disc-shaped objects of different sizes, both configurational parameters, which have opposite releasing values for prey capture, will interact, so that stimuli of certain diameter (5-10mm) are optimal prey. A large moving square object (e.g., 100mm diam.) elicits predator avoidance behavior.

The worm/antiworm discrimination of a moving stripe involves configurational perception. Configuration (Gestalt), here, is defined by the relationship between the stimulus extension in (xl(1)) and perpendicular to (xl(2)) the direction of movement, that is, stimulus geometry in relation to movement direction. The effects of these parameters are determined by certain spatio-temporal properties of a central "visual matrix". The visual prey stimulus thus is not defined by a highly specific "key", but characterized by a relationship between features, xl(1):xl(2), within certain ranges of variable area, xl(1)·xl(2) (Ewert 1968, see 1984). Neither "the" worm nor "the" beetle nor "the" antiworm seems to be represented by innate neural templates. But the related parameters xl(1) and xl(2) are abstracting behaviorally significant features of moving objects. Quantitative investigations on different visual patterns allow the general conclusion that continuous or discontinuous extensions in and perpendicular to the direction of stimulus movement, respectively, are the key features by which prey is distinguished from non-prey (Ewert et al. 1970). The corresponding perceptual operations precede any of the ballistic motor responses of the prey-catching sequence, such as orienting toward prey or snapping.

We have to consider that the "configuration concept", in this context, is not defined by relationships which are exclusively internal to the signal itself; the background has to be also taken into account. The following example illustrates this point. Suppose a toad is sitting in front of the

Fig.2. Example to illustrate signal-noise relationships. A match moving in direction of its long axis (see arrow) provides a prey signal for the toad.

contents of a match-box (Fig.2); how could a stimulus (match stick) with prey features emerge? It has to be different from the background (represented by the other match sticks) as far as certain spatio-temporal characteristics are concerned; i.e., it must move and be oriented parallel to the direction of movement. If several matches are moving simultaneously and irregularly, the toad's prey-capture will be inhibited, a phenomenon known in ethology as "swarm-effect". If all the matches move at the same speed in the same direction, they will be treated as a moving surround.

Although the worm vs. antiworm preference is largely independent of stationary background structures (Ewert 1984), there is evidence to show that the background, too, contributes to configurational properties: toads mainly orient and snap toward the leading edge of a black worm-like stripe that is moved against a white background. Obviously, a change from white to black, an "off"-effect, is the target for this "head preference" phenomenon (the head of the stripe being defined by the edge that points in the direction of movement). If the stimulus-background-contrast is reversed, i.e., a white stripe is moved against a black background, the "off"-effect occurs at the trailing edge of the stripe and the toad, as predicted, now responds mainly to this, namely that part of the contrast border that belongs to the background (Burghagen and Ewert 1982).

Discriminate Value, D(W,A)

Worm/antiworm discrimination is a powerful tool with which to explore in the visual system the effects of changes in configurational parameters xl(1) and xl(2). Let RW be the prey-catching activity in response to a black worm-like (W) and RA to a black antiworm-like (A) stripe, each presented separately and moving at the same velocity against a white background, the discriminate value $D(W,A)=(RW-RA)(RW+RA)^{-1}$. In behavioral experiments, D(W,A) is betweeen +1 and 0; furthermore, it can be shown that

$$D(W,A)=f(x)$$

which means that the acuity of W/A discrimination increases within limits

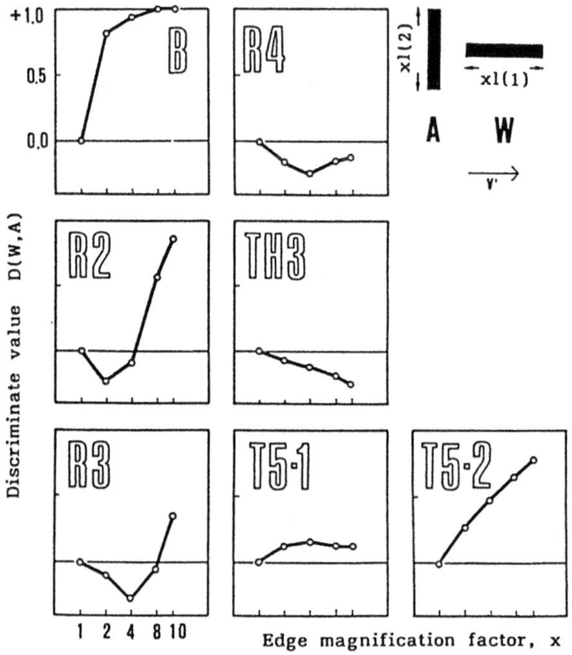

Fig.3. Discriminate value D(W,A) as a measure of discrimination between worm-like (W) and antiworm-like (A) moving stripes of constant width and different length parallel (xl(1)) or perpendicular to (xl(2)) the direction of movement; l(1)=l(2)=2.5mm, in behavioral experiments; l(1)=l(2)=2deg visual angle, in neurophysiological experiments; black stimuli were moved at constant angular velocity against a homogeneous white background. The D(W,A) values were calculated for the prey-catching orienting activity (B) and for the activation of various classes of retinal (R2, R3, R4), pretectal (TH3), and tectal (T5(1), T5(2)) neurons in the common toad. The receptive fields of retinal ganglion cell classes are in the range of 4deg (class R2) to 16deg (class R4), of tectal T5 neurons 20-30deg, and of TH3 neurons at about 45deg (From Ewert et al. 1978).

with edge magnification x of the stripe (Fig.3B). It has been demonstrated that the D(W,A) values also remain positive when other stimulus parameters are varied, such as the speed or the direction of stimulus movement (Ewert 1984).

Motivational State

Prey-catching activity depends upon motivational states of the animal (Ewert 1984), e.g., those related to hormonal changes. During the mating period, in spring, toads do not feed. In summer, during the hunting season, satiated toads will not respond to even optimal worm-like stimuli, which suggests a correlation with the nutritional state. Motivational states of the toad have an effect on D(W,A). The value decreases with increasing prey-catching motivation, but it remains positive, which means that the worm configuration of a moving stripe is preferred to the antiworm configuration.

Individual Experience

The worm vs. antiworm preference is determined by the properties of an innate releasing mechanism. This has been demonstrated by experiments (Traud 1983, cf. also Ewert 1985) in which toads (Bufo bufo) were raised singly from eggs and kept in homogeneous or differently structured environments. But within the relatively wide "prey schema" adaptations are possible in relation to certain individual experience. For example, in the presence of known prey odor (previously associated during feeding), D(W,A) becomes smaller (Ewert 1968 in 1984). The D(W,A) values are almost permanently decreased - but still positive - after conditioning of prey-capture to the antiworm stimulus (or to the experimenter's hand) presented a couple of times on successive days together with a natural prey object, such as a mealworm (For a review on "olfactory/visual" and "visual/visual" conditioning in toads see Ewert 1984).

Stimulus specific habituation is another example demonstrating "behavioral lability". Toads become habituated to a prey dummy if the same stimulus is repeatedly offered in a longterm stimulus series. But if certain cues (e.g., tip edges, dot or stripe patterns) of a dummy within the optimal prey-schema are slightly changed, the toad may immediately respond by orienting and snapping (For review see Ewert 1984).

FEATURE ANALYZING NEURONAL ASSEMBLIES

Sensory Maps

Moving from the behavioral to the neurophysiological level of analysis of visual perceptual operations, it is important to note that the x-y coordinates of visual space are represented in the visual system in a multiple manner (Fite and Scalia 1976), which refers to visual maps in the retina, the optic tectum, the pretectum, and the anterior thalamus (Fig.4). Within these, information is processed at different emphasis in functional units (assemblies) of cells, which is known for the retina (Hartline 1938; Barlow 1953; Lettvin et al. 1959; Grüsser and Grüsser-Cornehls 1976), and which has been suggested for the tectum (Ewert 1974; Székely and Lázár 1976; Arbib and Lara 1982), and the pretectum (Lara et al. 1982). Neurophysiological experiments have shown (Ewert 1984) that visual signals are analyzed in the corresponding assemblies, so that feature-coding neurons in various parts of the central visual system might be responsible for the "representation" of a visual object, some neuronal classes taking part in different types of assemblies and contributing to the characterization of certain objects. The important point I would like to emphasize here is the fact that central visual maps are mutually connected, so that they can talk to and influence each other, which has been demonstrated anatomically by Wilczynski and Northcutt (1977) and Weerasuriya and Ewert (1983) and physiologically by Ewert et al. (1974). Sort of "cross-talk" is essential for various perceptual operations.

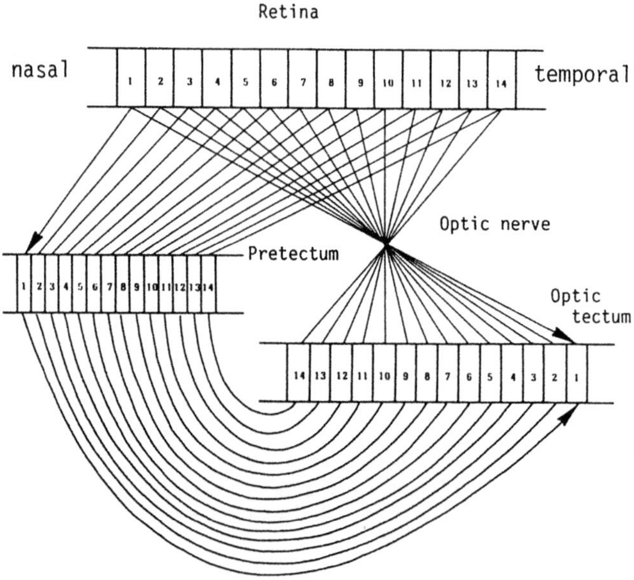

Fig.4. Visual maps. Schematic diagram illustrating retinal projection fields in the optic tectum and pretectum and their connections. The projection and connectivity patterns are not precisely "point-to-point" as regards overlap due to convergence and divergence, not considered here. (According to Fite and Scalia 1976; Wilczynski and Northcutt 1977).

Feature Analyzing Neurons

Toward a neurophysiological analysis of prey recognition we have taken advantage of the D(W,A) values and have tested against these stimulus criteria the properties of various classes of neurons: retinal ganglion cells (R), pretectal caudal thalamic (TH) neurons in the lateral posterodorsal Lpd and lateral posterior P nuclei, and tectal (T) neurons (Ewert et al. 1978, Ewert 1984). Among these, many neuronal classes displayed different sensitivities in response to the same sets of configurational stimulus parameters (Fig.3, Table 1). Class T5(1) and TH3

Neuronal class	Correlation (rA;rW)
R2	+0.6;+0.5
R3	-0.7;+0.2
R4	-0.9;+0.6
T5(1)	0.0;+0.8
T5(2)	+0.9;+0.7
TH3	-0.9;+0.9

Table 1: Correlation coefficient pair (rA;rW) for comparison between behavioral and neurophysiological responses to worm-like (W) and anti-worm-like (A) stripes (From H.-W. Borchers and J.-P. Ewert, 1979, Behavioral Processes 4:99-106).

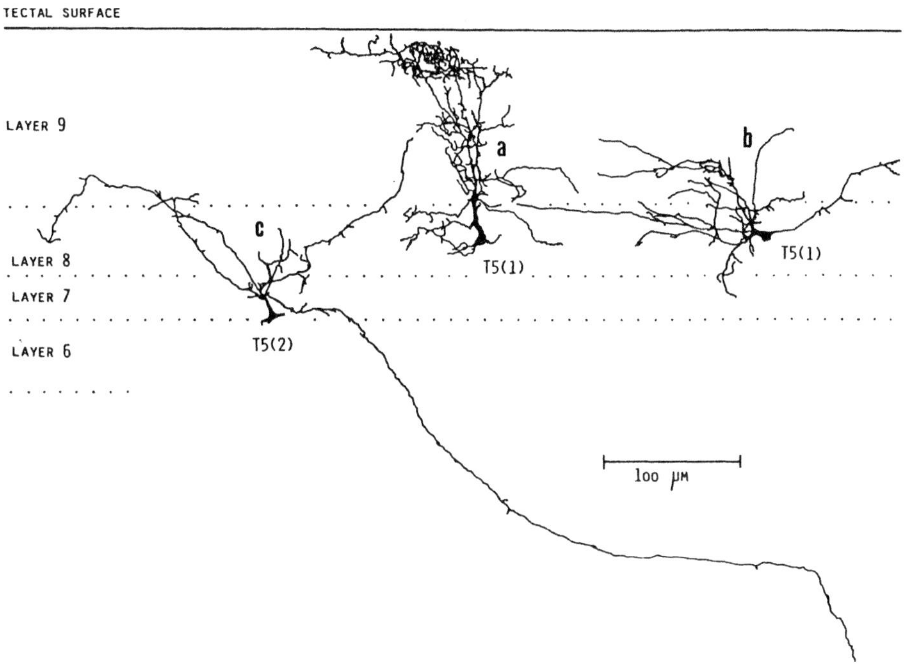

Fig.5. Identified tectal neurons from intracellular recording and iontophoretic staining with K+-citrate/Co3+-lysine filled micropipettes. Camera-lucida reconstructions of pear-shaped (a), large ganglionic (b), and pyramidal neurons (c) in the grass frog's optic tectum (From Ewert et al. 1985).

(or T5(3)) neurons are sensitive either to the parameters xl(1) or to xl(2). Class T5(2) neurons are sensitive to xl(1) and xl(2), in different ways. They show best positive correlation with the probability that a configurational visual stimulus fits the innate "prey-schema". Altogether these classes of neurons provide the neuronal basis of the assumed "visual matrix".

This is, of course, only a condensed result of extensive quantitative investigations on the variations in various stimulus parameters, indicating that T5(1), T5(3), and TH3 neurons are involved in the configurational perceptual operations related to different behavioral goals (for a review see Ewert 1984). Class T5(2) neurons, by contrast, are specific in that they are gating the perceptual processes of visual feature analysis toward the occurrence of prey capture.

The extracellular data are consistent with the results of intracellular recording and iontophoretic staining of tectal cells (Fig.5) with K+-citrate/Co3+-lysine-filled micropipettes, obtained recently at our laboratory in collaboration with N.Matsumoto from Osaka University (Japan). In these studies, class T5(1) cells were identified as pear-shaped or large

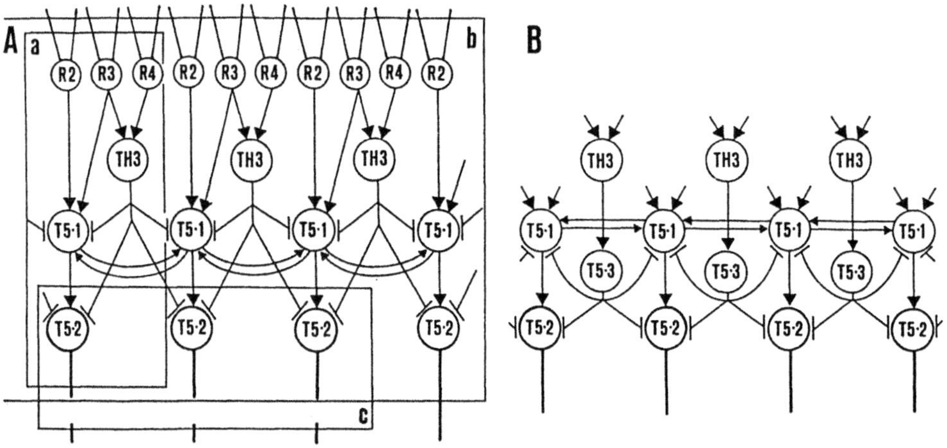

Fig.6. Global model of prey feature analyzing networks (A and B). Neural processes responsible for prey feature analysis of an IRM are based on central lateral excitation and inhibition. Note in (A) the definitions used for: (a) functional unit of cells as described by Ewert (1974) broadly corresponding to the "column concept" defined anatomically by Székely and Lázár (1976) and described mathematically by Lara et al. (1982), (b) neuronal network, and (c) neuronal population. Arrows indicate putative excitatory synapses and lines with cross bars putative inhibitory synapses; morphological features of neurons are not considered here (see Fig.5). The lateral excitatory and inhibitory connections are not restricted to immediately adjacent neurons. The networks shown in (A) and (B) are thought to be pieces of an integrated whole (After Ewert 1986).

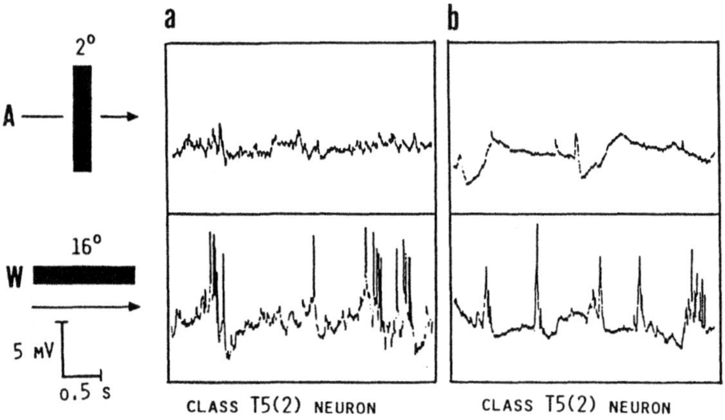

Fig.7. Intracellularly recorded membrane potentials of grass frog T5(2) cells in response to worm (W) and antiworm (A) configuration of a moving stripe; two neurons (a) and (b) (From Ewert et al. 1985).

ganglionic neurons the somata of which were located in tectal layer 6/7 or 8, class T5(3) as large ganglionic neurons in layer 8, and T5(2) as pyramidal neuron in layer 6/7 with its long axon coursing within layer 7 and projecting to the medulla oblongata (Ewert et al. 1985; Matsumoto et al. 1986).

Discrimination Through Inhibitory Interactions

Configurational prey-selection depending on the characteristic of the innate releasing mechanism might be a matter of information processing that takes place in assemblies of interacting cells (Fig.6Aa) inherent in a network (Fig.6Ab) (Ewert 1974, 1984). Such a functional unit consists of feature-sensitive cells (T5(1), T5(3), TH3), each of which receives different combinations of retinal input. It is suggested that class T5(2) neurons - through specific integrative excitatory and inhibitory synaptic contacts to these cells - are "weighting" the information. It seems as if these prey-selective cells, the property of which is governed by inhibition, make use of what Barlow (1985) calls "veto logic", or rather NAND logic (For a system-theoretical evaluation see Ewert and v.Seelen 1974). Intracellular records of T5(2) cells (Fig.7) have shown spike levels reaching EPSPs in response to a worm-like moving stripe but no change in the membrane potential, or IPSPs, in response to the antiworm-like moving stripe (Ewert et al. 1985; Matsumoto et al. 1986). The important result here is that change of the configuration of a stimulus (from worm to antiworm) may be accompanied in a T5(2) cell with a corresponding change of EPSPs to IPSPs, hence suggesting "configuration specific" excitatory and postsynaptic inhibitory inputs to these cells. The "veto-logic" can be also demonstrated if worm and antiworm stimuli are presented together, e.g., forming a cross; in this case, T5(2) neurons are silent (Ewert unpublished).

Various other studies have indicated that tectal intrinsic functions of visual processing are mainly excitatory (e.g., Ewert et al. 1970) and that feature analysis of tectal prey-selective neurons is determined by prosencephalic influences based on pretecto-tectal inhibitory influences (Ewert 1968 in 1984) which may also involve tectal inhibitory transmission (cf. Fig.6A,B). After a direct current- or a Kainic acid-induced lesion in the pretectal Lpd and lateral P (Lpd/P) region, both extracellularly recorded T5(2) neuron activity and the prey-catching activity become "disinhibited" in response to any visual stimulus; the D(W,A) values in this case were about zero, hence resembling a sort of "agnosia". This has been shown most convincingly by extracellularly recording the same T5(2) neuron (Fig.9A) in a paralyzed toad or frog pre and post lesion (Schürg-Pfeiffer, Finkenstädt, Cromarty, and Ewert in prep.) and by chronic recordings from a T5(2) cell in the behaving toad with an implanted second electrode for a pretectal DC lesion (Schürg-Pfeiffer and Ewert in prep.).

From a broad view, the inhibitory pretecto-tectal connections - through a process of lateral inhibition (Fig.6A,B) - may lead to various perceptual operations regarding: (i) configurational prey-selection (Ewert 1968 in 1984), (ii) distinction between object motion and self-induced motion (Ewert et al. 1983), and (iii) figure/ground discrimination. With respect to the latter function, in collaboration with H.-J.Tsai from Academia Sinica (Peking, VR China), we have noted that T5(2) neurons, the responses of which toward prey were normally inhibited by a simultaneously moving structured background (visual noise), discharged strongly in such a situation after a pretectal Lpd/P lesion that eliminates putative inhibitory input of

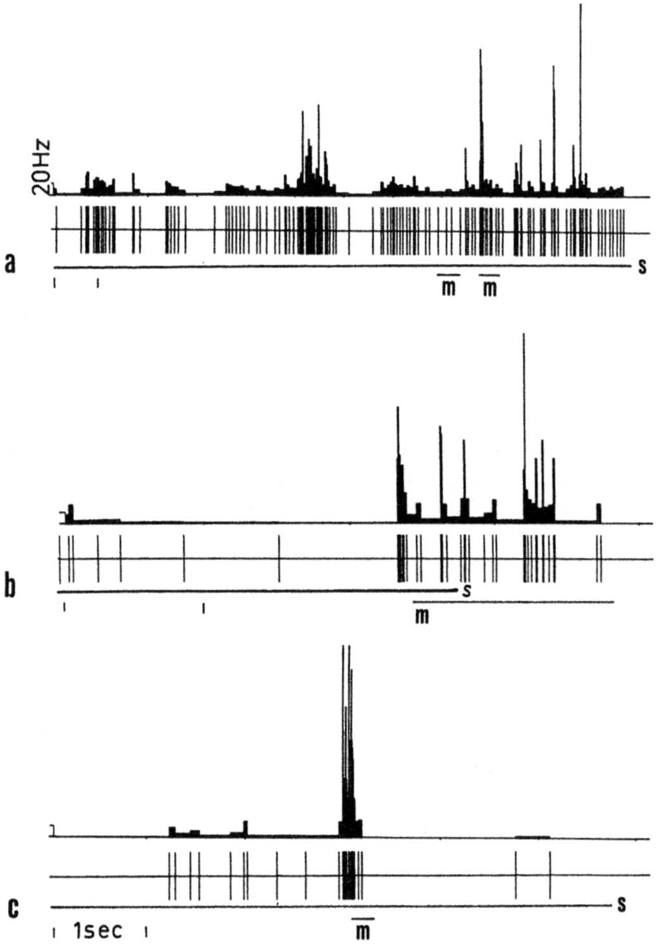

Fig.8. Activity of single tectal neurons in the behaving, disinhibited, thalamic-pretectally lesioned toad (representative examples). These animals showed stereotyped, predictable, non-habituating prey-catching responses to all moving visual stimuli, and therefore provide suitable subjects in which to examine the correspondence between visual stimulation, neuronal activity and motor pattern. Each computer print-out shows the pattern of action potentials and the corresponding dwell-histogram of spike frequency against time. The horizontal lines indicate stimulus (s) and behavior events (m). (a) Activity of a class T2(1) neuron during a traverse of its large frontal excitatory receptive field (ERF $\sim$ 90deg) by a moving prey stimulus (s); high discharge frequency did not necessarily coincide with an orienting movement (m). (b) Activity of a T5 neuron; increased activity during visual stimulation (s) always preceded and persisted during an orienting response (m). (c) Spontaneously active T8 neuron; the burst preceded all motor responses (m), here, snapping toward a moving mealworm; note the subsequent silent period (From Megela et al. 1983).

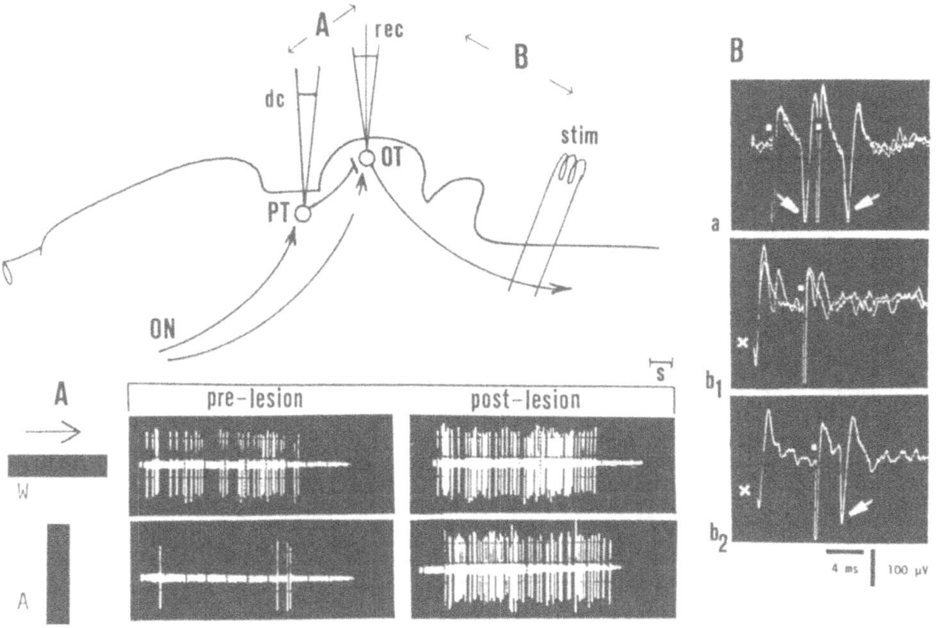

Fig.9. Neurophysiological steps (A and B) in the analysis of pretecto-tecto-bulbar/spinal pathways. A) Evidence of pretecto-tectal inhibitory influences provided by extracellular recordings from the same class T5(2) neuron pre and post a direct current induced ipsilateral pretectal (PT) lesion in a paralyzed toad. Representative example. Pre-lesion: strong activity of the T5(2) neuron in response to a worm-like moving black stripe that traversed the center of the excitatory receptive field (ERF diam. = 28.7deg) in horizontal direction at 7.6deg/sec against a white background; weak activity in response to the same stripe in antiworm configuration. Post-lesion: 5min after passing an anodal DC current of 60µA for 6sec through a coagulation electrode, the discharge rate of the same T5(2) neuron increased in response to the worm and, particularly, in response to the antiworm configuration, so that D(W,A)=0. (Schürg-Pfeiffer, Finkenstädt, Cromarty, and Ewert in prep.). - B) Specification of tecto-bulbar/spinal output. Antidromic activation ("backfiring") of a physiologically identified, extracellularly recorded tectal T5(2) neuron in response to electrical stimulation (negative 0.1 msec square wave impulse) applied through bipolar electrodes to the tecto-bulbar/spinal tracts contralaterally in the medulla oblongata (100-1000 µm rostrad to the obex) in a paralyzed common toad. a) Constant latency response at 2.8 msec to repeated electrical impulses of just suprathreshold intensity (several superimposed traces), and following ability with two successive impulses down to a stimulus interval of 3.5 msec (superimposed); the minimal interval is not shown, here. b) Collision test in which a visually elicited orthodromic spike (see "x") triggered the electrical stimulus to evoke an antidromic spike (see "arrow") at a variable spike-stimulus interval; b1: collision between the visually elicited spike and the electrically evoked antidromic spike occurred at a spike-stimulus interval (critical delay) of 5.1 msec (two superimposed traces), but not at larger intervals as shown in c2. The "point" refers to the onset of the electrical impulse followed by a stimulus artifact (From Satou and Ewert 1985).

TH3 neurons. The latter can be activated by moving background structures (Tsai and Ewert, in preparation).

In summary, from our neuroethological investigations the picture emerges that the "innate prey schema" in toads is represented by genetically labeled connectivity patterns in functional units (assemblies) of cells inherent in a network (as shown in Fig.6), i.e., the specific integrative synaptic contacts of prey-selective T5(2) neurons to feature-coding T5(1), T5(3), and TH3 neurons. A small population of T5(2) neurons (Fig.6Ac), with partly overlapping receptive fields, may transmit the result of information processing in the corresponding cell assemblies to appropriate bulbar/spinal motor systems. Recordings from a T5(2) neuron in the freely moving toad showed that a neuronal burst preceded, and "predicted", a subsequent prey orienting or snapping response (Borchers et al. 1983; Schürg-Pfeiffer and Ewert in prep.; see also Megela et al. 1983, Fig.8).

It is important to emphasize that the same or similar network may be responsible - through specific integrative connections by other output-neurons - for other recognition tasks, e.g., those related to a predator or a sexual partner (Ewert 1980).

Signal Localization

Stimulus discrimination, performed by "feature analyzing units" (Fig.6Aa), is one aspect of the IRMs that trigger the motor patterns of the prey-catching sequence. Stimulus localization could be achieved by various receptive field properties: (i) evaluation of the x-y coordinates of a stimulus according to a code transmitted by a small population of adjacent T5-type neurons with partly overlapping receptive fields in correspondence with the visual map, (ii) stimulus distance measurements on the z-axis related to information transmitted by distance sensitive T3 neurons, or by binocular T1-type neurons (cf. Grüsser and Grüsser-Cornehls 1976).

There are other neurons, class T4 and TH4, the excitatory receptive fields of which cover the entire visual field. Some of these may be involved in processes related to arousal.

COMMAND (RELEASING) SYSTEMS

Global Concept

Since activation of the various action patterns of the prey-catching sequence by IRMs requires basically the same recognition processes (Ewert et al. 1983), but involves different localization mechanisms, I suggest that certain combinations of various classes of neurons dealing specifically with prey-feature selection (T5(2) neurons), localization (e.g., T3, T1(3) neurons), and arousal (e.g., T4 neurons) - each acting as a command element - trigger corresponding bulbar/spinal motor pattern generating circuitries. We call such a combination of specified neurons a "command (releasing) system" (Fig.10) and postulate that coincidence of appropriate excitation in all these command elements is necessary to activate, in the manner of an AND gate, the motor pattern generator (Ewert 1980, 1985; for a general discussion of the "command system concept" see Kupfermann and Weiss 1978).

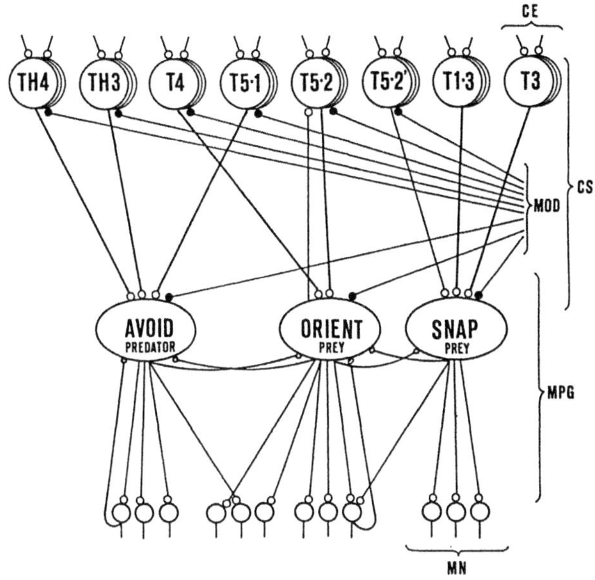

Fig.10. The "command system concept" of innate releasing mechanisms. Motor pattern generators for different action patterns require coincidence of different combinations of appropriate outputs of neurons (command elements) specified for feature selection (e.g., class T5(2)), localization (e.g., class T1(3)), arousal (e.g., class T4), respectively. Such a combination is called a command system. Modulatory influences (MOD) may enter a command system at the level of command elements and/or the motor pattern generators. (Each oval symbol stands for a neuronal premotor circuitry). CE, command element; CS, command system; MPG, motor pattern generating circuitry; MN, motoneuronal pools (According to Ewert 1980).

An important aspect of this concept is the fact that different command (releasing) systems may share a command element, which would invalidate a restrictive view of monopolized behavior-specific channels. Class T5(2) neurons - which as a class are shared by the various command systems of the prey-catching sequence (Fig.10) - are essential and decisive for prey-catching behavior: If their output is not appropriate, no prey-catching command can be executed.

According to this concept, IRMs are command systems. They can be regarded as a sensorimotor interface that fulfills tasks with respect to (i) signal recognition and localization, (ii) command functions by activating the corresponding motor pattern generation system, (iii) motor pattern generation through participation in the temporal sequence of the time program, and (iv) integration of modulatory inputs depending on attentional and motivational states and acquired experience (Ewert 1985).

Sensorimotor Aspects

The above mentioned concept postulates that the optic tectum contains neurons which transmit certain "command signals" to bulbar/spinal motor systems. As on a keyboard - to use an analogy by Graham Hoyle (1984) - certain combinations of these output neurons (command elements) activate corresponding motor pattern generators (Fig.10). Artificially, in the freely moving animal, they can be "called into play" by electrical stimulation of the tectum at sites that include the appropriate command elements (Ewert 1967 in 1984).

How are the putative tectal command elements connected with bulbar/spinal motor systems? In collaboration with A.Weerasuriya from the University of Colombo (Sri Lanka) we tackled this problem first from an anatomical view point, showing in toads that application of small amounts of horseradishperoxidase (HRP) into the medulla oblongata of toads led to backfilling of pear-shaped, large ganglionic, and pyramidal cells near the tectal output layer 7, a site from where the putative command elements with prey-selective and localization properties have been previously recorded (Weerasuriya and Ewert 1981; for comparable HRP studies in frogs, see Ingle 1983). In collaboration with M.Satou from Tokyo University (Japan) we have shown that tectal putative command elements (Fig.10) - such as T5(2), T5(1), T4, and T3 neurons - could be backfired antidromically in response to short electrical shocks applied to the medulla oblongata. Criteria of antidromic activation were: constant latency response (Fig.9Ba), following ability (Fig.9Ba), and collision test (Fig.9Bb) (Satou and Ewert 1985).

The exact destination of tectal output neurons in the bulbar/spinal motor systems is presently being analyzed. Satou et al. (1984), by means of intracellular recordings from tongue muscle motoneurons, have provided evidence of the existence of interneurons between the optic tectum and motoneuronal pools. This fits the conclusions drawn from neuroanatomical and neurophysiological studies by Weerasuriya and Ewert (1981, 1986 in prep.) and Ewert et al. (1984).

In a broad sense, a motor pattern generating circuitry (MPG) for a fixed action pattern of the prey-catching sequence would fulfill the criteria defined by Doty (1976): (i) sensory integration at the input side - via command systems - assures the activation of the MPG if, and only if, a specific combination of input occurs; (ii) an intrinsic pattern of neuronal connectivity, upon activation, is capable of generating a consistent spatio-temporal distribution of excitation and inhibition; (iii) the output of the internuncial network, mediated via pre-motor neurons, has privileged access to the required motoneuronal pools.

Our preliminary experimental data suggest that the pattern generating elements of an MPG are not localized in any one area, but rather distributed in subtectal, reticular, bulbar, and spinal systems. It is reasonable to assume that different MPGs may share certain pattern generating elements (Ewert et al. 1984). Furthermore, we have to consider that different command elements, which arrive from the tectal layers 6/7 (and/or the pretectum) do not need to impinge all on a single cell of the MPG, so that there must not necessarily be a particular neuron on which all the sensory messages converge.

MODULATION OF COMMAND SYSTEMS

The concept of "command (releasing) systems" forming the mediating circuitries between stimulus and response includes the possibility of "modulatory" influences, exerted by systems in which prosencephalic (diencephalic and telencephalic) structures participate. Stimulus specific habituation and learning processes have to be mentioned in this context. Furthermore, the "inner" state of the organism may influence activities of certain neurons. During the mating period, prey-selective T5(2) neurons cannot be recorded; at this time of the year the sexual partner (female) is the goal of orienting movements (by the male). In summer, during the hunting season, T5(2) neurons are weakly activated or appear to be silent in satiated animals.

Anatomical data provided by Wilczynski and Northcutt (1977), Neary and Northcutt (1983), and our own neurophysiological studies (Ewert 1984) suggest that various rostral thalamic and pretectal structures are involved in different internal loops (Fig.11A), by which pallial and striatal telencephalic nuclei may influence tectal cells, thus determining or modifying the output of the tectum to medullary/spinal pre-motor structures. In other words, the "mediating" command circuitry may be influenced by different "modulating" circuits. What can loop-operations do, and what may their advantage be? Suppose A is a neuronal substrate being modulated by a circuit involving other nuclei B, C, D, loop-operations may exert (Fig.11B):

(i) integration of external and internal input: "priming" a command system (Fig.11Ba),

(ii) disinhibition of an output system due to external and/or internal input: "gating" a command system (Fig.11Bb),

(iii) build-up of stimulus specific inhibition: "screening" a command system (Fig.11Bc),

(iv) associative disinhibition ("modifying" a command system, Fig.11Bd).

It may seem trivial to emphasize that the diagrams shown in Fig.11B represent a crude simplification of complex processes, but these "loop concepts" are suitable for experimental investigation. To illustrate this point, in the following I shall refer to operations (ii) and (iv).

With loop-operation (ii), the "striatal-thalamic-tectal" pathway may exert gating effects on tectal output systems that allow orientational head and/or body turning in response to appropriate visual signals. Autoradiographic studies with (14C)-2-deoxyglucose (2DG) have shown a statistically significant increase in glucose utilization in the ventral striatum during visually elicited prey-catching or escape behavior. Bilateral lesions to the striatum led to visual neglect of prey objects (Ewert 1967 see 1984, Finkenstädt in prep.).

With loop-operation (iv), there are theoretically different possibilities, by which innate prey-recognition can be modified or extended through conditioning: (a) The innate properties of T5(2) prey-selective command elements could be modified long-term if hypothetical "associative storing Ms neurons" (receiving excitatory inputs of both TH3 and T5(2)) depress the inhibitory influence from TH3 to T5(2). This may explain why

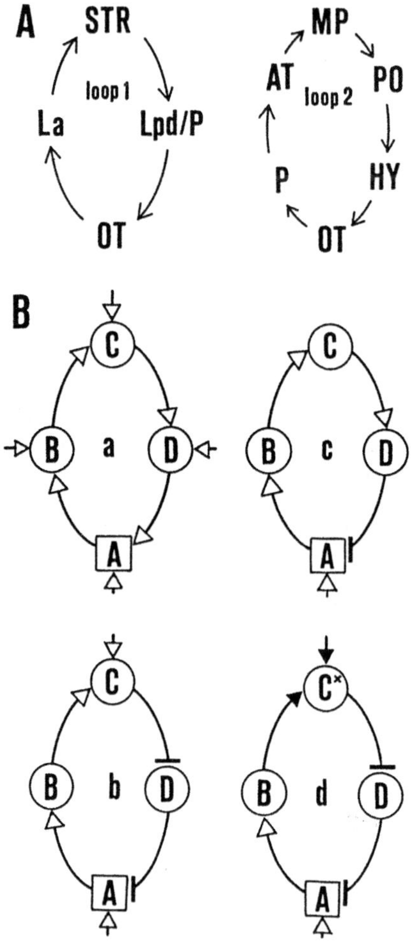

Fig.11. "Neural loops" and putative loop-operations, by which prosencephalic structures may determine the property of tectal neurons and/or modulate their respective activities. A) Anatomical substrates for various loops. AT, anterior thalamus; HY, hypothalamus; La, lateral anterior thalamic nucleus; Lpd, lateral posterodorsal thalamic nucleus; MP, medial pallium; OT, optic tectum; P, posterior thalamic nucleus; PO, preoptic area; STR, striatum. Arrows indicate descending pathways. (For anatomical data see Wilczynski and Northcutt 1977 and 1983; nomenclature of thalamic nuclei according to Neary and Northcutt 1983). – B) Ideas of loop-operations in hypothetical neural structures A,B,C,D exerting effects on A, a proposed output element: a) integration (e.g., "priming"), b) disinhibition (e.g., "gating"), c) build-up of inhibition (e.g., "screening"), d) associative disinhibition (e.g., "modifying"). Arrows: excitatory influences; solid black arrows: coincident associative inputs; lines with cross bars: inhibitory influences.

after "visual/visual" conditioning in the course of hand-feeding, non-prey stimuli, such as the experimenter's hand, elicit prey capture. (b) The T5(2) property might be changed temporarily by hypothetical "associative Ms/Mc units" (e.g., receiving excitatory visual T5(2) and olfactory input) if, and only if, the associated stimulus (prey odor) is present. (c) The innate properties of IRM might be extended by hypothetical "associative command element Ms/MT5 unit", of which MT5 command elements are activated only in the presence of the associated (e.g., olfactory) stimulus and, while inhibiting T5(2), might take over the function of the innate prey-selective command elements.

Preliminary results from our laboratory suggest that modification of the IRM for prey capture with regard to associative properties is mediated by a "pallial-thalamic-tectal" pathway. In a recent 2DG study my co-worker Th.Finkenstädt has shown that visual/visual conditioning after hand-feeding in toads is conspicuously accompanied by statistically significant changes in glucose utilization in certain areas of the posteroventral medial pallium. Bilateral lesions to this region prevent visual conditioning (Finkenstädt and Ewert in prep.).

Immunocytochemical studies by Kuljis and Karten (1982) and Reiner et al. (1980, 1982) as well as neuropharmacological investigations carried out by other authors (for a review see Freeman and Norden 1984), indicate that various prosencephalic influences to the tectum, mediated by pretectal (P and Lpd) and hypothalamic nuclei, may involve different neurochemical transmission patterns, throughout layers 3 to 9, so that a relatively complex intrinsic/extrinsic (inhibitory) framework may be postulated that shapes the properties of tectal neurons and modulates their respective activities.

CONCLUSIONS

The present study in toads sheds light on pretecto-tectal connectivities involved in perceptual operations as regards signal recognition and localization of prey-oriented action patterns. The stimulus-response mediating command circuitries are under the control of telencephalic (striatal and pallial) structures which in a broad sense may exert gating, shaping, or modulatory effects on prey-selective tecto-bulbar/spinal neurons in deciding as to whether or not a visual stimulus can be taken as a target for prey capture.

Despite the lack of homologies between the anuran "striato-pretecto-tectal" pathway and the mammalian "striato-nigro-tectal" pathway (Wilczynski and Northcutt 1983), the disinhibitory operational properties of both pathways (according to Fig.11Bb) exhibit some interesting similarities, which suggests comparable functional principles:

(i) Telencephalic ablations in the basal ganglionic area of anurans (Diebschlag 1935; Ewert 1967 see 1984) and in the striatum of mammals (Dray 1980; Pycock 1980) lead to a visual neglect as far as the "visual grasp reflex" (Akert 1949) is concerned. (ii) Behavioral studies on toads with telencephalic and/or pretectal lesions and recordings from rat nigro-tectal neurons in response to electrical stimulation of striatal areas suggest striato-pretectal (Ewert 1967 see 1984) and striato-nigral (Deniau et al. 1978) inhibitory connections, respectively. (iii) Surgical disruption of pretecto-tectal pathways in toads (Ewert 1967, 1968 see 1984; Ewert et al. 1974) and pharmacological blocking of nigro-tectal transmission in monkey (Wurtz and Hikosaka 1983) lead to "disinhibition" of visually oriented head or eye movements, respectively. (iv) Repetitive electrical stimulation (tetanization) in the pretectal Lpd/P area in toads (Ewert et al. 1974) and in substantia nigra in rats (Chevalier et al. 1984) decrease the sensory excitability of tecto-bulbar/spinal output-neurons, which are thought to carry "command signals" for orienting movements (Ewert 1974; Borchers et al. 1983; Satou and Ewert 1985).

The disynaptic basalganglio-tectal pathway operates along with other prosencephalo-tectal circuitries (involving the medial pallium as the "primordium hippocampi", Herrick 1933) which are relevant for attention, motivation, and learning (Pycock 1980; Ewert 1980; Rizzolatti 1983; Thinus-Blanc 1983; Kafetzopoulos 1983; Laming and Ewert 1984; Finkenstädt et al. 1986). Particularly in lower vertebrates - due to the lack of a laminated visual cortex - prosencephalo-tectal channel(s) have diverse functions which consist not only of gating the activities of certain tecto-bulbar/spinal neurons, but also of shaping their visual response properties and modifying them to fit individual experience.

In this context, the concept of IRM, that was for a long time - and partly still is - a controversial issue for various reasons, is not obsolete in ethology. It is again becoming up-to-date tool in neuroethological research, since it offers many advantages in the study of the associated neuronal processes involved in learning (see also Konishi 1985).

ACKNOWLEDGEMENTS. I wish to acknowledge the help of the late Professor Paul Ellen who commented on the manuscript and Mrs. Ursula Reichert for word processor type-setting. The work was supported by the Deutsche Forschungsgemeinschaft Ew 7/1-7/9 and the Fritz Thyssen Stiftung.

REFERENCES

Akert,K., 1949. Der visuelle Greifreflex. Helvetia Physiologica Acta 7:112-134.

Arbib,M.A. and Lara,R., 1982. A neural model of the role of the tectum in prey-catching behavior. Biological Cybernetics 44:185-196.

Barlow,H.B., 1953. Summation and inhibition in the frog's retina. Journal of Physiology (London) 173:377-407.

Barlow,H.B., 1985. The role of nature, nurture, and intelligence in pattern recognition, in "Pontificiae Academiae Scientiarum Scripta Varia 54". Ex Aedibus Academicis in Civitate Vaticana, Roma.

Borchers,H.-W., Schürg-Pfeiffer,E., Megela,A.L., and Ewert,J.-P., 1983. Single neuron activity in the optic tectum of intact and thalamic-pretectal (TP)-lesioned behaving toads. Neuroscience Letters Supplement

14:36.

Bullock,T.H., 1983. Implications for neuroethology from comparative neurophysiology, in "Advances in Vertebrate Neuroethology". J.-P.Ewert, R.R.Capranica and D.J.Ingle (Eds.) Plenum Press, New York.

Burghagen,H. and Ewert,J.-P., 1982. Question of 'head preference' in response to worm-like dummies during prey-capture of toads, Bufo bufo. Behavioural Processes 7:295-306.

Chevalier,G., Vacher,S., and Deniau,J.M., 1984. Inhibitory nigral influence on tectospinal neurons, a possible implication of basal ganglia in orienting behavior. Experimental Brain Research 53:320-326.

Deniau,J.M., Hammond,C., Riszk,A., and Feger,J., 1978. Electrophysiological properties of identified output neurons of the rat substantia nigra (pars compacta and pars reticulata): Evidence of the existence of branched neurons. Experimental Brain Research 32:409-422.

Diebschlag,E., 1935. Zur Kenntnis der Großhirnfunktionen einiger Urodelen und Anuren. Zeitschrift für vergleichende Physiologie 21:343-394.

Doty,R.W., 1976. The concept of neural centers, in "Simpler Networks and Behavior". J.C.Fentress (Ed.), Sinauer Ass., Sunderland, Mass.

Dray,A., 1980. The physiology and pharmacology of mammalian basal ganglia. Progress in Neurobiology 14:221-335.

Ewert,J.-P., 1974. The neural basis of visually guided behavior. Scientific American 230:34-42.

Ewert,J.-P., 1980. "Neuroethology. An Introduction to the Neurophysiological Fundamentals of Behavior". Springer-Verlag, Berlin - Heidelberg - New York.

Ewert,J.-P., 1984. Tectal mechanisms that underlie prey-catching and avoidance behaviors in toads, in "Comparative Neurology of the Optic Tectum". H.Vanegas (Ed.), Plenum Press, New York.

Ewert,J.-P., 1985. The Nico Tinbergen Lecture 1983: Concepts in vertebrate neuroethology. Animal Behaviour 33:1-29.

Ewert,J.-P., 1986. Measuring visual discrimination: Principles of configurational perception, in "Aims and Methods in Neuroethology". D.M. Guthrie (Ed.), Manchester Univ. Press, Manchester.

Ewert,J.-P. and Seelen,W.v., 1974. Neurobiologie and System-Theorie eines visuellen Muster-Erkennungsmechanismus bei Kröten. Kybernetik 14:167-183.

Ewert,J.-P., Speckhardt,I., and Amelang,W., 1970. Visuelle Inhibition und Exzitation im Beutefangverhalten der Erdköte (Bufo bufo L.). Zeitschrift für vergleichende Physiologie 68:84-110.

Ewert,J.-P., Hock,F.J., und Wietersheim,A.v., 1974. Thalamus/Praetectum/ Tectum: Retinale Topographie und physiologische Interaktionen bei der

Kröte (Bufo bufo L.). Journal of Comparative Physiology 92:343-356.

Ewert,J.-P., Borchers,H.-W., and Wietersheim,A.v., 1978. Question of prey feature detectors in the toad's Bufo bufo (L.) visual system: A correlation analysis. Journal of Comparative Physiology 126:43-47.

Ewert,J.-P., Burghagen,H., and Schürg-Pfeiffer,E., 1983. Neuroethological analysis of the innate releasing mechanism for prey-catching behavior in toads, in "Advances in Vertebrate Neuroethology". J.-P.Ewert, R.R.Capranica and D.J.Ingle (Eds.), Plenum Press, New York.

Ewert,J.-P., Schürg-Pfeiffer,E., and Weerasuriya,A., 1984. Neurophysiological data regarding motor pattern generation in the medulla oblongata of toads. Naturwissenschaften 71:590-591.

Ewert,J.-P., Matsumoto,N., and Schwippert,W.W., 1985. Morphological identification of prey-selective neurons in the grass frog's optic tectum. Naturwissenschaften 72:661-662.

Finkenstädt,Th., Adler,N.T., Allen,T.V., and Ewert,J.-P., 1986. Regional distribution of glucose utilization in the telencephalon of toads in response to configurational visual stimuli: A (14C)-2DG study. Journal of Comparative Physiology (in press)

Fite,K.V. and Scalia,F., 1976. Central visual pathways in the frog, in "The Amphibian Visual System: A Multidisciplinary Approach". K.V.Fite (Ed.), Academic Press, New York - San Francisco - London.

Freeman,J.A. and Norden,J.J., 1984. Neurotransmitters in the optic tectum of non mammalians, in "Comparative Neurology of the Optic Tectum". H.Vanegas (Ed.), Plenum Press, New York.

Grüsser,O.-J., and Grüsser-Cornehls,U., 1976. Neurophysiology of the anuran visual system, in "Frog Neurobiology". R.Llinás and W.Precht (Eds.), Springer-Verlag, Berlin - Heidelberg - New York.

Hartline,H.K., 1938. The response of single optic nerve fibers of the vertebrate eye to illumination of the retina. American Journal of Physiology 119:69-88.

Herrick,C.J., 1925. The amphibian forebrain. III: The optic tracts and centers of amblystoma and the frog. The Journal of Comparative Neurology 39:433-485.

Herrick,C.J., 1933. The amphibian forebrain. VIII: Cerebral hemispheres and pallial primordia. The Journal of Comparative Physiology 58:737-759.

Hoyle,G., 1984. The scope of neuroethology. The Behavioral and Brain Sciences 7:367-412.

Ingle,D.J., 1983. Brain mechanisms of visual localization by frogs and toads, in "Advances in Vertebrate Neuroethology". J.-P.Ewert, R.R.Capranica and D.J.Ingle (Eds.), Plenum Press, New York.

Kafetzopoulos,E., 1983. On the central dopaminergic mechanisms for motion and emotion, in "Advances in Vertebrate Neuroethology". J.-P.Ewert,

R.R.Capranica and D.J.Ingle (Eds.), Plenum Press, New York.

Konishi,M., 1985. Birdsong: From behavior to neuron. Annual Review of Neuroscience 8:125-170.

Krasne,F.B., Kandel,E.R., and Truman,J.W., 1979. Simple systems revisited. Neurosciences Research Program Bulletin 17:529-538.

Kuljis,R.O. and Karten,H.J., 1982. Laminar organization of peptide-like immunoreactivity in the anuran optic tectum. The Journal of Comparative Neurology 212:188-201.

Kupfermann,I. and Weiss,K.R., 1978. The command neuron concept. The Behavioral and Brain Sciences 1:3-39.

Laming,P.R. and Ewert,J.-P., 1983. The effects of pretectal lesions on neuronal, sustained potential shift, and electroencephalographic responses of the toad tectum to presentation of a visual stimulus. Comparative Biochemistry and Physiology 76:247-252.

Laming,P.R. and Ewert,J.-P., 1984. Visual unit, EEG, and sustained potential shift responses to biologically significant stimuli in the brain of toads (Bufo bufo). Journal of Comparative Physiology 154:89-101.

Lara,R., Cervantes,F., and Arbib,M.A., 1982. Two-dimensional model of retinal-tectal-pretectal interactions for the control of prey-predator recogition and size preference in amphibia, in "Competition and Cooperation in Neural Nets". S.Amari and M.A.Arbib (Eds.), Springer-Verlag, Berlin - Heidelberg - New York.

Lettvin,J.Y., Maturana,H.R., McCulloch,W.S. and Pitts,W.H., 1959. What the frog's eye tells the frog's brain. Proceedings of the Institute of Radio Engineers 47:1940-1951.

Lorenz,K., 1943. Die angeborenen Formen möglicher Erfahrung. Zeitschrift für Tierpsychologie 5:235-409.

Matsumoto,N., Schwippert,W.W., and Ewert,J.-P., 1986. Intracellular activity of morphologically identified neurons of grass frog's optic tectum in response to moving configurational visual stimuli (submitted).

Megela,A.L., Borchers,H.-W., and Ewert,J.-P., 1983. Relation between activity of tectal neurons and prey-catching behavior of toads, Bufo bufo. Naturwissenschaften 70:100-101.

Neary,T.J. and Northcutt,R.G., 1983. Nuclear organizatin of the bullfrog diencephalon. The Journal of Comparative Neurology 213:262-278.

Pycock,C.J., 1980. Turning behaviour in animals. Neuroscience 5:461-514.

Reiner,A., Brauth,S.E., Kitt,C.A., and Karten,H.J., 1980. Basal ganglionic pathways to the tectum: Studies in reptiles. The Journal of Comparative Neurology 193:565-589.

Reiner,A., Karten,H.J., and Brecha,N.C., 1982. Enkephalin-mediated basal ganglia influences over the optic tectum: Immunohistochemistry of the tectum and the lateral spiriform nucleus in pigeon. The Journal of Comparative Neurology 208:37-53.

Reiner,A., Brauth,S.E., and Karten,H.J., 1984. Evolution of the amniote basal ganglia. Trends in Neurosciences 7:320-325.

Rizzolatti,G., 1983. Mechanisms of selective attention in mammals, in "Advances in Vertebrate Neuroethology". J.-P.Ewert, R.R.Capranica and D.J.Ingle (Eds.), Plenum Press, New York.

Satou,M. and Ewert,J.-P., 1985. The antidromic activation of tectal neurons by electrical stimuli applied to the caudal medulla oblongata in the toad, Bufo bufo L. Journal of Comparative Physiology 157:739-748.

Satou,M., Matsushima,T., and Ueda,K., 1984. Neuronal pathways from the tectal "snapping-evoking area" to the tongue muscle controlling motoneurons in the Japanese toad: Evidence of the intervention of excitatory interneurons. Zoological Science 1:829-832.

Székely,G. and Lázár,G., 1976. Cellular and synaptic architecture of the optic tectum, in "Frog Neurobiology". R.Llinás and W.Precht (Eds.), Springer-Verlag, Berlin - Heidelberg - New York.

Thinus-Blanc,C., 1983. Localization, orienting responses and attention in the golden hamster, in "Advances in Vertebrate Neuroethology". J.-P.Ewert, R.R.Capranica and D.J.Ingle (Eds.), Plenum Press, New York.

Tinbergen,N., 1951. "The study of instinct". Clarendon Press, Oxford.

Traud,R., 1983. "Einfluß von visuellen Reizmustern auf die juvenile Erdkröte (Bufo bufo)". Doctoral Dissertation, University of Kassel.

Weerasuriya,A. and Ewert,J.-P., 1981. Prey-selective neurons in the toad's optic tectum and sensori-motor interfacing: HRP studies and recording experiments. Journal of Comparative Physiology 144:429-434.

Weerasuriya,A. and Ewert,J.-P., 1983. Afferents of some dorsal retino-recipient areas of the brain of Bufo bufo. Society for Neuroscience Abstract 9:536.

Wilczynski,W. and Northcutt,R.G., 1977. Afferents to the optic tectum of the leopard frog: An HRP study. The Journal of Comparative Neurology 173:219-229.

Wilczynski,W. and Northcutt,R.G., 1983. Connections of the bullfrog striatum: Efferent projections. The Journal of Comparative Neurology 214:333-343.

Wurtz,R.H. and Hikosaka,O., 1983. Deficits in eye movements after injection of GABA-related drugs in monkey superior colliculus. Society for Neuroscience Abstracts 9:750.

COGNITIVE MAPS AND NAVIGATION IN HOMING PIGEONS.

Wolfgang Wiltschko and Roswitha Wiltschko
Fachbereich Biologie der Universität, Zoologie, Siesmayerstraße 70, D-6000 Frankfurt a.M., FRG

It is an obvious necessity for all animals that have some kind of "home" to be able to orient within their home range, i.e. between their home, suitable food sources, water etc.. This allows fast and efficient movements which will minimize the time and energy expenditure and help to reduce the periods in which the animals are exposed to predation. Hence it is not surprising that orientational abilities have been found in many species of vertebrates, social insects and numerous other invertebrates (see Schmidt-Koenig 1975). But the most outstanding performances have been reported from birds. Migrating birds fly to their distant wintering areas and return to last year's breeding territory often after completing a journey of several thousand kilometers. Homing pigeons are able to return to their home after passive displacement of several hundreds of kilometers into completely unknown territory.

The orientation system of birds which we shall describe was derived from findings based on the behavior of free flying pigeons and displacement distances of up to hundreds of kilometers. The experiments took place in natural habitats, a highly complex environment, where we could not manipulate the ambient conditions. Thus, most experimental interferences involve the animals themselves, their sensory input and/or their experience.

THE "MAP AND COMPASS"-MODELL

The ability of pigeons to home has been known since antiquity. But the experimental analysis of this phenomenon was begun in this century. Some early authors - e.g. Heinroth and Heinroth (1941) - assumed that after displacement, the birds would search - either randomly or by applying some strategy - until they, by chance, recognized familiar terrain at the edge of their home range. From this point, they would follow sequences of familiar landmarks to reach home. Such a strategy, called "piloting", was taken to be the simplest way of homing and was designated as "type I" by Griffin (1952). However, this explanation did not consider the possible problems involved in recognizing landmarks when one does not know in advance from which side the home range will be approached and which of the landmarks one might meet.

Experimental evidence, however, did not support "piloting". Birds do not simply follow sequences of familiar landmarks. The nature of the homing process is best illustrated by clock-shift experiments (a type of test first performed by Schmidt-Koenig). Pigeons whose internal clock has been phase-shifted (for details, see Schmidt-Koenig 1961) depart in directions which deviate from those of controls in a predictable way. Fig. 1 gives an

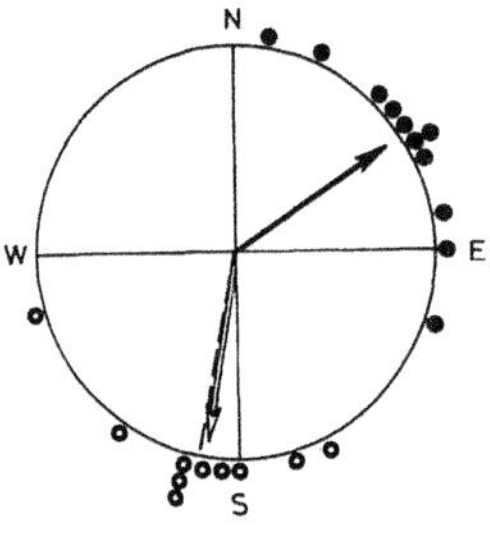

Fig. 1:
Pigeons whose internal clock has been shifted 6 h forward (solid symbols) show a characteristic counterclockwise deviation from their controls (open symbols) indicating the use of the sun compass. - The home direction 192° S is marked by a dashed line. The symbols at the periphery of the circle mark the vanishing bearings of individual birds, the arrows represent the mean vectors with their lengths drawn proportional to the radius of the circle.

example: After being displaced to the north, the birds determine their home direction. The navigational system does not indicate "fly there" (as Barlow's 1964 theory of inertial navigation assumed), but in our example something like "fly south". Shifting the internal clock does not affect this first step (the experimentals, too, intended to fly south), but it leads to a typical error when the birds afterwards use the sun compass to find out which direction is south. The experimentals whose internal clock had been shifted 6 h forward took the morning sun in the east to be the noon sun in the south and departed flying toward it.

These findings demonstrate that homing is a two step process. First the home direction is established as a compass heading (i.e. relative to an external reference system), and in a second step, a compass is used to localize this direction. Kramer (1953) was the first who recognized this very important fact; he described pigeon homing by his "map and compass"-model.

The characteristic deviations which indicate the use of the sun compass are also observed when clock-shifted pigeons are released from very familiar locations, e.g. within 2 km of their home loft at places which they should recognize from their daily exercise flights (cf. Keeton 1974), and at sites from which they have homed more than 60 times (Füller et al. 1983). These findings, while surprising at first, indicate that compass orientation is an intrinsic part of every homing process, enabling birds to cope with space in the range of hundreds of kilometers. Navigation without a compass has been discussed as a theoretical possibility, but there is no evidence that would support its existence.

The various compass systems used by birds are well known and relatively well understood: the sun compass, the magnetic compass and, in night-migrating birds, a star compass (for summary, see W. Wiltschko 1983). - In contrast mechanisms enabling the birds to determine their home direction (the "map"-step), in spite of considerable research efforts, remain enigmatic. We are just beginning to understand some of their basic properties, and these will be discussed in this paper.

DEVELOPMENT OF THE NAVIGATIONAL SYSTEM DURING ONTOGENY

Keeton (1971, 1974) was the first to become aware of the fact that the pigeons' navigational system undergoes a fundamental change during the early development. Today, we know that the mechanisms and strategies used by adult, experienced pigeons are based on information that had to be acquired in the first months of life, quite possibly as soon as the young birds become able to fly.

Initially, the only mechanism available to young birds is a "magnetic compass". Because of their ability to perceive the magnetic field (for details on possible mechanisms, see Semm et al. 1984), the space around them does not appear homogenous, they can recognize the various directions. This per se must not be thought of as a sophisticated mechanism. It simply provides them with an independent general reference system in the horizontal plane just as the perception of gravity provides such a system in the vertical. Discriminating "north" and "south" - or rather (since the magnetic compass is based on the inclination of the field lines and not on their polarity, W. Wiltschko a. Wiltschko 1972) "poleward" and "equatorward" is not more difficult than discriminating "up" and "down".

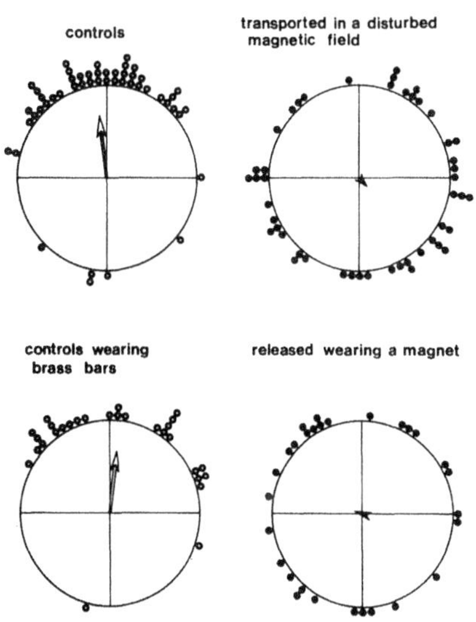

Fig. 2:
Young inexperienced pigeons need undisturbed magnetic information during displacement (upper diagrams) and during release (lower diagrams) to be oriented. - The bearings are plotted with home = 360° up. Symbols as in Fig. 1. (Upper diagrams: data from W. Wiltschko a. Wiltschko 1981; lower diagrams: data from Keeton 1971).

Yet when the young birds begin to extend their activity, this reference system allows them to keep track of their movements, since they are aware of the direction in which they are flying. When young, inexperienced pigeons were transported to the release site in a distorted magnetic field (being treated like their controls upon arrival, W. Wiltschko a. Wiltschko 1981) or when (after normal transportation) they were released carrying a small magnet (Keeton 1971), they were not able to orient (Fig. 2). This clearly shows that they require undisturbed magnetic information. - We believe that their strategy is to record the net direction of their displacement - integrating detours, if necessary - by the magnetic compass. They reverse this direction, and thus obtain the home direction by again consulting their magnetic compass to localize it in space. This strategy is called "route reversal", and it is in agreement with the "map and compass"-model insofar as the home direction is first determined as a compass direction.

Yet when the experimental treatments of deprivation of meaningful information during displacement or during release is given to experienced pigeons, it does not seem to affect the orientation. Older, experienced pigeons no longer rely solely on the magnetic compass. Their preferred compass is the sun compass; it is used whenever the sun is visible (comp. Fig. 1). The learning processes associating sun azimuth, time of day and geographic direction begin as soon as the birds obtain flying experience. Normally, the sun compass is established by the time pigeons are about 3 months old, but when this early flying experience is increased by training releases, they are able to use the sun compass at an even younger age (R. Wiltschko 1983).

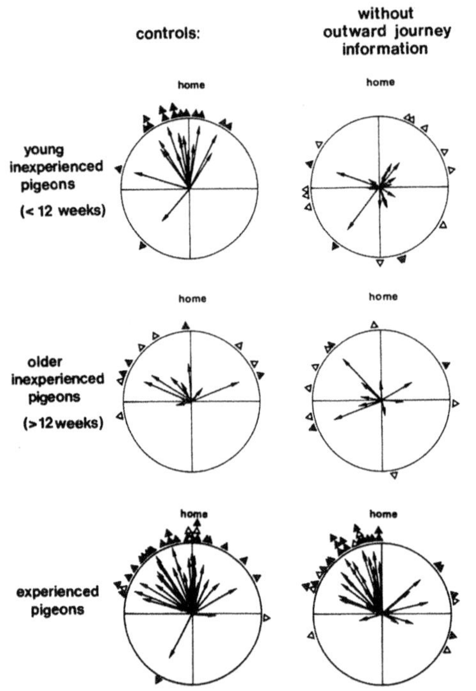

Fig. 3:
The importance of information collected during the outward journey decreases with increasing age and experience. - The data are pooled with respect to home = 360° up. The arrows represent the mean vectors of a sample; the triangles at the periphery mark the mean directions (solid: significantly oriented samples, open: non-significant samples). (Data from R. Wiltschko a. Wiltschko 1985).

At the same time, their navigational strategy changes radically. Route reversal is given up in favor of using a "map" of local factors, i.e. the birds now derive their home direction from information obtained at the release site. This fact is demonstrated by experiments in which pigeons of various ages and experience were prevented from collecting meaningful information during displacement (Fig. 3). In inexperienced birds younger than 12 weeks, such restraint causes disorientation (upper diagrams). In older inexperienced pigeons, the differences between experimentals and controls were much less noticable. In trained pigeons who had taken part in a series of releases away from their loft, we found hardly any effect at all (lower diagrams)(R. Wiltschko a. Wiltschko 1985). This indicates that experienced birds did not use outward journey information any longer.

The data presented in Fig. 3 seem to suggest that the switch from route reversal to the use of local information is correlated with age. But increasing age also means an increasing number of spontaneous flights around the loft and thus increasing flying experience. At the age of 2 to 3 months, young pigeons begin to venture farther from their loft; they sometimes stay out of sight for hours. The use of local information might, like the sun compass, depend on flying experience. - It must remain open whether the establishment of the sun compass and the establishment of a "map" of local information are somehow connected or whether they are two independent processes that happen to take place at the same time.

When a "map" of local information is available, it seems to be used preferentially. This is true even if this type of navigation has as yet not achieved high accuracy. During a stage of transition, its use may result in poorly homeward oriented bearings (comp. Fig. 3, middle diagrams; for a detailed discussion see W. Wiltschko a. Wiltschko 1982, R. Wiltschko a. Wiltschko 1985). Nevertheless the reason for the strong urge to use local information may be evident. As long as the birds rely solely on directional information collected during the outward journey, they are not able to check their course and thus they cannot correct mistakes, whereas the use of local information allows them to redetermine the home direction as often as necessary. Hence navigation by local cues reduces the risk of errors and adds safety, factors which become essential as soon as the birds venture farther and cover greater distances.

This development of the navigational system in young birds shows an interesting parallel to the development Acredolo (1986) describes for young humans. With the onset of self-produced locomotion, the formerly egocentric spatial reference system begins to change into an exocentric one that allows them to keep track of their own movements in space. Likewise, extended flights and increasing flying experience start the switch from outward journey information (which represents an home-centered system), to the independent, objective system of local information. In both cases, an added dimension in locomotion quickly leads to an adaptation in the spatial system which allows the individuals to master the new situation.

THE NAVIGATIONAL "MAP"

General characteristics

Any model describing how the navigational system based on local information works has to consider the following points:

First, it must include some explanation for the existence of release site bias, i.e. for the fact that pigeons rarely depart directly toward home, but show deviations from the true home direction which are characteristic for a given site (Fig. 4).

Second, it must be able to explain how pigeons are able to use local information in completely unknown territory.

Third, it must take into account that navigation is a fast process. A pigeon flies roughly to the direction in which it will finally vanish within less than a minute after release (Pratt a. Thouless 1955 a.o.).

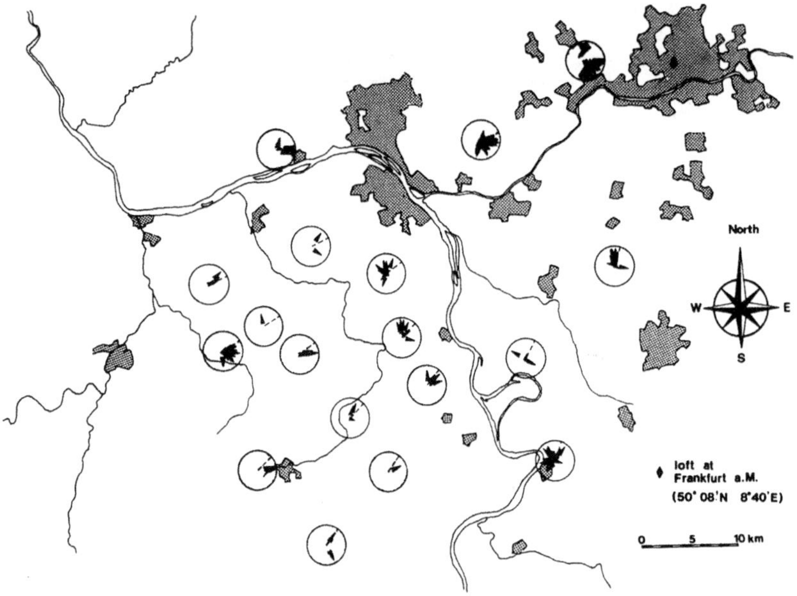

Fig. 4: Distribution of release site biases in an area south-west of the Frankfurt loft. The arrows within the circles indicate the mean vectors of the single releases proportional to the radius of the circles. Light circles: sites with predominantly counterclockwise deviations; shaded circles: sites with predominantly clockwise deviations.

The concepts presently under discussion may be characterized as spatial "map", which means a <u>mental picture of the spatial distribution</u> of the factors used. The idea of such "map" to be used for bicoordinate navigation dates back to the 40s and 50s when the global theories of navigation were published (e.g. Yeagley 1947, Matthews 1953; a detailed review was published by Wallraff 1974). Here we want to summarize our own ideas which are, in part, based on Wallraff's and which we have discussed in earlier papers (W. Wiltschko a. Wiltschko 1978, 1982).

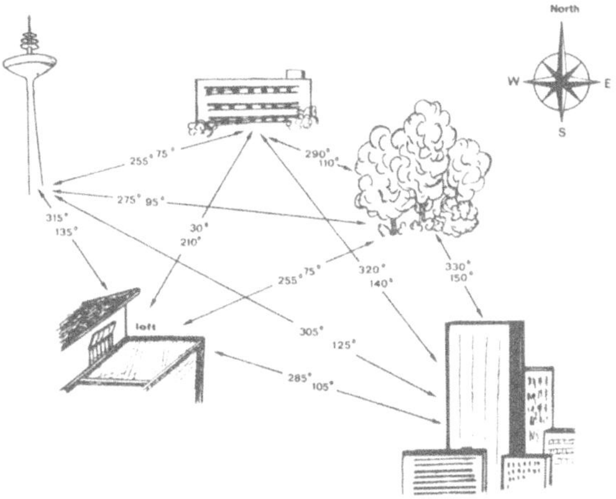

Fig. 5: The "mosaic map" - a directionally-oriented picture of the spatial distribution of landmarks around home (after W. Wiltschko a. Wiltschko 1982).

For orientation <u>within</u> the home range, i.e. in the very familiar area in the vicinity of the loft, the pigeons use a "mosaic map" (Fig. 5) which is assumed to be a <u>memorized</u>, <u>directionally-oriented</u> picture of the distribution of significant marks (which need not only be visual landmarks, but may involve other sensory qualities). We assume that the birds "know" the compass direction from any given mark to the other marks and to home (see W. Wiltschko a. Wiltschko 1978). Such a "map" would allow any birds direct routes and fast and efficient flights between goals within the home range. Since the landmarks are used to indicate position, this concept explains the use of compass orientation in familiar areas.

For orientation <u>outside</u> the home range in more or less unfamiliar territory, pigeons make use of a "navigational map" based on at least two environmental factors. These have the nature of gradients and intersect at an angle not too acute (Fig. 6). The birds <u>know</u> the directions of the gradients; in our example, they know that gradient A increases to the east. Upon release they measure the local value of A and compare it with the remembered home value. If the local value is smaller (P_1 in Fig. 6), they

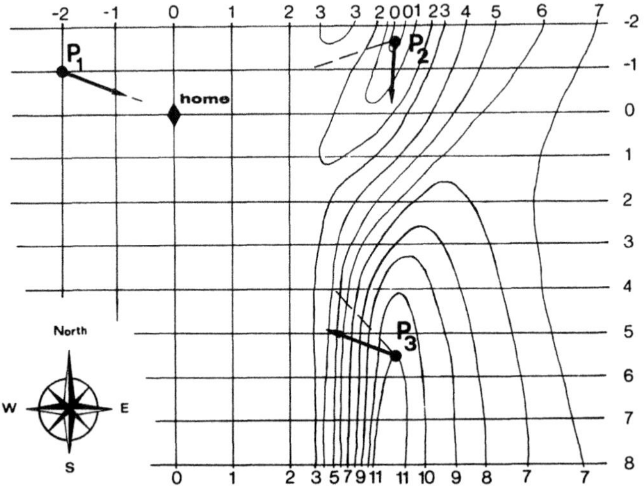

Fig. 6: The "navigational map" - a directionally-oriented picture of the spatial distribution of gradients. The gradient values are given relative to the home values. One gradient shows an anomaly in the eastern region (after W. Wiltschko a. Wiltschko 1982).

"conclude" that they must be somewhere west of home and thus have to fly in easterly direction. Another factor must give them a second direction, and they combine both to derive their (subjective) home direction (Fig. 6).

Since the course of the gradients can be extrapolated, the birds are able to use this mechanism of determining their home direction also outside the familiar range. Yet if the factors are not completely regularly distributed, extrapolation might cause the birds to "misjudge" their position and depart in directions deviating from the true homeward course (P_2 and P_3 in Fig. 6). Release site biases might thus be explained as the reaction to irregularities in the distribution of the "map" factors (Keeton 1973, W. Wiltschko a. Wiltschko 1982 - but see Wallraff 1974, 1982).

So far the described "navigational map" is an open system. The crucial questions to specify the concept are: What gradients form the co-ordinates of the "map", and how do the birds obtain their knowledge about the gradient directions and their spatial distributions? - Let us discuss the last question first. The authors of the early theories (e.g. Yeagley 1947, Matthews 1953) implied that this knowledge was innate. Today, however, most authors tend to assume that it is learned (i.e. acquired by experience and memorized. The circumstances of these learning processes become the essential problem, as they determine the nature of the "map".

Experience models the "map"

Wallraff (1974) assumes that pigeons learn the distribution of relevant factors during their long-term exposure at the home loft. His view is supported by the findings that the behavior of pigeons can be affected by differently shielded aviaries (e.g. Wallraff 1966, Ioalé 1982 a.o.) and by exposing them to various factors at the home loft (e.g. Baldaccini et al. 1975 a.o.). Yet all these pigeons had little, if any, opportunity to fly freely. Learning during long-term exposure, although certainly possible, is perhaps only of minor importance when the pigeons have normal access to the natural environment. It seems obvious that under normal conditions the young birds learn and memorize the direction of the gradients on their flights. During these flights, they experience directly how potential navigational factors change in the direction in which they are flying.

The literature contains many reports that flying experience (spontaneous flights as well as homing flights after training releases) affects the initial orientation of pigeons. - Since most authors used inexperienced pigeons that were older than ca. 3 months, the temporary decrease of homeward orientedness at the transition from route reversal to the use of the navigational map (comp. Fig. 3) has rarely been documented (W. Wiltschko a. Wiltschko 1982, R. Wiltschko a. Wiltschko 1985). Yet most authors agree that the pigeons' performance improves markedly when they have participated in a series of training releases. In general, trained pigeons are better oriented and return home in larger numbers than inexperienced pigeons (e.g. Matthews 1953b, Sonnberg a. Schmidt-Koenig 1953, Wallraff 1982 a.m.o.). Keeton (1974) pointed out that the very first training releases are the most effective. This is not surprising as they represent the most dramatic enlargement of the area known to the birds.

The "map", and the learning processes establishing it, are not directly accessible through experimental analysis. All our knowledge comes from indirect evidence. The development of the "navigational map" may take place in the following way: In the beginning, as long as the pigeons' experience is based on the spontaneous flights alone, their "map" strongly depends on the local distribution of the potential factors in the area around their loft. These might not be typical when a larger region is considered. As they can only extrapolate gradients on the basis of this knowledge, these inexperienced pigeons might misinterpret the local factors at distant sites. At the same time, the experience may vary greatly between individuals because of the different spontaneous tendencies to fly. Some birds venture farther away and explore larger areas than others who do not leave the immediate vicinity of their loft. These differences in experience may cause individually different assessment of their position. This is reflected in the frequently observed short mean vectors in samples of older inexperienced pigeons.

The effect of training releases is that pigeons get acquainted with the distribution of potential navigational factors in a larger area. They find out what factors form the most suitable coordinates, making the "map" a more realistic picture of the factors' true distribution. Apparently the "map", once established and in use, does not reach final form. It is continuously enlarged and improved over experience, at least in the early part of a bird's life. Over areas which the pigeons frequently cross, they might even learn to take irregularities of the "map" into account. At some locations where birds on their first release are disoriented or show very large biases, the homeward orientation improves on later flights (e.g. Kiepenheuer 1982, Kowalski a. Wiltschko 1984). Maybe odd local gradient values can be treated as anomalies whose positions are memorized. This may occur even if the birds in general still extrapolate the overall course of the gradients.

It is an interesting question as to whether or not the local distribution of "map" factors at the home loft continues to affect the pigeons' "map" in their later life. Birds from different lofts sometimes show very different orientation when released at the same sites, even if the home directions are similar (e.g. Schmidt-Koenig 1963, Wallraff 1970). Around some lofts, the biases form a somewhat regular pattern (cf. Wallraff 1982). Even if the birds continously enlarge their "map" and adapt it to their increasing experience, their interpretation of the navigational factors might remain to a certain extent "loft-specific". Perhaps the distribution of the gradients in the area of the loft itself, experienced every day and reinforced by every free flight, exerts a stronger influence on the pigeons' way of interpreting local cues than other types of distribution met only on occasional flights at greater distances from home. This might be the reason why experienced birds still show biases at familiar sites.

The learning processes establishing and improving the "map" seem to require that the birds have a genetic predisposition to recognize and memorize the spatial distribution of navigational factors in a quick and easy manner. Birds seem to possess some kind of mental compiler that arranges their experience in a spatial array to form the "navigational map". The "map" can be expanded beyond the area of immediate experience by extrapolation. This implies cognitive processes, as the birds must derive from the known courses of the gradients some rule about their farther course. Consequently, the experience obtained at one location might meaningfully modify the behavior at other locations in that it changes the way local factors are interpreted. A chance observation illustrates this point. In a group of young pigeons released for an exercise flight at their loft, about 2/3 ventured away and stayed out of sight for approximately 1:15 h. The remaining 1/3 stayed at the loft (their identity was recorded). When released the next day, the vanishing bearings of the birds who had explored on the previous day, differed significantly from those of the birds who had remained at the loft (Fig. 7). The experience in the area surrounding the loft had altered the reaction at a site 66 km distant. - In the above case, all birds were inexperienced, so the increase in experience through this one spontaneous flight was very large compared to the overall experience, effecting a marked change in orientation. Normally, when birds have a sufficient amount of flying experience, we would expect additional experience to modify the "map" only gradually.

The "map" continues to be adjusted to the existing conditions, at least during the first years. Yet there is evidence indicating a fundamental difference between the processes establishing and enlarging the "map" with increasing experience and the processes which later update the "map" within the familiar area (e.g. adjusting it to changing conditions). The first processes must be a general forming of the base for extrapolation, whereas the updating processes do not seem to be generalizing. This was suggested by a series of tests with adult experienced pigeons who were sub-

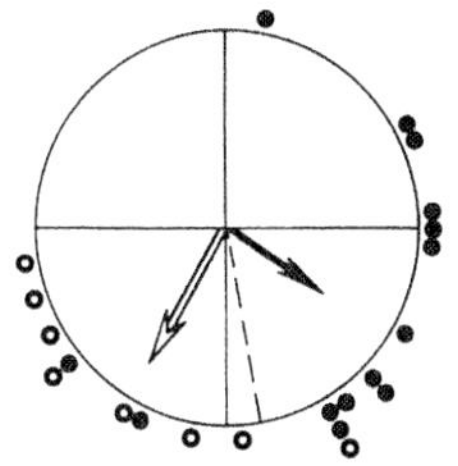

Fig. 7:
Orientation of inexperienced pigeons 66 km north of the loft. Open symbols: birds who had stayed near the loft, solid symbols: birds that went out of sight for more than 1 hour on the day before release. (symbols as in Fig. 1).

jected to a 6 h shift of their internal clock for an extended period of time. They were allowed to fly freely in the overlap time between the natural and their subjective day. Their sun compass was quickly adapted to the new situation, but they continued to deviate from their controls when they were released from any given site for the first time under shifted conditions. On later flights from the same sites, the deviation disappeared. A reverse deviation was observed when the birds were released there for the first time after their internal clock had been shifted back to normal (W. Wiltschko et al. 1984). We attribute these deviations to temporal variable "map" factors. Our findings indicate that the pigeons, while they were living in the shifted day, had somehow adapted this part of their "map" to the experimental situation. The adaptation was, however, closely limited to the location where they had had the experience contradicting their original "map". Although in the end of the test series the test birds had been living under the shifted conditions for more than 2 months and had altogether flown several hundred kilometers, they continued to deviate from their controls at every site where they were released for the first time under shifted conditions (W. Wiltschko et al. 1984). This seems to suggest that when the "map" is updated, any changes are effective only locally, otherwise the original "map" is maintained. This might be a safe strategy under natural conditions, since a general change in the environment is highly improbable. -

Regional differences in factors and strategies

Finally, we must turn to the question what factors form the gradients of the "map". Here our knowledge is also rather limited. Currently, mainly two factors are under discussion, namely olfactory information (see Wallraff 1980, Papi 1982 a.m.a.; it is not clear whether odors function as gradients of a "navigational map" or as local patches on an extended type of "mosaic map") and magnetic parameters (see Lednor 1982, Walcott 1982). These are certainly not all factors involved. We should be prepared that some unexpected cues may turn out to play a crucial role, as the pigeons' sensory world is much richer than our own (e.g. Kreithen 1978).

Some recent findings, however, seem to indicate that knowing the factors incorporated might not bring solution to the "map"-problem. Stimulated by the controversis about the significance of olfactory cues, we began a series of comparative tests in Tuscany, Italy, where Papi and his coworkers had obtained positive results, and at our own loft in Frankfurt,

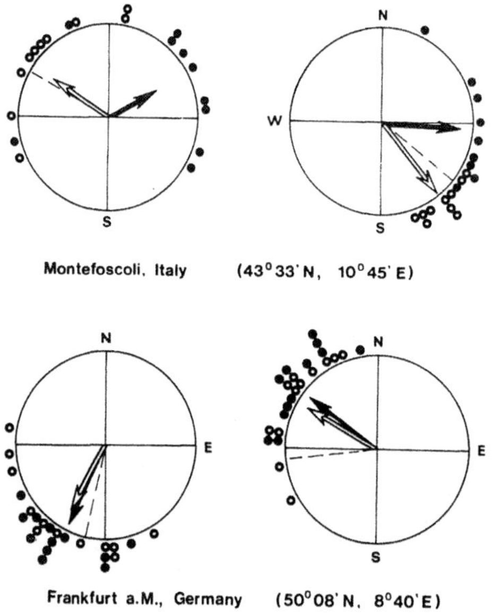

Fig. 8:
Identical experimental treatment effects different reactions.
Upper diagrams: In Italy, anosmic pigeons (solid symbols) differred from their controls (open symbols). (Data from W. Wiltschko et al. 1986).
Lower diagrams: In Germany, anosmic pigeons and controls were oriented alike. (Symbols as in Fig. 1).

Germany. The experimental treatment was identical, yet the results obtained in Italy and in Germany were very different (Fig. 8): In Italy, the anosmic pigeons differed significantly from their controls; they were not homeward oriented (W. Wiltschko et al. 1986). In Germany, we did not find a difference between the orientation of anosmic pigeons and controls. Either both groups were homeward oriented, or both showed the same bias (in prep.). - These results clearly demonstrate that pigeons in the two regions did not make use of the same navigational factors. Olfactory information was important in Tuscany, but not in Frankfurt.

Earlier findings had already suggested that not only the navigational factors involved in the "map", but also the navigational strategy itself might show regional variability. While treatments during the outward journey had, if any, only negligible effects on the orientation of experienced pigeons in Germany and Northeastern USA (e.g. Keeton 1974b, Wallraff 1980b, R. Wiltschko a. Wiltschko 1985, see Fig. 3), Papi and his coworkers consistently reported positive results. Their test birds in Italy show a decrease in homeward orientedness (e.g. Benvenuti et al. 1982 a.o.). At the same time, release site biases, if they occur at all, are rather small. Taken together, these findings indicate that in Italy outward journey information continues to be used by adult, experienced birds, and local information does not play the same dominant role as found elsewhere.

It is clear that any considerations on bird navigation have to take such regional differences into account. The orientation system has turned out to be extremely variable. Its final formation depends on a variety of factors the role of which we do not completely understand. Apparently pigeon's mental compiler that establishes and memorizes the "map" is not prepared to take any special type of navigational cues. Instead, it must be expected to process multi-sensory spatial information and be able to select from all available environmental factors those which are the most reliable in the region where the bird lives. If suitable local factors are not easily available, information obtained during the outward journey continues to be used.

All this makes the orientation system highly adaptive. It will always be perfectly tuned to the birds' local situation making optimal use of what information is available. - The experimenter, however, can no longer hope to find a universally valid answer to the question of how the birds orient. He can only describe how the orientation system will develop under given circumstances.

REFERENCES

Acredolo, L.P., 1986. Early development of spatial orientation in humans. in "Cognitive Processes and Spatial Orientation in Animal and Man", Vol. 2. P. Ellen and C. Thinus-Blanc (Eds.), Martinus Nijhoff Publ., Dordrecht, The Netherlands.

Baldaccini, N.E., Benvenuti, S., Fiaschi, V., Ioale, P. and Papi, F., 1974. Pigeon homing: Effects of manipulation of sensory experience at home site. Journal of Comparative Physiology, 94: 85-96.

Barlow, J.S., 1964. Inertial navigation as a basis for animal navigation. Journal of Theoretical Biology, 6: 76-117.

Benvenuti, S., Baldaccini, N.E. and Ioalé, P., 1982. Pigeon homing: Effect of altered magnetic field during displacement on initial orientation. in "Avian Navigation". F. Papi and H.G. Wallraff (Eds.), Springer Verlag, Berlin, Heidelberg, New York.

Füller, E., Kowalski, U. and Wiltschko, R., 1983. Orientation of homing pigeons: compass orientation vs. piloting by familiar landmarks. Journal of Comparative Physiology, 153: 55-58.

Griffin, D.R., 1952. Bird navigation. Biological Review of the Cambridge Philosophical Society, 27: 359-400.

Heinroth, O. and Heinroth, K., 1941. Das Heimfinde-Vermögen der Brieftauben. Journal für Ornithologie, 89: 213-256.

Ioalé, P., 1982. Pigeon homing: effects of differential shielding of home cages. in "Avian Navigation". F. Papi and H.G. Wallraff (Eds.), Springer Verlag, Berlin, Heidelberg, New York.

Keeton, W.T., 1971. Magnets interfere with pigeon homing. Proceedings of the National Academy of Sciences of the USA, 68: 102-106.

Keeton, W.T., 1973. Release-site bias as a possible guide to the "map" component in pigeon homing. Journal of Comparative Physiology, 86, 1-16.

Keeton, W.T., 1974. The orientational and navigational basis of homing in birds. in "Advances in the Study of Behavior, Vol. 5. Academic Press, New York, San Francisco, London.

Keeton, W.T., 1974b. Pigeon homing: No influence of outward-journey detours on initial orientation. Monitore zoologico italiano, 8: 227-234.

Kiepenheuer, J., 1982. The effect of magnetic anomalies on the homing behavior of pigeons. in "Avian Navigation". F. Papi and H.G. Wallraff (Eds.), Springer Verlag, Berlin, Heidelberg, New York.

Kowalski, U. and Wiltschko, R., 1984. Heimorientierung der Brieftauben: Ortserfahrung verkleinert den Ortseffekt. Verhandlungen der Deutschen Zoologischen Gesellschaft Gießen, 1984: 251.

Kramer, G., 1953. Wird die Sonnenhöhe bei der Heimfindeorientierung verwertet? Journal für Ornithologie, 94: 201-219.

Kreithen, M.L., 1978. Sensory mechanisms for animal orientation - can any new ones be discovered? in "Animal Migration, Navigation, and Homing". K. Schmidt-Koenig and W.T. Keeton (Eds.), Springer Verlag, Berlin, Heidelberg, New York.

Lednor, A.J., 1982. Magnetic navigation in pigeons: possibilities and problems. in "Avian Navigation". F. Papi and H.G. Wallraff (Eds.), Springer Verlag, Berlin, Heidelberg, New York.

Matthews, G.V.T., 1953. Sun navigation in homing pigeons. Journal of Experimental Biology, 30: 243-267.

Matthews, G.V.T., 1953b. The orientation of untrained pigeons: a dichotomy in the homing process. Journal of Experimental Biology, 30 268-276.

Papi, F., 1982. Olfaction and homing in pigeons: ten years of experiments. in "Avian Navigation". F. Papi and H.G. Wallraff (Eds.), Springer Verlag, Berlin, Heidelberg, New York.

Pratt, J.G. and Thouless, R.H., 1955. Homing orientation in pigeons in relation to opportunity to observe the sun before release. Journal of Experimental Biology, 32: 104-157.

Schmidt-Koenig, K., 1961. Die Sonne als Kompaß im Heim-Orientierungssystem der Brieftauben. Zeitschrift für Tierpsychologie, 18: 221-244.

Schmidt-Koenig, K., 1963. On the role of the loft, the distance and the site of release on pigeon homing (the "cross loft experiment"). Biological Bulletin, 125: 154-164.

Schmidt-Koenig, K., 1975. "Migration and Homing in Animals". Zoophysiology and Ecology, 6. Springer Verlag, Berlin, Heidelberg, New York.

Semm, P., Nohr, D., Demaine, C. and Wiltschko, W., 1984. Neural basis of the magnetic compass: interaction of visual, magnetic and vestibular inputs in the pigeon's brain. Journal of Comparative Physiology, 155: 283-288.

Sonnberg, A. and Schmidt-Koenig, K., 1970. Zur Auslese qualifizierter Brieftauben durch Übungsflüge. Zeitschrift für Tierpsychologie, 27: 622-625.

Walcott, Ch., 1982. Evidence for a magnetic map in homing pigeons? in "Avian Navigation". F. Papi and H.G. Wallraff (Eds.), Springer Verlag, Berlin, Heidelberg, New York.

Wallraff, H.G., 1966. Über die Heimfindeleistungen von Brieftauben nach Haltung in verschiedenartig abgeschirmten Volieren. Zeitschrift für vergleichende Physiologie, 52: 215-259.

Wallraff, H.G., 1970. Über die Flugrichtung verfrachteter Brieftauben in Abhängigkeit vom Heimatort und vom Ort der Freilassung. Zeitschrift für Tierpsychologie, 27: 303-351.

Wallraff, H.G., 1974. "Das Navigationssystem der Vögel". Schriftenreihe Kybernetik, R. Oldenbourg Verlag, München, Wien.

Wallraff, H.G., 1980. Olfaction and homing in pigeons: nerve-section experiments, critique, hypotheses. Journal of Comparative Physiology, 139: 209-224.

Wallraff, H.G., 1980b. Does pigeon homing depend on stimuli perceived during displacement? I. Experiments in Germany. Journal of Comparative Physiology, 139: 193-201.

Wallraff, H.G., 1982. Homing to Würzburg: an interim report on long-term analyses of pigeon navigation. in "Avian Navigation". F. Papi and H.G. Wallraff (Eds.), Springer Verlag, Berlin, Heidelberg, New York.

Wiltschko, R., 1983. The ontogeny of orientation in young pigeons. Comparative Biochemistry and Physiology A, 76: 701-708.

Wiltschko, R. and Wiltschko, W., 1985. Pigeon homing: change in navigational strategy during ontogeny. Animal Behavior, 33: 583-590.

Wiltschko, W., 1983. Compasses used by birds. Comparative Biochemistry and Physiology A, 76: 709-718.

Wiltschko, W. and Wiltschko, R., 1972. Magnetic compass of European Robins. Science, 176: 62-64.

Wiltschko, W. and Wiltschko, R., 1978. A theoretical model for migratory orientation and homing in birds. Oikos, 30: 177-187.

Wiltschko, W. and Wiltschko, R., 1981. Disorientation of inexperienced young pigeons after transportation in total darkness. Nature (London), 291: 433-434.

Wiltschko, W. and Wiltschko, R., 1982. The role of outward journey information in the orientation of homing pigeons. in "Avian Navigation". F. Papi and H.G. Wallraff (Eds.), Springer Verlag, Berlin, Heidelberg, New York.

Wiltschko, W., Wiltschko, R., Foà, A. and Benvenuti, S., 1986. Orientation behaviour of pigeons deprived of olfactory information during the outward journey and at the release site. Monitore zoologico italiano, 20: in press.

Wiltschko, W., Wiltschko, R. and Keeton, W.T., 1984. The effect of a "permanent clock-shift" on the orientation of experienced homing pigeons. Behavioral Ecology and Sociobiology, 15: 263-272.

Yeagley, H.L., 1947. A preliminary study of a physical basis of bird navigation. Journal of Applied Physics, 18: 1035-1063.

ACCURACY OF MAP-BUILDING AND NAVIGATION BY HUMANS DURING 'NATURAL' EXPLORATION: RELATIVE ROLES OF MAGNETORECEPTION AND VISION

R. Robin Baker

Department of Zoology, University of Manchester, M13 9PL, UK

INTRODUCTION

Humans, in common with perhaps all vertebrates (Baker 1978, 1982, 1985c), organise their lives and movements within an area of familiarity. Each individual learns not only when and where to go to obtain this or that resource but also which routes are the most economical. Such behaviour is only possible if information concerning sites and routes is stored in the central nervous system, presumably as some form of 'mental map'. Of course, at birth an individual has no familiar area and no mental map. Yet, eventually, its map may span up to thousands of square kilometres. Areas that were previously unfamiliar have become familiar, a process of familiarisation that relies primarily on 'exploration'.

Exploration involves travel to sites beyond the previous limits of the familiar area. Its function is to encounter and assess novel sites that may provide needed resources. The most useful sites are adopted and placed on the mental map. The nature of exploration, however, is such (Baker 1978) that rarely will the route by which a new site is first discovered just happen to be the most economical link with other sites. In the wake of exploration and discovery must come the pioneering of economic routes, feats of path-finding that require an ability to 'navigate'. Ultimately, the efficiency with which an individual exploits its environment is a function of its ability at map-building and navigation (Baker 1985c).

Studies of fish (Smith 1984), birds (Baker 1984b), and humans (Gatty 1958; Lewis 1972; Baker 1981) show that, as exploration proceeds, an armoury of senses detect a wide range of environmental information. The 'least navigation' hypothesis (Baker 1978) predicts that any given problem of map-building or navigation during exploration will be solved using the most economical subset of this array of information. One aim of the study of animal navigation is to evaluate the relative roles and importance of different senses under different conditions.

This paper describes experiments on humans engaged in walks across totally unfamiliar terrain; an attempt to study the accuracy of map-building and navigation during 'natural' exploration. Treks were a few kilometres, lasted about an hour, and took place under different conditions of daylight, cloud cover, and height of sun. The primary aim was to identify the role of magnetoreception in subconscious map-building and navigation and to evaluate how this role might change in the presence of other, visually mediated, cues. Experimental design, however, rests heavily on current ideas on the physiology of magnetoreception.

PHYSIOLOGY OF MAGNETORECEPTION

A wide range of animals (e.g. molluscs, insects, fish, amphibians, reptiles, birds, mammals) are now known to be sensitive to the geomagnetic field in a way that may contribute to orientation and navigation during exploration (Kirschvink et al. 1985). However, the organ responsible for this sensitivity, the magnetoreceptor, has not yet been identified for any animal.

Many authors (see collected papers in Kirschvink et al. 1985) favour the hypothesis that magnetoreceptors consist of deposits of magnetite in specialised tissue, ramified by nerves. As the animal moves within the geomagnetic field, electric, magnetic or even pressure events take place around the magnetite. Nerves fire and directional information is carried to the central nervous system for interpretation and integration with input from other sensory modalities. Magnetic deposits, often identified as magnetite, have now been described in a wide range of animals. In vertebrates (fish, reptiles, mammals and perhaps birds), these deposits are often localised around the region of the ethmoid sinus, leading to speculation that there may be a magnetoreceptor common to the vertebrate stock in this region (Baker 1984a; Walker et al. 1984). However, as stressed elsewhere (Baker 1985d), despite the interest and excitement that has centred on the magnetite hypothesis, the first real evidence for any link between magnetoreception and magnetite remains to be found.

Alternatively, Leask (1977) proposed that magnetoreception takes place in the vertebrate retina, as a by-product of the normal visual process. Some support for this hypothesis may derive from the discovery that there is single unit electrical activity in the nucleus of the basal optic root of homing pigeons under earth-strength magnetic stimulation (Semm et al. 1984) but only if there is simultaneous stimulation of the vestibular organ and not at all in total darkness.

Whether the vertebrate magnetoreceptor is located in the sinuses, the retina, or elsewhere, there is evidence that it is based on components that can be aligned and realigned by an applied magnetic field. Strong ac fields applied to the head of live pigeons disrupt the pigeons' subsequent ability to navigate when released after the artificial field is removed (Walcott 1982). Similarly, strong dc fields may actually improve subsequent performance. Walcott suggests that: 1) accuracy of magnetoreception depends on degree of common alignment of an array of particles (perhaps magnetite); 2) alignment of these particles may be changed by applied magnetic fields; and 3) the new alignment persists after removal of the applied field.

Bar magnets influence human magnetoreception when placed on the forehead (Gould 1980; Adler and Pelkie 1985; see Baker 1985b) and on the right temple (Baker 1984a,b,1985a,c,d) but less so, or not at all, when placed further back than the ear (Baker 1985c). This influence persists when the magnets are removed (Baker 1984a,b,1985d) implying, as for pigeons, that the human magnetoreceptor is based on components that are normally aligned but which may be realigned by applied magnetic fields. So how does normal alignment originate?

Gould et al. (1978) demonstrate that the front of the abdomen of honey bees possesses a natural magnetic remanence the alignment of which may be attributable to the orientation of the bee in the geomagnetic field when immobile as a pupa. The alignment of this remanence can be altered by an applied magnetic field. The influence of magnets on compass orientation by humans is a function of a person's normal (i.e. >14 days) bed orientation. Details of these effects and the 'bed orientation hypothesis' to which they led are given in Baker (1984a,b,1985a). This hypothesis postulates that, as for bees, a person's natural alignment of magnetoreceptor components is influenced by the action of the geomagnetic field during periods of relative immobility, in this case while asleep.

The magnetoreceptor in the head of birds does not 'read' the polarity of the geomagnetic field in the same way as a person reading the familiar hand-held compass (Wiltschko and Wiltschko 1972). Instead, the avian magnetoreceptor functions as an 'inclination' compass, detecting the direction in which the lines of force dip most steeply into the ground. This direction is that of the nearest geomagnetic pole. As such, therefore, the avian compass does not detect 'north' and 'south', but rather 'poleward' and 'equatorward'.

Experiments on humans on Grand Bahama (Baker 1986) suggest that the human magnetoreceptor also functions as an inclination compass. Subjects from the Southern Hemisphere (Australia, New Zealand) were transported to the Caribbean across the equator and compared in their ability to judge compass direction, both when sighted and blindfolded, with subjects displaced similar distances but without crossing the equator (from Britain, Japan, Oman, and Hong Kong). Tests began within 36 h of arrival on Grand Bahama and continued for 4 weeks. Southern Hemisphere subjects were significantly worse at judging compass direction than subjects from the Northern Hemisphere, but the two groups were not significantly different in their judgement of poleward and equatorward. Similarly, magnets influenced the two groups differently in their judgement of compass direction, but similarly in their judgement of pole direction. The results are as expected if humans, like birds, respond to the inclination of the geomagnetic field, not its polarity.

These various experiments have led to formulation of the following hypothesis of the physiology of human magnetoreception (Baker 1986). Somewhere in the front half of the human head is a magnetoreceptor with components having a natural alignment. This alignment varies between individuals and is a function of each individual's normal bed alignment and sleeping habits. During exploration, as the individual changes bearing, the configuration changes between these aligned components within the magnetoreceptor and alignment of the lines of force of the geomagnetic field. The result is a signal, related to direction, that is then carried to the central nervous system. Applied magnetic fields that induce change in the alignment of receptor components may have one of a number of effects. If the applied field simply strengthens or weakens the normal alignment of the magnetoreceptor, and thus strengthens or weakens the neural signal, it may simply improve or impair magnetoreception. However, induced realignment of receptor components, if the result is a change in the geographic direction in which magnetoreceptor and geomagnetic field interact in a particular way, should result in misinterpretation of direction. Any of these influences of applied magnetic fields will persist for a time after the applied field is removed.

This hypothesis of the physiology of human magnetoreception has dictated the form of magnetic manipulation in recent walkabout experiments.

WALKABOUT EXPERIMENTS

Experiments to study the role of magnetoreception in human orientation and navigation have been of three major types: 1) chair experiments (Baker 1984a,b,1985d); 2) bus experiments (Baker 1980, 1981, 1985b); and 3) walkabout experiments (Baker 1985c,d). All three produce data for recognition of compass direction; chair experiments test only this ability. Bus experiments test for navigational ability over distances of tens of kilometres. An impromptu bus experiment over 109.5 km was carried out on the participants of this NATO ASI and the results are given in an appendix to this paper. The main body of the paper, however, is concerned only with walkabout experiments, designed to test human navigation and orientation under conditions that approach as closely as possible the process of 'natural' exploration. In particular, the aim is to study the role of magnetoreception when all other sensory modalities are in operation.

Experimental Protocol

Walkabout experiments have now been carried out at four separate sites in England and the Bahamas (table 1). Groups of subjects without blindfolds are led by experienced guides through unfamiliar woodland. Routed so as not to provide a clear view of distant landmarks, journeys have been 2-4 km long ending at a test site 0.31-1.75 km from the starting point of the journey (table 1). Routes, vegetation and topography are described and illustrated elsewhere (Baker 1986).

'Celestial conditions' during walkabout experiments divide so far into four main categories: 1) at night (with no Moon); 2) by day under total overcast but good visibility; 3) with the sun visible but high (45-63°) in the sky, providing a weak directional cue (when overhead the sun has no value for horizontal orientation); and 4) with the sun visible and low (10-40°) in the sky, providing a strong directional cue. Conditions for each experiment are given in table 2.

Most (21) of the groups of subjects were aged 19-22 y and were mainly University students. However, two groups on Grand Bahama were aged 16-19y and five groups at Woodchester Park were aged 7-10y. Sex ratios approached unity. Until they arrived to be tested at the test-site, subjects did not know either the questions to be asked or even the aims of the experiment (but see CONCLUDING REMARKS). Indeed, most groups engaged in other tasks (e.g. collecting insects) during the walk. It is a safe assumption for most subjects that any navigation was more or less entirely subconscious.

Table 1. Distance of walkabout experiments at different sites

		distance (m) from start to test site	
site	location	outward journey	air-line
Woodchester Park	SW England	2300	670
Delamere Forest	NW England	2150	1200
Burnt Wood	mid-England	2500	308
Lucaya	Grand Bahama	4000	1750

Measurements of Navigational Performance

At the test site that marks the end of a walkabout journey, subjects are asked to: 1) draw an arrow pointing toward 'home' (the starting point of the journey); 2) estimate the compass direction of the arrow they have just drawn; and 3) estimate the air-line distance between the test site and 'home'. Only estimates 1) and 3) are used in this paper. Full details of experimental protocol are given elsewhere (Baker 1985c, 1986).

The most sensitive measure of navigational accuracy during exploration is of ability to pinpoint position in 2-dimensional space, as on a map. Such a measure reflects ability to monitor both distance and direction in combination. In walkabout experiments, people estimate both the direction and air-line distance of home from the test site. These estimates are in effect a vector that allows the coordinates of the person's estimated location in space to be calculated by simple trigonometry. The distance between this point and the actual location of the test site may then also be calculated. Expressed as a percentage of the distance walked, this 'distance-error' measures accuracy of map-building. It is shown in the appendix to this paper that errors in judging location (expressed as a percentage of distance travelled), after a bus journey of 109.5 km, are not significantly different from similarly expressed errors in judging location after walkabout journeys of only 2-4 km under similar celestial conditions. The implication is that relative accuracy of map-building does not change with increase in distance.

If a new site is to be located on a mental map with sufficient accuracy not to distort its position relative to sites already on the map, it is important that both direction and distance are judged accurately. Less accuracy is needed simply to return home after exploration. Any, even a gross, overestimation of the distance to home is relatively unimportant. An explorer needs simply to identify home direction with sufficient accuracy to pass near enough to recognise familiar landmarks. Naturally, the nearer to home the person passes, the greater the chance of recognising such landmarks and the shorter the total distance of the return journey. A functional measure of navigational accuracy, therefore, is to measure or calculate the nearest distance to home that will be achieved by travelling a particular distance in a particular direction.

In a walkabout experiment, each subject walks a tortuous route through unfamiliar terrain to a test site, then estimates the direction of 'home' (i.e. the starting point of the journey). Suppose the subject then attempts to return home by walking in the direction he or she has pointed. Let e° be the angular error (i.e. the difference between the direction the person points and the true direction of home). When e>90°, the subject never gets nearer to home than the distance (T) of the test site. If, on the other hand, e<90°, at first the subject approaches nearer to home and, after a certain distance (D), reaches the nearest distance (ND). Thereafter, if the subject continues in a straight line, he or she will pass further and further away. Given that T and e° are both known, the distances D and ND may be calculated by simple trigonometry ($D = T(\cos e°)$; $ND = T(\sin e°)$). When $e \geq 90°$, $ND=T$.

In all of the walkabout experiments listed in table 2, subjects have been asked to 'point' (draw an arrow) towards home. In all but five, they have also been asked to estimate the air-line distance (A) to home. If A<D

then a person attempting to return home by walking the estimated distance in the estimated direction would stop before being as near to home as if they had continued further. For such people, the nearest distance to home is given by:

$$ND = ((D-A)^2 + (D^*\tan e°)^2)^{0.5}.$$

The data collected during walkabout experiments thus allow analysis of the accuracy of map-building (using error in estimating location), and the accuracy of returning home (using nearest distance). Moreover, judgement of direction and distance may be analysed separately. As experiments are performed under different conditions of daylight, cloud-cover and height of sun, there is the opportunity to begin to evaluate the potential roles of distance-vision and celestial orientation. Finally, by suitable magnetic manipulation, it is possible to study how these visual mechanisms interact with magnetoreception.

Testing for the Role of Magnetoreception

As in experiments on other animals, a role for magnetoreception in human orientation and navigation may be studied by comparing the performance of subjects exposed to different ambient magnetic fields. If magnetoreception is involved, experimental manipulation of the magnetic field should influence the strength of homeward orientation. In the experiments reported here, magnetic treatment takes the form of pre-treatment. It thus relies on manipulation of the magnetoreceptor rather than on manipulation of the magnetic field to be detected during the experiment. Subjects wear bars aligned vertically on the right temple for 15 minutes, then remove them before setting out on the walkabout. Pre-treatment has the advantage that subjects are not encumbered by bars while walking.

The bar magnets used for pre-treatment have been of two types: type-A magnets measure 77x15x6 mm and have a pole strength of about 20mT; type-B magnets measure 75x8x3 mm and have a pole strength of about 30mT. Three treatments have been used: some subjects wear magnets with the N-pole uppermost (Nup); some wear magnets with the S-pole uppermost (Sup); a third group wear brass bars of the same dimensions as the type of magnet in use. All walkabout experiments are double-blind. Bars are sealed inside opaque brown-paper envelopes and neither subjects nor experimenters know which subjects receive which treatment. Double-blind protocol persists throughout both the experiment and the processing and tabulation of results.

Since 1983 when the technique of pre-treatment with vertical bars was first adopted, 28 such walkabout experiments have been carried out at sites in Britain and the Bahamas. Dates, conditions, locations and sample sizes are summarized in table 2.

The influence of type-A and -B magnets on accuracy of map-building, the most sensitive measure of navigational performance, is summarized in table 3. Nup and Sup pre-treatments produce significantly different estimates, both for type-A ($z = -2.218$; P(2-tailed) = 0.026; Mann-Whitney U-test) and for type-B ($z = 3.260$; $P = 0.001$) magnets. The influence of type-A and -B magnets, however, is significantly different (combined z-values = $-2.218 - 3.260 = -5.478$ which, when divided by $\sqrt{2}$ to restore unit degrees of freedom, gives: $z = -3.874$; P(2-tailed) = 0.0001).

Table 2. Walkabout experiments: dates, conditions and number of subjects exposed to different magnetic pre-treatments with vertical bars

TYPE-A MAGNETS					TYPE-B MAGNETS				
		no. subjects					no. subjects		
date	cond.	B	N	S	date	cond.	B	N	S
Woodchester Park					Lucaya				
21.06.83	h.sun	10	11	11	20.12.84	l.sun	5	3	4
08.07.83	h.sun	8	8	8	22.12.84	l.sun	18	8	8
04.09.83	black	13	14	12	22.12.84	h.sun	13	6	9
18.09.83	l.sun	10	15	12	22.12.84	stars	18	9	10
12.04.84	l.sun	4	4	4	23.12.84	l.sun	5	3	2
02.09.84	stars	30	10	11	24.12.84	stars	2	0	0
23.09.84	l.sun	11	8	7	14.01.85	h.sun	10	10	10
08.10.84	o'cast	16	11	10	15.01.85	h.sun	8	10	10
Delamere Forest					Woodchester Park				
14.03.84	l.sun	4	4	3	13.04.85	l.sun	7	7	7
05.11.84	l.sun	8	8	7	25.06.85	o'cast	12	10	12
Burnt Wood					12.07.85	o'cast	11	11	11
21.11.84	o'cast	21	7	8	08.09.85	stars	10	8	8
22.11.84	o'cast	16	12	12	16.09.85	o'cast	13	13	15
					22.09.85	l.sun	8	10	10
					Burnt Wood				
					20.11.85	o'cast	14	15	14
					21.11.85	o'cast	20	15	14
TOTAL		151	112	105	TOTAL		174	138	144

B=brass; N=Nup magnet; S=Sup magnet

l.sun = low sun (altitude 10-40°); h.sun = high sun (altitude 45-63°)
black = overcast night

Table 3. Walkabout experiments: influence of 15-minute pre-treatment with type-A and -B vertical magnets on accuracy of map-building

LOCATION ERROR (distance as % of distance walked)

	medians			inter-quartiles			no. subjects		
	magnets		brass	magnets		brass	magnets		brass
	Nup	Sup		Nup	Sup		Nup	Sup	
type-A	27	44	36	18-47	24-81	23-54	58	57	78
type-B	49	32	41	25-88	20-72	26-71	138	144	174
	+ve	-ve		+ve	-ve		+ve	-ve	
Total	31	46	39	19-56	25-85	25-67	202	195	252

+ve = 'enhancing' (Nup type-A; Sup type-B) magnets
-ve = 'disruptive' (Sup type-A; Nup type-B) magnets

Pre-treatment with Nup type-A magnets (Baker 1985c) and Sup type-B magnets promotes an improvement in map-building accuracy. The estimates made are significantly better (z = 2.137; P(2-tailed) = 0.033; Mann-Whitney U-test) than those made by subjects pre-treated with brass bars. Conversely, the combined estimates of location made by subjects pre-treated with either Sup type-A magnets or Nup type-B magnets are significantly worse (z = -2.426; P = 0.015) than those made after treatment with brass bars. In view of these results, Nup type-A and Sup type-B will henceforth, for simplicity, be described as 'enhancing' magnets. Sup type-A and Nup type-B will be described as 'disruptive' magnets. The discovery that magnetoreception can be enhanced or disrupted by suitable magnetic manipulation should perhaps be no more surprising than the discovery that vision may be enhanced or disrupted by spectacles.

In the analyses that follow, the role of magnetoreception is tested under different visual conditions by comparing the levels of performance of subjects pre-treated with enhancing and disruptive magnets. Subjects treated with enhancing magnets are assumed to have their capacity for magnetoreception intact; those treated with disruptive magnets to have their capacity impaired. The questions asked are: 1) does impairment of magnetoreception effect all aspects of the navigational mechanism equally; and 2) can any other cue in the environment substitute for magnetoreception when the latter is impaired?

Levels of Performance and Influence of Magnets

Errors in estimating the air-line distance of home may be expressed as a percentage of the true distance. Thus, from a test-site 1 km from home, estimates of 500 m and 1500 m would both qualify as 50% errors. If, for the moment, underestimates are given negative values (and overestimates, positive), the 252 subjects pre-treated with brass bars in walkabout experiments produced a median error of +60% with an inter-quartile range of -31% to +174%. There is a clear tendency for people to overestimate the distance to home. Pre-treatment with enhancing magnets (N = 202) reduces this tendency to overestimate (median = +37%); pre-treatment with disruptive magnets (N = 195) exaggerates the tendency (median = +71%). The difference is significant (z = 1.701; P(1-tailed) = 0.045; Mann-Whitney U-test). However, disruptive magnets also increase the range of errors (IQR for: enhanced = -9 to +139%; disrupted = -18 to +258%). Apparently, therefore, impaired magnetoreception leads to increased error per se, rather than simply to increased overestimation. Consequently, all further analyses of distance estimates ignore sign and deal only with percentage error.

All measures of performance that involve distance are summarized in table 4 using medians and inter-quartile ranges (IQR). To illustrate performance from these data, consider a walkabout journey of 4 km that ends at a test-site 1 km from home. Suppose that, throughout the walk, the sun shines low in the sky. Under such circumstances, table 4 shows that 25% (i.e. most accurate quartile) of people pre-treated with enhancing magnets judge their location to within 840 m (i.e. <21% of 4 km), the air-line distance to home to within 190 m (i.e. <19% of 1 km) and, in trying to return home, would pass within 350m (i.e. <35% of 1 km).

Table 4. Walkabout experiments: influence of magnets and visual conditions on accuracy of map-building, returning home and judgement of air-line distance to home

ERROR IN JUDGING LOCATION (distance as % of distance walked)

	medians			inter-quartiles			no. subjects		
	magnets		brass	magnets		brass	magnets		brass
	+ve	-ve		+ve	-ve		+ve	-ve	
night									
starlit	45	45	43	32-120	34-182	33-128	18	17	30
day									
overcast	23	48	38	15- 46	22-105	21- 71	85	84	107
high sun	41	61	49	22- 74	42-108	31- 87	48	45	49
low sun	31	38	35	21- 46	21- 59	25- 44	51	49	66
Total	31	46	39	19- 56	25- 85	25- 67	202	195	252

ERROR IN RETURNING HOME (nearest distance as % of distance from test site)

night									
starlit	73	81	75	44- 93	72-100	44- 84	18	17	30
day									
overcast	76	88	76	37- 99	50-100	37- 99	85	84	107
high sun	61	93	69	24- 91	56-100	47-100	48	45	49
low sun	59	67	60	35- 83	39- 94	34- 80	51	49	66
Total	67	85	68	36- 91	49-100	40- 97	202	195	252

ERROR IN JUDGING DISTANCE OF HOME (estimate as % of true distance)

night									
starlit	77	85	77	35-378	54-616	54-378	18	17	30
day									
overcast	63	152	145	39-167	49-378	40-258	85	84	107
high sun	52	83	58	19-106	19-199	37-139	48	45	49
low sun	33	43	54	19- 83	19-124	37- 79	51	49	66
Total	63	83	77	19-139	31-258	40-174	202	195	252

+ve, expected to enhance; -ve, expected to disrupt
high sun, altitude 45-63°; low sun, altitude 10-40°

Ability to point towards home is expressed (table 5) as the homeward component (h; Batschelet 1981). h takes a value between +1 (when all subjects are 0° in error at pointing) to -1 (when all are 180° in error). Guessing, or a mean error of 90°, gives a homeward component of 0.0. Homeward components are tested for significance (i.e. probability that observed value is greater than 0.0 by chance) by Rayleigh's V-test (Batschelet 1981). In table 5, all judgements of home direction are significantly better than chance, though three categories for subjects pre-treated with brass (total; overcast; high sun) have mean errors significantly different from 0° as indicated by 95% confidence intervals (Batschelet 1981). The most striking feature of judgement of direction is that, even with the sun shining (if it is high in the sky), subjects with impaired magnetoreception do not perform as well (h = 0.29) as subjects with magnetoreception intact under starlit skies at night (h = 0.50).

Table 5. Walkabout experiments: influence of magnets and visual conditions on accuracy of pointing to start of journey

	homeward component			mean error ± 95% CI			no. subjects		
	magnets		brass	magnets		brass	magnets		brass
	+ve	-ve		+ve	-ve		+ve	-ve	
night									
starlit	0.50	0.24	0.41	-18±30°	-39±60°	1±26°	28	28	60
day									
overcast	0.46	0.27	0.39	13±16°	-15±28°	29±16°	96	94	123
high sun	0.54	0.29	0.44	15±18°	-12±40°	25±22°	48	45	49
low sun	0.72	0.55	0.64	0±12°	4±18°	11±12°	70	64	80
Total	0.57	0.35	0.47	5± 9°	-9±14°	18± 9°	242	231	312

+ve magnets, expected to enhance; -ve magnets, expected to disrupt
Negative mean errors = anticlockwise errors; positive = clockwise
High sun, altitude 45-63°; low sun, altitude 10-40°
All homeward components are significant (P<0.05; V-test)

Differences in the estimates made by subjects pre-treated with enhancing and disruptive magnets are tested for significance by the Mann-Whitney U-test (Siegel 1956; Batschelet 1981). Results are presented in table 6 and show that, generally, magnetoreception is involved in all aspects of navigation (map-building; returning home; judgement of direction and distance) but not equally so under all conditions. A significant role in judging direction is apparent both by day (if the sky is overcast or the sun is shining high in the sky) and by night (under starlit skies). By contrast, only under overcast skies, by day, does magnetoreception seem to play a significant role in the judgement of distance. However, there is no evidence that magnetoreception plays any significant role in the judgement of either direction or distance if the sun is shining low in the sky.

Table 6. Walkabout experiments: influence of 15-minute pre-treatment with enhancing and disruptive magnets on accuracy of map-building and navigation

Mann-Whitney U-test of hypothesis that enhanced performance is more accurate than disrupted (data from tables 4 and 5)

	accuracy of				judgement of			
	map-building		return home		direction		distance	
	z	P	z	P	z	P	z	P
night								
starlit	0.297	ns	1.756	*	1.926	*	0.729	ns
day								
overcast	3.041	***	1.953	*	1.897	*	2.768	***
high sun	2.714	***	3.187	****	3.068	***	1.622	ns
low sun	0.593	ns	1.214	ns	1.147	ns	0.568	ns
Total	3.902	****	3.570	****	3.819	****	2.929	***

high sun, altitude 45-63°; low sun, altitude 10-40°
P(1-tailed): ns, P>0.05; *, <0.05; **, <0.02; ***, <0.01; ****, <0.001

The influence of distance-vision and celestial orientation

The influence of daylight, cloud-cover and height of sun on the navigational performance of subjects with enhanced and disrupted magnetoreception is summarized in table 7. Increased opportunity visually to assess bends and distances (compare tests under overcast daytime skies with those under starlit night skies) does not lead to a significant improvement in the judgement of direction or distance. Opportunity to see the sun shining high in the sky (compare tests under high sun with tests under overcast daytime skies) leads to a significant improvement only in the ability to judge distance, not direction. The overwhelming factor to emerge from table 7, however, is the dichotomy in benefit associated with opportunity to see a strongly directional sun shining low in the sky (compare tests under low sun with tests under high sun). Subjects with magnetoreception intact gain no significant benefit from this extra information. By contrast, subjects with impaired magnetoreception gain a highly significant benefit.

Table 7. Walkabout experiments: influence of daylight, cloud-cover and height of sun on accuracy of map-building and navigation by humans with ability at magnetoreception enhanced and disrupted

	low sun v high sun		high sun v overcast day		overcast day v starlit night	
	z	P	z	P	z	P
MAGNETORECEPTION:						
ENHANCED						
map-building	1.190	ns	-1.820	ns	2.870	***
return home	0.466	ns	1.355	ns	-0.555	ns
direction	0.186	ns	1.541	ns	-0.108	ns
distance	0.446	ns	1.787	*	0.487	ns
DISRUPTED						
map-building	3.792	****	-1.527	ns	1.466	ns
return home	2.354	**	-0.091	ns	-0.149	ns
direction	2.653	***	-0.167	ns	0.636	ns
distance	1.695	*	2.589	***	-0.050	ns

high sun, altitude 45-63°; low sun altitude 10-40°

Hypotheses tested: accuracy greater with low sun than high, high sun than overcast, and on overcast day than starlit night.

z-values calculated by Mann-Whitney U-test from data in tables 4 and 5. Negative z-values indicate accuracy runs counter to hypothesis (e.g. data shows trend for smaller map-building errors under overcast than high sun).

P(1-tailed): ns, P>0.05; *, <0.05; **, <0.02; ***, <0.01; ****, <0.001

CONCLUDING REMARKS

Walkabout experiments help to evaluate the processes involved in map-building and navigation as humans explore unfamiliar terrain. So far, the distances involved are relatively short (walks, 2-4 km; test sites, 0.31-1.75 km from home) and this limit to the data should be borne in mind. Other mechanisms may become important over longer distances and/or the relative roles of the different senses may change. It may also be important to note that the data in tables 4-7 relate almost certainly to subconscious navigation. Subjects were unaware of their task and the questions they were to be asked. Two recent experiments at Burnt Wood divided test groups into two, half being told in advance the questions to be asked, the other half not. Subjects with magnetoreception intact gained no advantage from this information. Subjects with impaired magnetoreception gained a significant advantage. These experiments, which seem to indicate a primarily subconscious role for magnetoreception, are to continue.

Two strong patterns for subconscious navigation emerge from the data in tables 4 to 7. First, magnetoreception seems to be of major importance in judging direction, even by day. The only apparent substitute for the role normally played by magnetoreception seems to be another strongly directional cue, the low sun (tables 5 and 7). Without such a substitute, subjects with impaired magnetoreception perform no better by day than normal subjects perform under starlit skies at night (table 5). Secondly, and in contrast, magnetoreception seems to play less part in maintaining an awareness of the distance of home, except under overcast (table 6). Instead, by far the most important factor seems to be a visible sun, of any height (table 7).

The importance of magnetoreception or sight of a low sun to accuracy in judging direction, suggests that their major role in exploration and navigation is subconsciously to measure angles of turn during the outward journey. The importance of the sun in the judgement of distance is less clear. One intriguing possibility is that the sun's influence may in some way involve subconscious assessment of time.

ACKNOWLEDGEMENTS

Experiments on Grand Bahama were carried out in collaboration with Operation Raleigh and the Scientific Exploration Society. My participation was funded by the Royal Society. Research on magnetoreception at Manchester University is supported by SERC research grant GR/B74337.

REFERENCES

Adler, K. and Pelkie, C.R., 1985. Human homing orientation: critique and alternative hypotheses. in "Magnetite Biomineralization and Magnetoreception in Organisms: A new Magnetism". J. L. Kirschvink, D. S. Jones and B. J. MacFadden (Eds.), Plenum, New York.

Baker, R. R., 1978. "The Evolutionary Ecology of Animal Migration". Hodder and Stoughton, London.

Baker, R. R., 1980. Goal orientation by blindfolded humans after long-distance displacement: possible involvement of a magnetic sense. Science, 210:555-557.

Baker, R. R., 1981. "Human Navigation and the Sixth Sense". Hodder and Stoughton, London.

Baker, R. R., 1982. "Migration: paths through time and space". Hodder and Stoughton, London.

Baker, R. R., 1984a. Sinal magnetite and direction finding. Physics in Technology, 15: 30-36.

Baker, R. R., 1984b. "Bird Navigation: Solution of a Mystery?" Hodder and Stoughton, London.

Baker, R. R., 1985a. Magnetoreception by man and other primates. in "Magnetite Biomineralization and Magnetoreception in Organisms: A new Magnetism". J. L. Kirschvink, D. S. Jones and B. J. MacFadden (Eds.), Plenum, New York.

Baker, R. R., 1985b. Human navigation: a summary of American data and interpretations. in "Magnetite Biomineralization and Magnetoreception in Organisms: A new Magnetism". J. L. Kirschvink, D. S. Jones and B. J. MacFadden (Eds.), Plenum, New York.

Baker, R. R., 1985c. Exploration and navigation: the foundation of vertebrate migration. in "Migration: Mechanisms and Adaptive Significance". M. A. Rankin (Ed.), Port Aransas Marine Laboratory.

Baker, R. R., 1985d. Human magnetoreception for navigation. in "Electromagnetic Waves and Neurobehavioral Function". M. E. O'Connor and R. Lovely (Eds.), Liss, New York.

Baker, R. R., 1986. "Human Navigation and Magnetoreception". University Press, Manchester.

Batschelet, E. 1981."Circular Statistics in Biology" Academic Press, London

Gatty, H., 1958. "Nature is your Guide". Collins, London.

Gould, J. L., 1980. Homing in on the home front. Psychology Today, 14:62-71

Gould, J. L., Kirschvink, J. L. and Deffeyes, K. S., 1978. Bees have magnetic remanence. Science, 201:1026-1028.

Kirschvink, J. L., Jones, D. S., MacFadden, B. J. (Eds.), 1985a. "Magnetite Biomineralization and Magnetoreception in Organisms: A new Magnetism". Plenum, New York.

Leask, M. J. M., 1977. A physico-chemical mechanism for magnetic field detection by migratory birds and homing pigeons. Nature, London, 267:144-146.

Lewis, D., 1972. "We, the Navigators". Australian National University Press, Canberra.

Semm, P., Nohr, D., Demaine, C. and Wiltschko, W., 1984. Neural basis of the magnetic compass: interactions of visual, magnetic and vestibular inputs in the pigeon's brain. Journal of Comparative Physiology A, 155:1-6.

Siegel, S., 1956. "Non-parametric Statistics for the Behavioral Sciences". McGraw-Hill Kogakusha, Tokyo.

Smith, R. J. F., 1984. "The Control of Fish Migration". Springer, Heidelberg.

Walcott, C., 1982. Is there evidence for a magnetic map in homing pigeons? in "Avian Navigation". F. Papi and H. G. Wallraff (Eds.), Springer, Heidelberg.

Walker, M. M., Kirschvink, J. L., Chang, S-B. R. and Dizon, A. E., 1984. A candidate magnetic sense organ in the yellowfin tuna,Thunnus albacares. Science, 224: 751-753.

Wiltschko, W. and Wiltschko, R., 1972. Magnetic compass of European robins. Science, 178:62-64.

APPENDIX: AN IMPROMPTU NAVIGATION EXPERIMENT ON CONFERENCE PARTICIPANTS

R.R. Baker, V. Bingman, J. Bovet, A. Etienne, C. Thinus-Blanc, R. Wiltschko, W. Wiltschko

On Tuesday, 2 July 1985, many participants of the NATO ASI conference on Spatial Orientation spent their free afternoon on an organised tour of the Lubéron. Circumstances led to the tour being changed as the journey progressed. The route eventually taken, starting from the conference centre at La Baume-Lès-Aix (Aix-en-Provence, France), was first north and east to Ansouis, then south-west, west, and finally north again to Gordes. The total length of the journey was 109.5 km.

Participants were transported in two coaches. En route to Ansouis, ethologists in the smaller coach planned an impromptu navigation experiment to be carried out on the occupants of the larger coach. On paper provided as the bus stopped at Gordes, subjects were asked to: 1) draw an arrow pointing towards La Baume; 2) write down the compass direction in which their arrow was pointing; and 3) write down the air-line distance from the bus to La Baume.

The long-axis of the bus was estimated (by RRB, RW and WW) to be oriented along compass bearing 22°-202° (North=0°), with subjects facing 22°. La Baume lies 52 km to the SSE (154° from grid north) from Gordes. These data allow calculation, for this much longer journey, of all the measures of accuracy at map-building and navigation already described for walkabout experiments. The data also allow calculation of subject's errors at judging both the compass direction of their own arrows and that of La Baume from Gordes.

Table A1. Bus experiment: estimates of the direction of La Baume from Gordes and estimates of compass direction at Gordes

homeward component (h) and mean error (e°±95%CI) at judging direction

	'natural' navigators				map-readers				Mann-Whitney
	n	h	e°±CI	V-test	n	h	e°	V-test	P
Pointing to La Baume									
males	17	0.28	57±41	0.054	3	0.97	-2	(0.013)	0.017
females	10	0.27	8	0.114					
combined probability				0.037					
Compass Direction of La Baume									
males	17	0.71	3±24	<0.001	3	0.64	35	(0.055)	0.825
females	10	0.46	-1±75	0.022					
combined probability				<0.001					
Compass Direction of Arrow									
males	17	0.39	-45±36	0.012	3	0.54	40	(0.085)	0.417
females	10	0.64	-23±35	0.003					
combined probability				<0.001					

n = number of subjects
-ve mean errors = anticlockwise errors; +ve = clockwise
Bracketed probabilities for V-test are less reliable due to sample size.
Combined probability = Fisher's combined probability test

Table A2. Bus experiment:comparison of 'natural' navigators and map-readers for accuracy of map-building, returning home and judgement of air-line distance to home

	'natural' navigators		map-readers		
	n	median+IQ	n	median+IQ	P(Mann-Whitney)
Map-location (km error)					
males	13	59(36-81)	2	7(5- 9)	(0.018)
females	10	51(39-75)			
Nearest distance to home (km)					
males	17	51(23-52)	3	9(6-18)	0.015
females	10	41(26-52)			
Air-line distance (km error)					
males	13	17(5-27)	2	3(2- 4)	(0.045)
females	10	21(2-28)			

n = number of subjects
Bracketed probabilities for Mann-Whitney test are less reliable due to sample size.

Throughout the journey there was continuous sunshine from a high sun (altitude app.60-70°). None of the 30 subjects knew they were to be tested and none were blindfolded or treated magnetically. Three males followed their journey on a road map. Five males, including one map-user, omitted to estimate distance. Tables A1 and A2 present the results separately for male and female 'natural' navigators and male map-users. Methods of analysis are as for walkabout experiments.

None of the sex differences are significant (P(2-tailed) > 0.05). For example, median errors at judging distance were 17 km for male natural navigators, 21 km for females (table A2). If underestimates are scored as negative errors and overestimates as positive, median and inter-quartile distances were -12 (-20 to +3) km for males and -2 (-2 to +28) km for females.

Compass orientation by both sexes of natural navigators is significant (V-test, table A1), though males show a significant anti-clockwise deviation (i.e. 95% confidence intervals do not include 0°) in judging the compass direction of their arrows. Map-readers gained no significant, or even apparent, advantage (compare h-values, table A1).

In contrast to judgement of compass direction, map-readers were significantly better at pointing to La Baume (table A1), judging air-line distance (table A2), and thus at estimating location (table 2) and potential for returning to La Baume (table A2). For example, 50% of male natural navigators, if they had travelled the distance and direction estimated after this journey of 109.5 km, would have passed no nearer to La Baume than 51 km. Only 25% would have passed within 23km. Corresponding distances for the very small sample of map-readers were 9 km and 6 km.

Accuracy at judging location in this bus experiment after passive displacement for 109.5 km may be compared with accuracy during walkabout experiments under similar conditions (i.e. high, continuous sun). Expressed as a percentage of the distance travelled, median error for the 23 'natural' navigators in this experiment was 52% (IQR = 33-68%). This compares with 49% (31-87%) for 49 subjects on walkabout experiments (table 4). The difference is not significant (z = 0.151; P = 0.881; Mann-Whitney U-test) indicating no change in accuracy of map-building with distance.

THE CONTROL OF SHORT-DISTANCE HOMING IN THE GOLDEN HAMSTER

A.S. ETIENNE

Laboratory of Ethology, F.P.S.E.
University of Geneva
CH-1211 Geneva 4
Switzerland

The aim of spatial orientation studies is to determine what information controls the subject's orientation in space and how this information is integrated and organized. In the ethological approach of this field of research, the first step always consists of defining the cues which a particular species uses in a given situation on the basis of its sensory equipment and with respect to the requirements it has to fulfill in its natural way of life. At this point in our investigations we have not only to ask what sensory modalities are involved, but, more fundamentally, whether our subjects rely on information which is registered "on site", i.e. at particular locations or whether they use information which has been generated "en route", i.e. during a preceding phase of locomotion (Baker, 1981;1984; Mittelstaedt and Mittelstaedt, 1982). Adopting the terminology which R. Robin Baker has coined with respect to navigation (usually defined in the ethological literature as goal-directed orientation across unfamiliar space), we shall refer to these two categories of spatial information as "location-based" and as "route-based" information.

Location-based information always refers to exteroceptive cues, which either consist of particular features or landmarks of the familiar spatial environment, or represent a general directional reference such as provided by an astronomical compass or a geomagnetic gradient. Whereas the capacity to read a general reference system may be genetically coded or individually acquired (Wiltschko and Wiltschko, 1976, 1981; Baker, 1984), the reliance on landmarks always involves preliminary learning processes which lead to the association between particular cues and specific places or directions. These associatons can be stored in a long term memory to allow the subject to organize familiar space in a more stable manner. At higher zoological levels, the long term storage of a variety of location-based cues forms the basic condition for structuring the spatial environment as a system of interconnected places, which the psychological literature designs as cognitive map (for references see O'Keefe and Nadel, 1978; Olton, 1979; Thinus-Blanc, this Nato book, Vol. 1; Vauclair, this Nato book, Vol. 1). As suggested by experimental data and theoretical considerations on bird navigation such a map could be used in a most economical manner if it were directionally orientated in relation to a compass system (Graue, 1963; Wallraff, 1974).

Route-based information allows a subject to orientate its itinerary independently of any preliminary familiarisation with a particular spatial environment. It consists in the registration and storage of variables which are generated during an (immediately) preceding phase of locomotion; therefore the information content of these variables derives from the feedback of locomotion rather than from external spatial elements <u>per se</u>.

Depending on the subject's motivational state, the path-dependent variables will be submitted to specific transformation rules and determine a "Sollwert" (or command value) which controls the next phase of locomotion.

Let us illustrate these conditions by an example which anticipates our own experiments on homing behaviour: A small mammal proceeds at night in an exploratory manner from its nest to a feeding place where it collects food, and then returns along a direct path to its nest to store the food in its granary. We assume that the orientation of the direct return derives from the integration of the angular and linear components of the outward journey; further, the vector which results from this integration has to be inverted by 180°, either in a continuous manner during the outward journey, or at a particular moment, such as the end of the outward journey or the beginning of the return; finally, the variable (or variables) which controls the choice of a particular return direction has to be stored at least until the return itinerary has been initiated. Evidently, these descriptions originate from our own conceptions of vector navigation, and a different type of formalism may be more adequate for describing the animal's behaviour.

Two important distinctions have to be made with regard to route-based information. First of all, this type of information can be registered with or without the help of an external frame of reference. In our example of homing behaviour, the different components of the outgoing journey could for instance be integrated with regard to a geomagnetic compass, or exclusively on the basis of internal selfgenerated cues which result directly and uniquely from the subject's locomotor activity. Secondly, the storage of route-dependent information may occur according to two modes. On the one hand, it may involve the retention of particular cues, or of an orderly sequence of such cues. Coming back to our homing example, this means that a subject which stores a sequence of detailed landmarks on the way out would follow the same landmarks on its way back, and thus proceed twice along the same route. On the other hand, route-dependent information can be processed and stored through the continuous updating of a limited set of variables, as is the case for vector navigation (Wehner, 1982) or path integration (Mittelstaedt and Mittelstaedt, 1982). The explicit description given earlier of homing by a small mammal shows that the subject need not follow the same route twice in the case of vector navigation, but may return to its destination along short cuts and therefore follow the shortest route.

Let us briefly consider the functional, ontogenetic and phylogenetic relationship between the two basic forms of spatial information we have been describing. Any individual or social group which exploits a home range and explores adjacent areas in an efficient manner will rely conjointly on a) location-based and b) on route-based information (Baker, 1984). In the first instance (a), in order to satisfy various functional needs, the subject will return to familiar places which are connected by an olfactory and/or tactile network of routes and trails, or, depending on the species' level of cognitive differentiation, by a more abstract mapping system. In the second case (b), during exploratory excursions, and also when certain categories of location-based cues are not available (e.g. visual landmarks at night), the subject will switch from location-based to route-based orientation which does not require any preliminary knowledge of particular spatial relationships.

At the ontogenetic level, young animals start to explore their familiar environment without, of course, the help of known, location-based landmarks. The inexperienced individuals may therefore rely primarily on route-based information. We may thus consider the possibility that the ontogenetic development of a location-based orientation system derives, among other things, from the consolidation of spatial cues which have at first been recorded " en route" on a short term basis. For instance, an inexperienced bird may pick up a sequence of visual cues on the way out from A to B, and follow the same landmarks in a reversed order to return from B to A. At a later stage, these landmarks could be combined with other features from the environment to form a system of location-based cues that would allow the subject to orientate freely from one place to another (Wiltschko and Wiltschko, 1978).

From a phylogenetical point of view, invertebrates can attain a high level of spatial orientation on the basis of route-dependent information. Spiders from the family Agelenidae "home" toward the retreat on their web through the joint integration of proprioceptive and optical inputs (Mittelstaedt, 1978), and various hymenopterous insects are capable of precise vector navigation by using the sun as optical reference. These performances may be attributed to the fact that route-based cues a) are integrated in an "egocentric" manner, i.e. with respect to the organism's axis of locomotion and therefore in relation to a fixed body axis, and b) that these cues are most probably submitted to a limited and invariant sequence of computation rules.

On the other hand, these species use "exocentric" or "objective" location-based information to a lesser extent and/or in a more rudimentary manner. Ants of the genus Cataglyphsis can for instance adjust their foraging excursions to stable landmarks. However, they are unable to extrapolate general spatial relationships between these landmarks or between the landmarks and the nest location. Rather, they store a sequence of landmarks as a fixed series of two-dimensional images of a constant angular size and expect to match these images precisely with the actually perceived view of the landscape (Wehner, 1982). Among other things, the insects have therefore to learn separately a given sequence of visual cues on their way out to the foraging place and on their return to the nest (Wehner and Flatt, 1972).

In short, at the invertebrate level, space is not organized according to stable criteria, which imply the extrapolation of objective relationships between general external reference systems, landmarks and places. Such an organization may only be expected to emerge in vertebrates which show a progressively increasing propensity for learning and more flexible forms of information processing.

The research which is presented here concerns the orientation of the golden hamster (Mesocricetus auratus W.) to its nest site in the hoarding situation. Our data refer exclusively to short-distance homing. However, the initial experiments (Etienne, 1980) of the research were designed to simulate long distance homing in rodents (see Bovet, 1978; 1984; this Nato book, Vol.1) through the combination of a passive outward journey to the food source and the simultaneous elimination of various exteroceptive cues. It is shown that the hamster is capable of integrating path-dependent information both during an active and a passive outward journey and in conditions which exclude the reliance on an external reference system. Special importance is given to the integration of a passive outward journey which raises the question whether animals can home on the basis of an inertial navigation system (Barlow, 1964).

Two series of experiments will illustrate specific limitations which affect path integration: Without the conjoint intervention of location-based cues, the control of homing by self-generated information is open to cumulative errors. In this respect it will be shown that the hamster compensates active and passive rotations only within certain limits. On the other hand, route-dependent homing behaviour does not lead the animal back to a given point in space, but to the ("egocentrically" defined) location where the registration of the outward journey has been initiated. As suggested by a further series of experiments, the hamster can be induced to initiate the integration of the outward journey at the wrong place and moment and in these conditions return systematically in a wrong direction.

A final section of the paper illustrates how the animal switches from route-based to location-based information as soon as peripheral visual cues become available, and how the two types of information interact in situations where they are presented in a conflicting manner.

PROCEDURES

Throughout the experimental period, each subject lived in its own arena (Ø: 2.20m) where it was tested repeatedly in particular experimental situations. The daily testing session took place at the beginning of the dark phase, when the dusk-and night active animals tend to be most active. A nest box was attached to the outside of the arena's peripheral wall and gave, through a small door, free access into the arena. With the exception of experimental trials in which the arena was rotated, the nest box was maintained at a standard or 0° position in space.

During one trial the animal 1) left the nest exit (its longitudinal or x-axis pointing towards the center of the arena), 2) performed an active or a passive outward journey to the centre of the arena where 3) it collected food items, and 4) returned from the centre to the periphery of the arena. At the periphery, it found the entrance of the nest box either directly, or after exploring the peripheral zone of the arena.

The animal's return itinerary was coded in terms of its location at the moment it crossed each of six concentric and equidistant circles within any one of twelve 30°-sectors which subdivided the arena's floor. Special importance was given to the crossing of the last (largest) circle located at a distance of 15 cm from the arena's peripheral wall and giving access to the arena's peripheral annular zone (for more details, see Etienne et.al, 1986).

During an active outward journey to the centre the animal followed a baited spoon which was lit by a dim light; during a passive outward journey, the animal entered from the nest exit into a transportation box (16x18x25cm) in which it was transferred to the centre. Care was taken to ensure that the subject entered the central zone of the arena with its head pointing in one of four different directions, making angles of 0°, ± 90° or 180° with the radius vector pointing to the nest.

The following procedures were applied to eliminate a variety of exteroceptive cues from the intra- and extra-arenal environment. 1) The entire trial was video-recorded under infra-red (IR) light with wavelengths well beyond the limit of the hamster's responsiveness to red and near IR light. The experimenter worked with the help of an IR viewer. 2) Four mobile loudspeakers diffused pink noise (45-20'000 Hz) into the experimental room, the sound-reflecting properties of which were altered by panels covered with two heavy layers of acoustically absorbent material. Both the loudspeakers and the panels were displaced at regular intervals. 3) Tactile and olfactory cues on the floor of the arena were erased by stirring a thick layer of sawdust which covered the floor and then flattening the sawdust out again. 4) The spatial orientation of all kinds of cues within the arena was changed by turning the arena (and therefore also the nest box) by ± 90° or 180° with respect to its standard or 0° position in space. In trials with an active outward journey the arena was always rotated before the subject had left the nest exit. In trials with a passive outward journey, the arena was turned before or after the animal had left the nest exit; in the second case, the transportation box was lifted above the arena's floor during the rotation. 5) The total intensity of the earth magnetic field (EMF) was reduced by a factor of 10 by a system of Helmholtz coils which surrounded the arena.- In certain experiments the five above mentioned procedures were carried out simultaneously and combined with a passive outward journey.

In general, the statistical evaluations of the results were carried out in two steps through circular statistics (see footnote for description of the statistical tests). 1) First-order circular statistics (Batschelet, 1981 and pers. com.) were applied to the population of itineraries which were performed by one individual in particular experimental conditions. The Rayleigh test was used to examine whether a given subject was significantly orientated when it crossed the six concentric circles on the arena's floor during its repeated itineraries from the centre to the periphery of the arena. Significant vectors pertaining to the subject's orientation in a

FOOTNOTE : The Rayleigh z-test indicates whether first-order data show a unimodal departure from a uniform circular distribution. Moore's test, a second-order non-parametric test, shows whether the first-order vectors of a given sample of animals show any directionality or whether they are uniformly distributed. This test takes into account both the direction and the dispersion of first-order vectors. The Mardia-Watson-Wheeler uniform scores test is used to compare two or several independent samples of first-order vectors in order to test whether the samples differ significantly from each other with respect to their mean angle and/or their angular dispersion.

particular situation were compared to the reference direction by the test of Stephens. 2) Second-order circular statistics (Batschelet, 1981) were applied to the mean values which had been obtained from particular individuals. The mean vector for the animals' orientation when they entered the peripheral annular zone was calculated for each experimental group and then submitted to Moore's test. The distribution of the first-order results for separate experimental groups were compared by the Mardia-Watson-Wheeler uniform scores test. Unless mentioned differently, the reference direction for statistical evaluations corresponded to the radius vector pointing to the nest box in its standard or 0° position (see above).

THE MAINTENANCE OF ORIENTATED RETURN ITINERARIES DURING THE ELIMINATION OF VARIOUS LOCATION-BASED CUES COMBINED WITH A PASSIVE OUTWARD JOURNEY

The initial results of this research (Etienne, 1980) showed that after a passive outward journey and during the simultaneous elimination of visual, acoustical, tactile and olfactory cues the hamster continued to orientate from the centre of the arena to the standard location of its nest

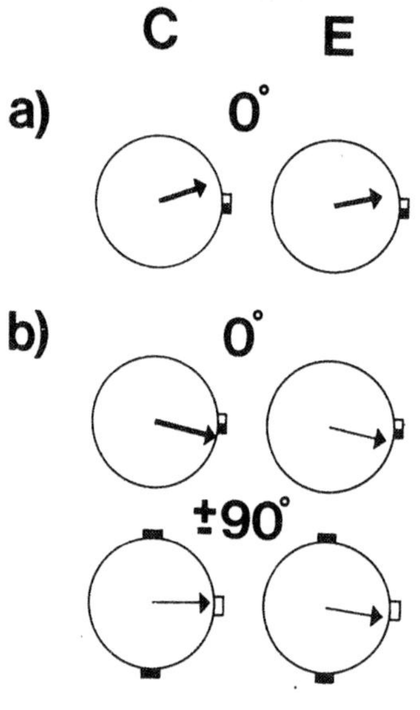

Fig. 1. The orientation in a weakened EMF combined with the elimination of other exteroceptive cues and a passive outward journey. All trials involved a passive outward journey and the elimination of visual, acoustical, tactile and olfactory cues. The white rectangles indicate the location of the nest box when the subject left it to enter the transportation box; the black rectangles indicate the position of the nest box when the subject was relased at the centre of the arena and during its return to the periphery. (These two locations were identical if the arena was not rotated). The vector on each circle indicates the orientation of the subjects when they entered the peripheral annular zone of the arena. C: Control trials in the natural EMF (total intensity: 0.45 G). E: Experimental trials in an artificial EMF (total intensity at centre of arena's floor: 0.03 G). Graph a) represents the mean orientation of 10 animals which were tested without rotation of the arena (0°), the arena being positioned on antivibratory supports. Each animal underwent 24 experiments (r=0.84 for the control and 0.72 for the experimental trials, P<0.01, Moore's test, second-order statistics). Graph b) represents the orientation of 4 animals in experiments where the arena was rotated by ± 90° in every second trial. Each subject underwent 23 experiments (0.92 r 0.98; solid line: P<0.05; heavy line: P<0.01; Moore's test, second-order statistics). There were no significant differences between the results which were obtained in the different experimental situations of experiment a) and b) (Mardia-Watson-Wheeler test).

box, even when the latter had been rotated and was therefore pointing in another direction during the animal's return to the periphery. On the basis of additional data (which will be presented later on) it was thought that in order to choose a constant direction, the animal could still use a category of location-based cues which had not been eliminated as yet. Conjecturing that these directional cues might be provided by a biologically meaningful reference system, namely the earth magnetic field (Mather and Baker, 1981), we repeated our initial experiments in an artificial magnetic field with a strongly reduced total intensity.

Fig.1 summarizes the majority of experiments which have been carried out in altered geomagnetic conditions. It can be seen that our subjects continued to return towards the standard location of the nest box, irrespective of the tenfold reduction of the total intensity of the EMF (fig. 1a and b) and of whether the arena had been rotated or not after the animal's exit from the nest box (Fig. 1b).

Thus, the hamster is able to return from the center of the arena to its nest site in conditions which involve not only a passive outward journey, but also the simultaneous elimination, neutralization or reduction of optical, acoustical, olfactory, tactile and geomagnetic information. This unexpected result led us to question whether the animal really uses location-based cues to return to the standard location of the nest, or whether it was able to integrate route-based information in spite of the passive modality of the outgoing trip.

TESTING THE ALTERNATIVE LOCATION-BASED INFORMATION VERSUS ROUTE-BASED INFORMATION

Location-based cues are collected in function of a given location in space, independently of the subject's (immediately) preceding itinerary. Route-based information, on the contrary, depends on the registration of information during the subject's preceding displacements and therefore also on the point in space and time where this registration has been initiated. By changing the point of departure of a complete hoarding trip we may consequently test whether the animal really chooses a constant direction and therefore uses location-based information, or whether it returns to the point where it initiated the outgoing trip, thus relying on route-dependent information.

For the sake of greater clarity, fig. 2 represents the orientation of one particular animal which had to start its outgoing journey at different points of departure. In the experimental trials of experiment a) and b), the subject left the nest exit after the rotation of the nest box and either walked (a), or was carried passively (b) to the centre of the arena. It can be seen that in these conditions the animal always returned to the point where it had left the nest exit during the preceding outward journey, independently of the latter's active or passive modality. The same animal was also submitted to a small number of trials which involved our usual technique: 1) The subject leaves the nest box at its standard location and enters the transportation box, 2) which is then lifted above the floor of the arena; 3) the arena is rotated, and 4) the subject is deposited at the centrally located food source. As shown on fig.2c), under these conditions the hamster tends to return in the usual 0° reference direction, i.e. to the standard location of the nest box. According to the results of experiments a) and b) the animal seems, however, only to return to a constant point in space because it happened to have always left the nest box at the same location.

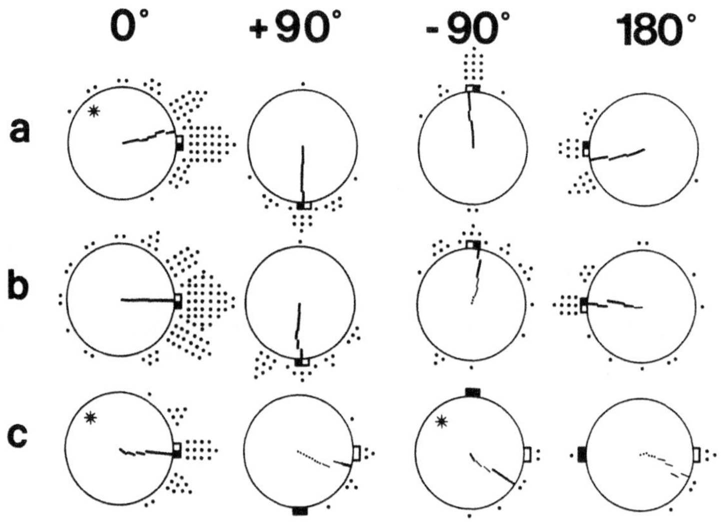

Fig. 2. The orientation of one subject in experiments with a changing point of departure. All trials took place with the elimination of visual, olfactory and tactile cues. In the control trials, the arena remained in position (0°); in the experimental trials, it was rotated by ± 90° or 180°. Experiment a) involved an active outward journey; experiments b) and c) a passive outward journey. In the experimental trials the arena was rotated before the subject had left the nest exit in experiments a) and b), and after it had left the nest exit in experiment c). The position of the nest box at the moment of the subject's departure from it is represented by a white rectangle, its position during the animal's return by a black rectangle. The six vectors on each circle represent the animal's mean orientation when it crossed the six circles on the arena's floor (dotted line: $P>0.05$; solid line: $P<0.05$; heavy line: $P<0.01$, Rayleigh z-test, first order statistics). In these experiments, the experimental vectors were compared to a new reference direction, namely that corresponding to the radius vector pointing to the nest box when the animal left it on its outward journey. In the circles designated by a star, certain vectors differ significantly from this reference direction ($P<0.05$, Stephens' test). The dots around the circles represent the subject's location with respect to the twelve 30°-sectors on the arena's floor when it entered the peripheral zone of the arena in each particular trial.

The repetition of experiments a) and b) with seven further subjects confirmed the interpretation that the hamsters do not chose a constant homing direction with the help of location-based cues, but that they return to the point of departure of the complete hoarding trip, and therfore rely on route-based information.

The nature of this information can be specified with regard to the two alternatives which have been mentioned befbrehand (see Introduction):

1) To proceed from the center of the arena to its periphery the animals followed in general in 2.5 to 5 sec a fairly direct path. After an active outward journey, the return itinerary only exceptionnaly matched the outgoing path, and after a passive outward journey, this was excluded anyway. The hamsters' homing behaviour can therefore be explained by analogy to vector integration.

2) As our subjects continued to orientate towards the point of departure in conditions which excluded the use of intra-arenal cues and made the availability of extra-arenal cues very unlikely (see fig. 1b), it seems that they integrate route-based information without relying on an external reference system.

This interpretation is straightforward with respect to the hamsters' homing attempts after an active outward journey which provides the subjects with different categories of self generated information (see Discussion). On the other hand, path integration during a passive outward journey has to rely mainly on vestibular information which stems from the combination of the subject's passive displacements and its own movements within the (rather large) transportation box. Before discussing this point any further, let us analyse what components of the outward journey the hamster actually assesses during its transportation.

THE INTEGRATION OF A PASSIVE OUTWARD JOURNEY

At the very beginning of this research, the following experiment was carried out in order to establish whether the hamster orientates with respect to route-based or location-based cues in conditions of a passive outward journey combined with the elimination of visual (and other exteroceptive) cues: From its nest exit which leads into its own arena A, the hamster enters the transportation box and is transferred to an adjacent arena B, where it has never been before. It was assumed that if the hamster continued to perform systematically orientated return itineraries in the unfamiliar arena B, it would 1) either choose to go in the same direction as in its own, familiar arena with the help of a location-based directional reference system, or 2) return in the opposite direction, i.e. towards the true location of its nest box, on the basis of inertial navigation (Etienne, 1980).

As shown by fig. 3a, the hamsters' response in this experiment justifies the first hypothesis: 11 out of 12 animals continued to head in the same direction in the unfamiliar arena B as in their own arena A. At first sight, this result would seem to contradict the conclusions reached from the above mentioned data according to which the animal does not choose a constant direction, but returns consistently to the place where it initiated the preceding outward journey. However, our experimental material can be interpreted in a coherent way on the basis of the following interpretation (Etienne <u>et al.</u>, 1985). Let us assume that the hamster assesses the angular, but not the linear component of a passive outward

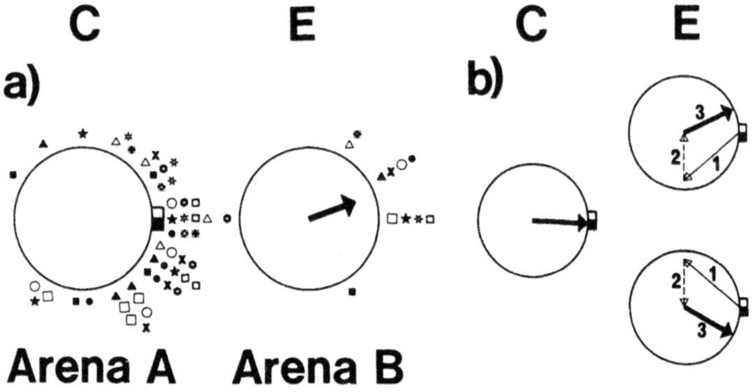

Fig. 3. The non-compensation of passively induced translations. a) The homing direction after a transfer into an unfamiliar arena. All trials involved a passive outward journey and the elimination of visual, olfactory and tactile cues. The subjects performed at first four control trials (C) in their own arena A. In the fifth experimental trial (E) they were transferred from their own nest exit (which leads into arena A) to the centre of an adjacent, unfamiliar arena B. On circle C each category of symbols shows in which 30°- sector each subject entered the peripheral zone of arena A in the four control trials. On circle E, these symbols indicate in the same way the subjects' orientation in the experimental trial in arena B; the vector ($r=0.71$, $P<0.001$, Rayleigh z-test, first-order statistics) represents the subjects' mean orientation in arena B. b) The homing direction after an outward journey with an active and a passive linear component. All trials involved the elimination of visual, acoustical, olfactory and tactile cues. In the control trials (C) the subjects walked from their nest exit to a centrally located platform, collected food there and returned to the periphery of the arena. The vector on circle C indicates the average orientation of 8 subjects when they entered the peripheral zone of the arena in 30 control trials ($r=0.98$, $P<0.001$). In the experimental trials (E), the animals walked actively from their nest exit to the periphery of the arena (arrow 1), where they stepped on a platform. While they collected food on the platform, the latter was pulled to the centre of the arena (arrow 2). Thereafter, the animals returned from the centre to the periphery of the arena. The vector (arrow 3) represents the average orientation of the subjects when they entered the peripheral zone of the arena in 15 trials of each experimental situation (upper circle: $r=0.96$, $P<0.001$, lower circle: $r=0.97$; $P<0.001$; Moore's test, second- order statistics).

journey. During its transportation it therefore always knows the angular amount by which it deviates from its initial orientation at the nest exit without, however, being aware of its actual location in space. If we assume that in order to return to its nest, the hamster compensates its final angular deviation from its orientation at the point of departure and inverts the resultant direction by 180°, it will be orientated a) towards its nest at the centre of its own arena A, and b) in exactly the same direction at the centre of arena B (as well as at any other point in space, see Discussion). This is indeed what happened in experiment 3a.

However, if our assumption, that passive translations are not compensated, is taken literally, our subjects should not move away from a particular hoarding place, but only orientate their longitudinal axis in a specific direction. The fact that they actually walk to the periphery of the arena may therefore depend on a conditioning effect: they have learned that the nest entrance is located at some distance from the hoarding place. They should then move towards the periphery of the arena even if their transfer to the food source did not include any translation at all. This is indeed the case: experiments in progress show that the hamster returns in an identical manner towards its nest when, after leaving its nest exit, it has been deposited with, or without translation to the centre of the arena. (In the second case, the arena is displaced linearly, while the transportation box is only rotated).

The results which are presented on fig. 3b confirm that passively induced translations are not compensated by our subjects, or at best, only insufficiently so: If the outgoing journey started with an active leg, and was then followed by a passive translation, the animal tended to return to its nest in a direction opposite to that of the first leg. Thus our subjects behaved again as if they were not aware of their passive linear displacement from the periphery to the centre of the arena. Comparable results have been obtained on gerbils in a testing situation very similar to that of our own research (Mittelstaedt and Mittelstaedt, 1982).

THE LIMITATIONS OF COMPENSATING ACTIVE AND PASSIVE ROTATIONS

Our data suggest that in our experimental situation the subjects are capable of integrating the angular component of an active as well as a passive outward journey. Path integration has, however, its limitations. It is in fact open to errors which affect in a cumulative manner the integration of path-dependent information, and which may also intervene in its storage and in the adjustment of the next phase of orientateded displacement to a path-dependent "Sollwert". It is therefore to be expected that the homing performance decreases in proportion to the amount of the animal's preceding angular displacement. Furthermore, passive rotations should be compensated to a lesser extent than actively performed ones, as the former lead mainly to vestibular information, while the latter generate a variety of self generated cues (see Discussion).

These predictions were fully confirmed by the two sets of experiments which are presented on fig. 4. The same subject was tested by a procedure which combined an active outward journey to the centre of the arena with a given amount of a) active or b) passive rotations. Whereas the animal was able to compensate up to 5 actively performed rotations of 360°, it compensated only 2 full rotations on a rotating platform where it was collecting food. The repetition of these experiments showed that the limit of compensating the angular component of the outward journey is reached after 3 or 5 (n=6) active rotations, and after 1 or 2 (n=6) passive rotations of 360°.

These results also confirm that in these and the previously described testing conditions, the animals do not complement route-based information by location-based cues and therefore rely exclusively on path integration.

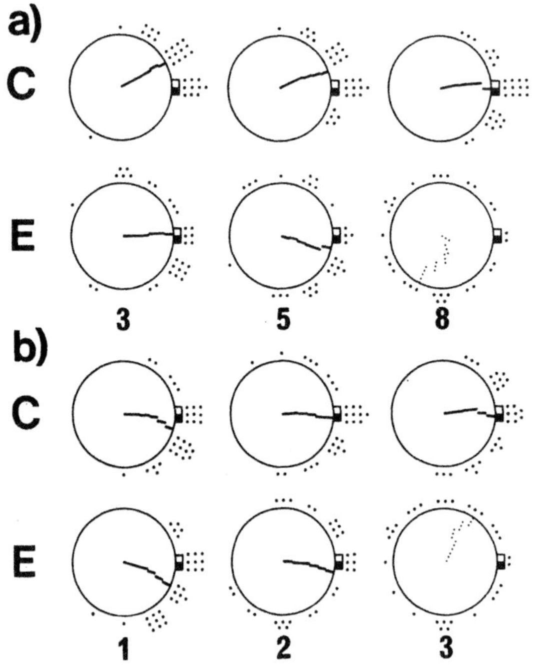

Fig. 4. The limitations of compensating active and passive rotations. All trials involved the elimination of visual, acoustical, tactile and olfactory cues. a) Active rotations. C, control trials: After an active outward journey to the centre of the arena the hamster collected food and returned to the nest. E, experimental trials: After an active outward journey to the centre of the arena the animal continued to follow a dimly lit bait which described 3, 5 or 8 unidirectional circles of 360° around the centre of the arena, and was only thereafter presented with food. b) Passive rotations. C, Control trials: the animal walked actively to the centre of the arena where it collected food on a circular platform, then it returned to the nest. E, experimental trials: The animal walked actively to the centre, stepped on the platform and was rotated in a constant direction by 1,2 or 3x360° while it collected the food items. Additional rotations of ±90° guaranteed that the final orientation of the platform changed from one trial to the nest. For further explanations, see fig. 2.

THE INITIALISATION OF PATH INTEGRATION

The link between path dependent information which is generated within the organism and the (absolute) space in which the animal actually moves occurs through the conditions in which path integration is initialised (Schöne, 1980): In our experimental situation the animal has to start integrating route-based information at the point (and moment) of departure, i.e. at the nest exit, when it has a particular location, orientation and speed with respect to the surrounding space. If path integration is initialized at the wrong place and time the animal will commit systematic errors in its homing itinerary which reflect the position and instant at which its accelerometers are set to zero.

Our subjects were induced to commit such an error in the following conditions: Instead of being transferred to the centre of the arena in the

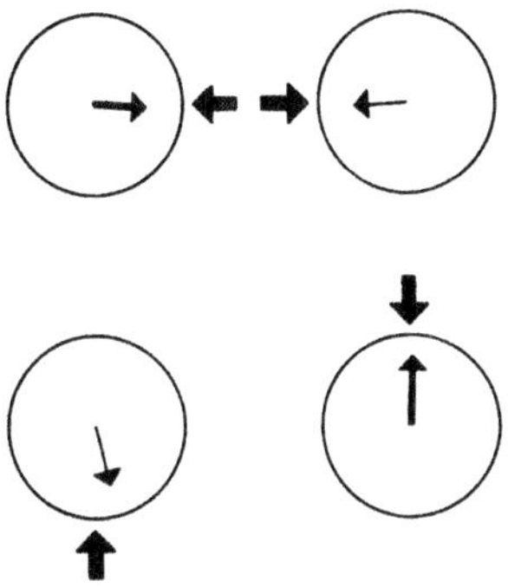

Fig. 5. Errors in the initialization of path integration. All trials involved the elimination of visual, acoustical, tactile and olfactory cues. The subjects entered from the nest exit into a narrow transportation tube in which they could not move. At the centre of the arena, they left the tube through its opposite end. In one particular trial, the tube was deposited at the centre of the arena in one of four different directions (see heavy arrows) and the subject therefore left its vehicle and approached the food source in the same direction. The vectors (0.6 r 0.77; thin line: P<0.05; heavy line: P<0.01; Moore's test; second-order statistics) represent the mean orientation of 13 subjects which underwent 7 trials in each of the four experimental situations defined by the positioning of the tube at the centre of the arena. The presentation and the statistical evaluation of the results is based on a new reference direction which is defined by the orientation of the transportation tube at the moment the subjects left it.

usual, rather large transportation box (see Procedures), the nest exit led to a narrow transportation tube, in which they were prevented from moving or turning in any direction. Instead of facilitating the integration of the angular component of the passive outward journey, this procedure seemed to confuse our subjects: they no longer showed any directional preference in space. However, a closer analysis of their behaviour revealed that the animals' orientation was correlated with the direction in which the transportation tube had been deposited near the food source and in which the animals had therefore left the tube in order to approach the central zone of the arena: All subjects returned to the periphery of the arena inverting by 180° the direction in which they had entered the centre of the arena (see fig. 5). In other words, the subjects behaved as if they were initiating path integration at the moment they left the transportation tube. Thus, their return itinerary reflected a compensation of their (mostly angular) movements around the food source, but not of the angular component of the passive outward journey from the nest exit to the centre of the arena. This interpretation is supported by the fact that the transportation tube resembles the channel shaped nest exit of a natural hamster dwelling.

THE ROLE OF LOCATION-BASED VISUAL CUES AND THEIR INTERACTION WITH PATH INTEGRATION

In a complementary line of research in our laboratory, E. Teroni presents the hamsters with various location-based cues instead of withdrawing the latter from their intra-and extra-arenal environment. So far, only visual cues seem to exert a major influence on the hamster's homing behaviour. Fig. 6 a) illustrates the general role of visual cues from outside the arena, which the hamster can perceive under ordinary room lights: Whereas the animals orientate by path integration under IR light, they return towards the standard location of the nest box under conditions

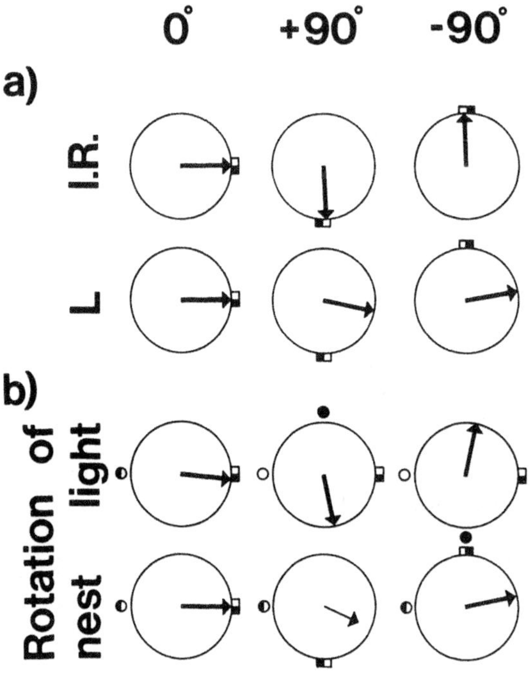

Fig. 6. The role of location-based visual cues. a) The orientation with a changing point of departure under infra-red and white light. All trials involved an active outward journey. In the control trials, the arena remained in position (0°); in the experimental trials, it was rotated by $\pm 90^\circ$. Upper row, IR: the experiments took place under IR light (n=7). Lower row: the experiments took place under the ordinary room light (n=11). b) The orientation with respect to a light spot. During the entire experimental period, a weak light source was located outside of the arena, opposite to the nest box. All trials involved an active outward journey. Upper row: rotation of the light spot (n=4). In the control trials, the light source remained in position (0°); in the experimental trials, it was rotated by $\pm 90^\circ$ when the subject had reached the centre of the arena and started to collect food. Lower row: rotation of the nest box (n=5). In the control trials, the arena remained in position (0°); in the experimental trials, the arena was rotated by $\pm 90^\circ$ before the animal had left the nest box. On a) and b) the rectangles indicate the position of the nest box during the total length of the trials; on b) the white and black circles indicate the position of the light source before and after the animal had started to collect food at the centre of the arena. For each subject, one particular experiment involved 20 control and 20 experimental trials. The vectors represent the animals' mean orientation when they entered the peripheral zone of the arena (0.74 r 1.0; thin line: P<0.05; heavy line: P<0.01; Moore's test, second -order statistics).

in which path dependent and location-based visual cues are presented in a conflicting manner. Only a slight influence of path integration is expressed by the small deviations of the vectors towards the location of the (rotated) nest box. Thus, the animals rely predominantly on location-based visual cues, from which they can deduce the standard location of the nest box.

Fig. 6 b) represents two initial series of experiments which are carried out at present to test the role and interaction of 1) location-based visual cues, 2) route-based visual cues, and 3) path integration without an external reference system (see Introduction). In these experiments, a very weak light beam is directed from a constant location outside of the arena to its centre throughout the experimental period. The experiments consist either of rotating this spot at the end of the outward journey to the centre, or of rotating the arena before the beginning of each trial.

In the first case (see upper row of fig. 6 b), the hamster should proceed in a direction away from the new position of the light if it uses the latter as location-based and/or route-based reference, but towards the standard location of the nest box if it relies on path integration without external reference system. As shown by our results, the orientation of the animals is mainly controlled by the visual cue, and only a slight influence of (non-visual) path integration can be observed. In conditions where the nest box is rotated (see lower row of fig. 6 b), the spot as location-based cue is set in conflict with path integration, both with and without visual reference. Here, the spot continues to play a predominant role as a location-based cue, whereas path integration plays a noticeable, but only secondary role.

Further experiments are carried out at the present time to create a conflict between each of the three above mentioned categories of information. Indications so far are that location-based visual cues seem to have a major influence on the animal's behaviour.

DISCUSSION

The visual conditions in which our subjects were tested exerted a major influence on their orientation. Under ordinary room light, the animals used preferentially location-based optical cues which were provided by the familiar environment around the arena. These cues play a predominant role after an active outward journey (see fig. 6 a) and completely control the subject's return itineraries after a passive outward journey (Teroni, unpubl. results). Further, in otherwise complete darkness, the animals relied mainly on a very weak light spot which had a constant location with regard to the standard position of the nest box and could therfore be used as directional location-based cue (see fig. 6 b).

Thus, location-based visual information controls to a large extent short distance homing in the golden hamster. This result is in agreement with data from psychological and ethological research concerning short and long distance orientation in rodents. Evidently, constant features of the visual environment allow a subject to organize its familiar surroundings in a stable manner and therefore seem to have a major functional significance even for dusk and night active species. So far, we do not know how the gain of location-based information and the long term organization of familiar space is related to route- based orientation in the adult and in the young animal (see Introduction).

In the absence of visual cues, the hamsters continued to orientate in a specific manner within their own arena : They returned along a direct path to the peripheral locations where they had left their nest at the beginning of each particular hoarding trip (see fig. 2). This result suggested that, both after an active and a passive outward journey, the animal relies on a dead-reckoning system or on vector navigation which has also been designed as path integration in the recent literature (Mittelstaedt and Mittelstaedt, 1982). Since its capacity of exhibiting significantly orientated return itineraries resisted the simultaneous elimination of various sensory cues combined with a passive (fig. 1) or an active (Etienne et al., 1986) outward journey, the hamster must integrate path dependent information without relying on any external reference system.

Within our limited experimental space, an active outward journey provides the subjects with all the information necessary to compensate the linear and angular components of the outgoing journey; in fact, the animals are capable of orientating adequately from any place in the arena to its periphery and not only from its center (unpubl. res.), as it is the case after a passive outward journey (see below). This self-generated information consists mainly of proprioceptive and vestibular cues and may also include variables which derive from the activation of neural circuits controlling locomotion as well as metabolic correlates of the subject's locomotor activity.

As expected on theoretical grounds (see Introduction), the assessment of the resulting values of all path dependent variables reflects in a cumulative manner the margin of errors that affect their integration. Fig. 4 a) illustrates these limitations of path integration with respect to the compensation of the angular component of an active outward journey: The longer the subject had to follow a unidirectional circular pathway before being presented with food, the more its return itineraries were dispersed, until they reached a completely homogenous distribution. However, it is quite remarkable that the hamster can compensate up to five actively performed full rotations.

The return itineraries of our subjects after a passive outward journey suggest that in these conditions their orientation depends on information from the vestibular system, whose functional principles follow those of an inertial guidance system (Barlow, 1964; Mayne, 1974). Let us explore this assumption with regard to the absence of compensation of passive translations (see fig. 3) and to the assessment of the angular displacements which the animal experiences within the transportation box.

Inertial forces due to accelerations act on the two types of organs of the vestibular system, namely the semicircular canals and the otoliths. The semicircular canals function as angular accelerometers and are therefore only stimulated during angular displacements of the head. The otoliths respond to linear accelerations, but also to changes (caused by altering head tilt) of the effective shear force due to gravity on the hair cells. It is therefore possible that without the availability of additional, non-vestibular information, the subject can interpret the stimulation of the semicircular canals in an unambiguous manner, but may confuse linear acceleration with changes of effective gravitational shear due to head tilt. This hypothesis, which has recently been reformulated by Mittelstaedt and Mittelstaedt (1982) with respect to the non-compensation of passive translations by gerbils finds partial support in the literature of information - processing by the vestibular system (for references and further discussion, see Etienne et al., 1986; see also Potegal, 1982 and this Nato book, Vol. 2). If our interpretations are correct, they would represent a further argument against the assumption that after passive displacement over long distances pigeons (and other species) home on the basis of inertial navigation (Gould, 1982; Wiltschko, 1978).

If our results suggest that passive translations are not interpreted as such by our subjects, they also confirm the animals' outstanding capacity for compensating angular displacements. The hamsters' orientation towards the point of departure of a passive outgoing trip can in fact be explained in the following manner. Through the stimulation of the semicircular canals, the hamster integrates passive rotations from the very moment it has left the nest exit and entered the transportation box. As the subject can move in this rather large vehicle, the vestibular assessment of changes in the orientation of the head yields the resultant of actively and passively induced rotations. After having left the transportation box at the centre of the arena, the animal moves actively around the food source and thereby continues to integrate self generated rotations. To return from the centre to the peripheral location where it left the nest exit, the subject might for instance compensate for the final deviation of its x-axis with respect to its initial orientation at the nest exit, and invert the resulting direction by 180°. According to our assumption that the hamster compensates only the angular component of its passive outward journey, it will return within its own arena towards the peripheral location where it left the nest exit, provided the animal was deposited at a point located on a line coinciding with its x-axis as it was orientated at the moment of the initiation of the outward journey. This is for instance the case when the food source is located at the centre of the arena. From any other place in space, the animal will orientate in the same compass direction, provided it has initiated the outward journey at its own nest exit.

The animals' remarkable performance in compensating rotations is also illustrated by figure 4 b: Hamsters can actively move around a food source on a rotating platform and at the same time integrate up to two full passive rotations.

Vector navigation (or related systems) imply precise rules of computation which are applied not only to the particular output of angular and linear sensors, but to the combined information yielded by both of these receptor systems. Furthermore, figure 5 illustrates the importance of a correct initialisation of the measurement of path dependent variables. Any error concerning the conditions in which path integration is initiated misleads the subject in an irreversible manner, unless correction procedures can intervene through the additional use of location-based information. Homing by path integration <u>per se</u> leads in fact not to a goal which is defined through its objective relationships with other points in space, but back to the subjectively conceived point of departure of the subject's total itinerary.

To conclude, let us speculate on the strategies our dusk and night active golden hamster may use to return to its dwelling in its natural habitat. It is very likely that a small mammal uses not only visual landmarks and path integration, but also a safe network of trails to orientate within its home range. Most of the time, the animal would thus benefit from a redundant and consistent system of information. On the other hand, the animal may have to leave its familiar pathways in order to explore the environment in specific functional contexts. It could then rely on location-based visual cues so long as they are available, and otherwise use path integration. In total darkness the hamster could therefore orientate over short distances without depending on any category of exteroceptive information.

ACKNOWLEDGMENTS

This research was supported by the "Fonds national suisse de la recherche scientifique", grants No.3.349.0.74. and 3.753.0.80. I wish to thank A. Kfouri and R. Maurer for critical suggestions, J. Bovet for pertinent comments concerning a first draft of the paper, R. Schumacher for drawing the figures and M. Mounir for preparing the print ready manuscript.

REFERENCES

Baker, R.R., 1981. "Human navigation and the sixth sense". Hodder and Stoughton, London, Sidney, Auckland, Toronto.

Baker, R.R., 1984. "Bird Navigation". Hodder and Stoughton, London, Sidney, Auckland, Toronto.

Barlow, J.S., 1964. Inertial Navigation as a Basis for Animal Navigation J. Theor. Biol. 6: 76-117.

Batschelet, E., 1981. " Circular Statistics in Biology ". Academic Press, New York.

Bovet, J., 1978. Homing in Wild Myomorph Rodents: Current Problems. In: "Animal Migration, Navigation and Homing". K. Schmidt-Koenig and W.T. Keeton (Eds.), Springer Verlag, Berlin, Heidelberg, New York.

Bovet, J., 1984. Strategies of homing behavior in the red squirrel, Tamiasciurus hudsonicus. Behavioral Ecology and Sociobiology, 16: 81-88.

Etienne, A.S., 1980. The orientation of the golden hamster to its nest-site after the elimination of various sensory cues. Experientia, 36, 1048-1050.

Etienne, A.S., Teroni, E., Maurer, R., Portenier, V., Saucy, F., 1985 Short-distance homing in a small mammal: the role of exteroceptive cues and path integration. Experientia, 41: 122-125.

Etienne, A.S., Maurer, R., Saucy, F., Teroni, E., 1986. Short-distance homing in the golden hamster after a passive outward journey. Animal Behavior. In press.

Gould, J.L., 1982. The map sense of pigeons. Nature, 296: 205-211.

Graue, L.C., 1963. The effect of phase shifts in the day-night cycle on pigeon homing at distances of less than one mile. Ohio J. Science, 63: 214-217.

Mather, J.F., Baker, R.R., 1981. Magnetic sense of direction in woodmice for route-based navigation. Nature, 291: 152-155.

Mayne, R., 1974. A Systems Concept of the Vestibular Organs. In : "Handbook of Sensory Physiology VI/2, Vestibular System, Part 2: Psychophysics, Applied Aspects and General Interpretations". H.H. Kornhuber (Ed.), Springer, Berlin, Heidelberg, New York.

Mittelstaedt, H., 1978. Analyse von Orientierungsleistungen. In "Kybernetik 1977". G. Hauske und E. Butenandt (Eds.), R. Oldenbourg, München, Wien.

Mittelstaedt, H. and Mittelstaedt, M.-L., 1982. Homing by Path Integration. In: " Avian Navigation", F. Papi and H.G. Wallraff (Eds.), Springer, Berlin, Heidelberg, New-York.

Olton, D.S., 1979. Mazes, Maps and Memory. American Psychologist, 34: 583-596.

O'Keefe, J. and Nadel, L., 1978. "The Hippocampus as a Cognitive Map". Clarendon Press, Oxford.

Potegal, M., 1982. Vestibular and Neostriatal Contributions to Spatial Orientation. In : " Spatial Abilities. Development and Physiological Foundations". M. Potegal (Ed.). Academic Press, London, New York.

Schöne, H., 1980. "Orientierung im Raum". Wissenschaftliche Verlagsgesellschaft MBH, Stuttgart.

Wallraff, H.G., 1974. "Das Navigationssystem der Vögel". Schriftenreihe Kybernetik. H. Marko and H. Mittelstaedt (Eds.). R. Oldenbourg, München, Wien.

Wehner, R., 1982. " Himmelsnavigation bei Insekten ". Neujahrsblatt, Naturforschende Gesellschaft in Zürich (Ed.) Orell Füssli, Zürich.

Wehner, R. and Flatt, I., 1972. The Visual Orientation of Desert Ants, Cataglyphis bicolor, by Means of Terrestrial Cues. In: "Information Processing in the Visual Systems of Arthropods". Wehner, R. (Ed.) Springer, Berlin, Heidelberg, New York.

Wiltschko, W. and Wiltschko, R., 1976. Interrelation of Magnetic Compass and Star Orientation in Night-migrating Birds. Journal Comparative Physiology, 109:91-99.

Wiltschko, W. and Wiltschko, R., 1978. A theoretical model for migratory orientation and homing in birds. Oikos, 30: 177-187.

Wiltschko, R. and Wiltschko, W., 1981. The Development of Sun Compass Orientation in Young Homing Pigeons. Behavioral Ecology and Sociobiology, 9: 135-141.

COGNITIVE MAP SIZE AND HOMING BEHAVIOR

Jacques Bovet

Département de biologie, Université Laval, Québec (Qué.), Canada, G1K 7P4

A MODEL OF NAVIGATION WHICH LAYS STRESS ON DISTANCE

A map provides information on the spatial relationships that exist between any two points X and Y that are on it. The simplest of these relationships can be expressed as $\overline{XY}$, the straight line segment between X and Y, characterized by a certain length, and a certain angular deviation from an axis of reference (e.g., the South-North axis). If the sense of this relationship is specified, e.g. from X (as an initial point) to Y (as a terminal point), then the segment $\overline{XY}$ becomes the vector $\overrightarrow{XY}$ of the same length, with a certain direction with respect to a bearing of reference (e.g., North). Common sense indicates that if you have to travel from X to Y along the most direct route, it is not necessary to make use of, or even to know the length of vector $\overrightarrow{XY}$: if the map tells you the orientation of the vector with respect to North and if you are able to transpose this orientation to the real world situation (e.g., using a compass), then you have just to follow the due course until you inevitably hit Y. If you use this procedure and want to stop at Y, however, you obviously need to be able to detect a signal at Y that tells you that you have reached your goal. The important point here is that this signal is totally independent of the distance between X and Y.

Most studies carried out to date on navigation in animals have focused on the problem of direction finding (reviews in Schmidt-Koenig, 1975; Baker, 1984) and have neglected or ignored the question of whether and how animals make use of an evaluation of distance to be covered (1). This I believe is because the most common procedure for studying navigation has been by means of some form of homing experiments, in which animals are tested for their ability to travel along $\overrightarrow{XY}$ vectors where Y is their regular home site (e.g., the loft in homing pigeons). It is implicitely taken for granted in these studies that the homing animals stop travelling when reaching Y not primarily because they know that they have covered the right distance, but much more because they know that they are at home (i.e., they recognize local landmarks that, in their memory, are linked with home).

---

(1) Notable exceptions are the papers on navigation by path integration by Mittelstaedt and Mittelstaedt (1980, 1982) and von Saint-Paul (1982) where distance estimation is claimed to be a *necessary* element of the navigation process.

The point I wish to make in this paper is about the usefulness, for navigating animals, to include an estimation of the distance to be covered in their navigational strategy, in addition to direction finding. Admittedly, as explained above, it is possible for an animal to navigate from X to Y using directional criteria only. If it knows, for instance, that X is West of Y, it just needs to go eastwards until it hits Y back, without really caring for the distance. I argue, however, that this type of directional-only-strategy will be efficient only for animals able of high directional accuracy, and will be catastrophic for animals with only moderate directional accuracy. I further argue that animals with moderate accuracy will greatly enhance their chances of reaching their goal by adding an estimate of the $|\overline{XY}|$ distance in their overall strategy. This is explained in Fig. 1. Let us first examine the case of a

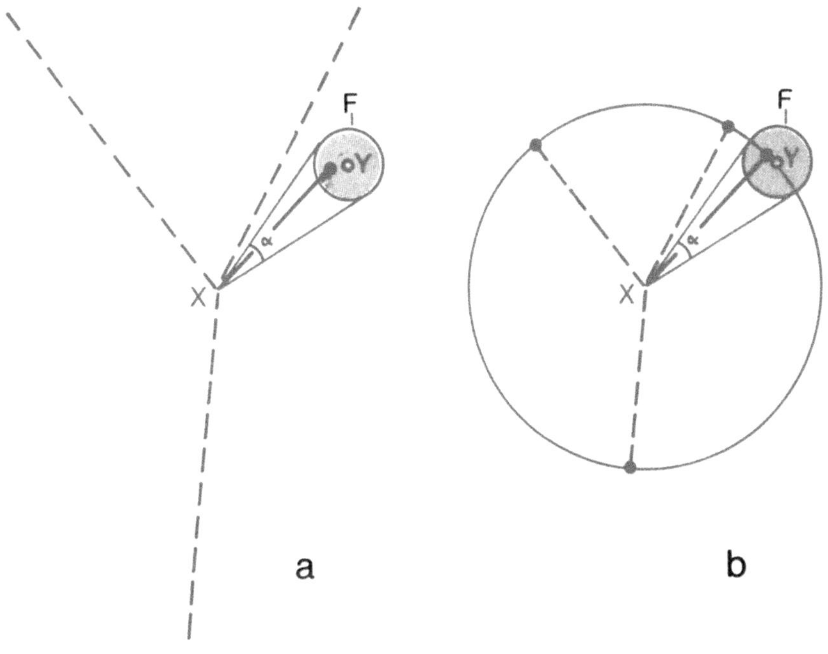

Fig. 1. Strategies of homing. F = area of familiarity. X = starting place of homing trip. Y = home. α = angle under which F is seen from X. Full thick lines: trips of animals which hit F. Dashed thick lines: trips of animals which miss F. Black dots = stops. a) The "directional-only-strategy": the animals which hit F stop because they recognize home-linked signs; the animals which miss F have no reason to stop and stick to the wrong direction. b) The "having-covered-the-$|\overline{XY}|$-distance" strategy: all the animals stop after covering bee-line distance $|\overline{XY}|$; the ones which have hit F recognize home-linked signs and resume regular home range activities; the ones which have missed F perceive the lack of home-linked signs as evidence for a mistake in orientation; they then search again, but restrict their search to the circle shown (center: X; radius: XY).

directional-only-strategy (Fig. 1a). Suppose a limited area of familiarity (F) around home site Y, which contains landmarks that are linked to home in the animal's memory. The homing trip will be successful only for those animals which happened to travel along the direction of $\overrightarrow{XY} \pm 0.5\ \alpha$. In the framework of the directional-only-strategy, all the other animals will be irremediably lost: since their chances of hiting a home-linked landmark are nil, they have no "navigational" reason to stop. Let us now introduce the estimation of the $|\overline{XY}|$ distance into the strategy (Fig. 1b), and suppose that "having-covered-that-distance" is in itself a sign for the animal that it should be at or near Y and, therefore, should watch for familiar landmarks. If the animal is now in its area of familiarity, it sees (or hears, or smells, etc.) the proper landmarks and does not search further. If the animal is not within its area of familiarity, the double circumstance of "having-covered-the-distance" *and* "not-seeing-(etc.)-familiar-landmarks" should be interpreted as evidence for an orientation mistake. From then on, several tactics can be imagined for the animal to reach Y at a minimal physiological or ecological cost, that do or do not include renewed attempts to make use of directional abilities. Most economically, all of them should be spatially limited to a circle of radius $|\overline{XY}|$ around X. The major advantage of remaining within this circle is the confidence of never being far from Y by more than twice the distance $|\overline{XY}|$. This lends to location X a major importance as a potential point of reference for the animal.

THE CASE OF RED SQUIRRELS

Recent field experiments I have carried out with free-living North-American red squirrels (*Tamiasciurus hudsonicus*) provide the basis to the idea of a navigational strategy based on moderately accurate orientation as well as on distance estimates. The reader is referred to the original publication (Bovet, 1984) for details, especially on methodology. Essentially, 17 squirrels were captured at their home site and, using a cage with no view on the surroundings, displaced to a location about either 600 m or 2 km away from home. They were then set free at this distant *release site*. Only one individual was displaced and released on any given day. Its behavior was then monitored, particular attention being paid to accurately surveying its travels during a few hours following release, for eventual transcription onto a map. The general scheme of the experiment was thus typical of standard homing experiments in which distances from home to release site are deliberately set much larger than the normal range of movements of the animals. This is done to eliminate (or at least minimize) the probability that any homeward oriented behavior observed at release time could be due to previous familiarity with the release site area and, presumably, with the route to be travelled from release site to home. This leaves only some form of navigation as a possible cause for homeward oriented behavior (e.g., route-based navigation, or location-based navigation: see Baker, 1984).

Typically, all 17 squirrels left the release site area in running a distance of at least 100 m along a fairly straight course. While 4 animals did not markedly change their course before cessation of monitoring (at 116, 188, 627 or 740 m from release site, respectively), the remaining 13 made a very sharp "U-turn" at an average 252 m bee-line distance from release site (range: 101-544 m). What happened next varied among

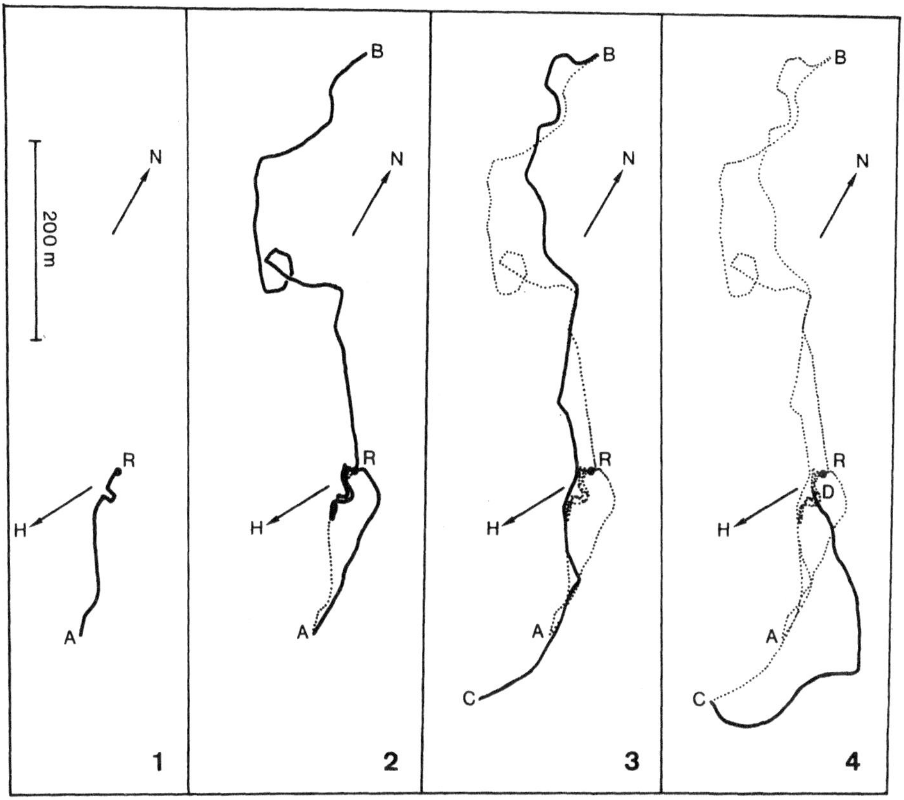

Fig. 2. Route travelled by squirrel # 12, decomposed for clarity into four consecutive stages, 1 to 4. Arrows N point to North, arrows H to the home site which is 560 m from release site R. Thick lines symbolize routes travelled during any one stage; dotted lines symbolize routes travelled during previous stage(s), if any. Scale given for stage 1 applies to other stages as well. Stage 1: from R to A, in 6 min. Stage 2: from A to B, in 57 min. Stage 3: from B to C, in 46 min. Stage 4: from C to D, in 11 min. Monitoring discontinued at D, when squirrel started midday napping.

individuals, the general pattern being that the squirrels returned to the release site or at least closer to it, made then another few hundred meter long fairly straight foray in another direction, made a new U-turn, came back to the release site area, made a new foray, and so on, until they stopped for a nap around noon, at which time we used to stop monitoring. Fig. 2 shows an almost caricatural example of this pattern. Fig. 3 shows a case that fits the pattern more loosely. The total distance travelled by a squirrel between leaving the release site area for a first time and end of monitoring was 2281 m on average, covered at a cruising speed of about 1 km/h.

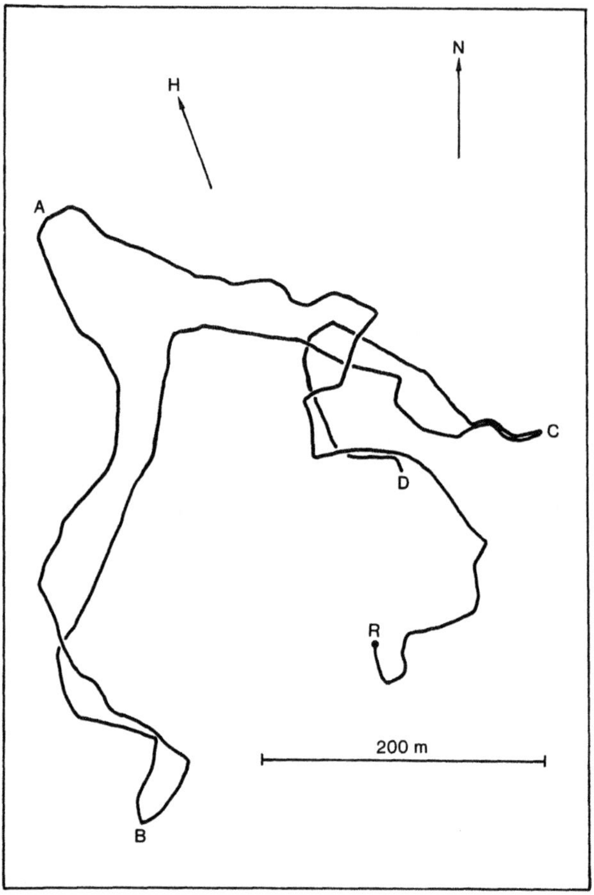

Fig. 3. Route travelled by squirrel # 4. Arrow N points to North, arrow H to the home site which is 2050 m from release site R. Travel from R to A in 57 min.; from A to B in 31 min.; from B to C in 44 min.; from C to D in 16 min. Monitoring discontinued at D, when squirrel started midday napping.

The orientation of the initial straight courses was non-random with respect to the home direction, although the relationship was far from simple. This indicates that the squirrels had some information on the home direction, and made use of it in the orientation of their initial straight courses (for details and justification of the statement, see Bovet, 1984). On the other hand, the fact that most squirrels did not proceed further than a few hundred meters along the direction of their initial straight courses, irrespective of displacement distance, suggests that they had no information on the distance to their home site, or that, for some reason, they "believed" that distance to be a few hundred meters only. This line of evidence points to the possibility that the squirrels were using information collected during the outbound trip, which enabled them to approximately assess the direction of their (passive) displacement, but not its length.

Of course, the normal life of a squirrel never includes an episode in which it is transported passively and set free away from home beyond normal range of movement, and I should now discuss how these "unnatural" experiments might relate to natural behavior. The area travelled over by an animal while performing its normal daily activities is called *home range*. In a parallel study carried out in the same area and same type of habitat, my student Hélène Lair found that a standard red squirrel home range is about 1 ha in surface area (see Lair, 1984, 1985). Occasionally, however, squirrels were found to engage into round-trip "excursions" outside their home range. The function of these excursions could be foraging on temporary, narrowly localized food sources or, in the case of males, searching for a mate in oestrus. They were performed along fairly straight courses over 100-400 m from the outskirts of the home range. Their homing leg was usually not a "reverse image" of their outbound leg (examples in Bovet, 1984, Fig. 5).

Forays of a few hundred meters outside home range are thus part of the *normal* life of a red squirrel, even though they are not part of its regular daily routine. Let us assume that squirrels develop and use some form of a cognitive map, and that their round-trip excursions are built into it. On this map, the vectors from any X (the "goal" of an excursion, from which a squirrel has to return) to any Y (a point within the home range) can have any direction, but not any length: they are at most "a few hundred meter" long. The size of the map is therefore limited, to the extent that the length of the excursions is limited. Thus, according to its map, there is no location in the normal range of movements of a squirrel which is further than a *critical distance* of a few hundred meters from its home range. My hypothesis is that this critical distance is the explanatory key to the fact that, in the homing experiments described above, 75% of the squirrels aborted their homing trip at an average 252 m from release site. More specifically, the hypothesis is that the squirrels, totally unprepared to the unnatural situation of a release in an unfamiliar area after passive displacement, behave "as if" they were due to return from a regular spontaneous exploratory trip, i.e., they apply to the unnatural situation a strategy that is based on past experience in natural situations. Just as they should do when returning from a spontaneous excursion, they interpret a failure to home after covering the critical distance as an indication that they made an orientation mistake. In these conditions, returning to the starting place of the presumably ill-oriented homing trip is a sensible thing to do: it is the only place recently visited which, in a normal situation, could be safely assumed to be within critical distance from the home site.

Close monitoring of the squirrels was interrupted a few hours after release. However, routine checks were made later throughout the study area and provided information on their eventual fate. Among the 13 squirrels that made U-turns, all 4 that had been displaced about 600 m eventually homed (within maximally 1, 1, 2 and 3 days from release, respectively), while only 3 of the 9 displaced 2 km did so (within maximally 1, 1, and 5 days from release, respectively). Of the 6 non-homers, all displaced 2 km, one was found dead 670 m from release site; another was spotted one day after release, 990 m from release site, and not relocated thereafter; the remaining 4 established residence in new home ranges at 380, 450, 530 and 530 m from release site, respectively. Globally, these figures appear to be about 2-3 times larger than the assumed critical distance. In line with the reasoning behind the hypothesis, this suggests that, after systematic search around release site within critical distance, the squirrels interpreted their lack of success as an indication that home was beyond critical distance, or that they underestimated the latter. They would then initiate a second round of longer radial forays. Attempts to home appeared to cease after a few days.

SUPPORT TO THE MODEL FROM OTHER STUDIES BY OTHER AUTHORS ON OTHER SPECIES

At this stage, it should be explicitly emphasized that the hypothetical model presented here was generated by the results of the squirrels' experiments. The latter therefore do not prove the validity of the model. *Ad hoc* tests are planed for the near future. But before these tests are performed, we might have a look at works published by other authors on other species and see how this hypothetical model could fit in the general picture of homing behavior.

Estimation of distance

First, I shall mention three totally independent field studies of homing behavior after experimental displacement, the results of which suggest a strategy that involves distance estimation.

(1) Early in this century, the International Council for the Exploration of the Seas sponsored an enormous research program on the biology of the plaice (*Pleuronectes platessa*), in which nearly 13000 tagged individuals were displaced and released at between 11 and 512 km from their original place of capture. De Veen (1970) has analyzed the distribution of the locations where some 1800 of these fish were eventually recaptured, These locations are clearly clustered within 100 km from release site, a distance which corresponds to the maximal intercapture distances found for control plaices that had been recaptured after release at their original site of capture.

(2) In a study of movements of free-living bank voles (*Clethrionomys glareolus*), which I happened to get acquainted with only after completion of my squirrels' study, Mironov and Kozhevnikov (1982) describe a pattern of successive radial forays performed around the release site in voles displaced and released outside their home range. The length of these forays was some 50-100 m. On the other hand, the authors observed occasional exploratory trips performed spontaneously by these voles. Although the paper does not give specific details on the length of these trips, the incidental evidence produced suggests that they were of the same order of magnitude, i.e., 100 m at the most.

(3) Mittelstaedt's and Mittelstaedt's (1980, 1982) model of "homing by path integration" includes distance estimation of the outbound trip as a necessary element of navigation. In an experiment specifically designed to test this model, von Saint-Paul (1982) manipulated the distance information available to displaced geese in such a way that they would underestimate

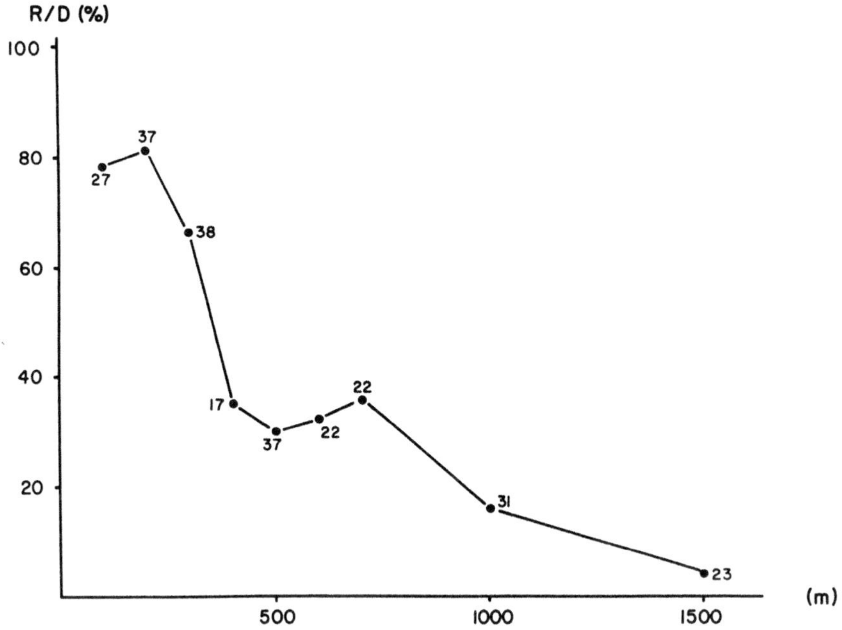

Fig. 4. Relationship between homing success (ordinate R/D, where D = number of animals displaced, R = number of animals which homed) and displacement distance (abscissa) in inexperienced cotton rats (*Sigmodon hispidus*). Figures along the curve correspond to the values of D for any one distance. (Reworked from DeBusk and Kennerly, 1975).

the distance actually covered during displacement. In six trials where the geese had real opportunity to get away from the release site over large distances, the birds made an "initial straight course" the length of which corresponded remarkably well to the experimentally underestimated displacement distance. After covering that distance, the geese either "stayed there", or "returned to starting place" (von Saint-Paul, 1982, p. 305).

Homing success

Second, I shall discuss how my hypothetical model could account for a fact well documented in homing studies and which is still in need of a satisfactory explanation. In virtually any set of homing experiments performed to date in any species with inexperienced animals (i.e., animals that are used in a homing experiment for the first time of their life), the proportion of animals that actually home (called *homing success*) decreases with increasing displacement distances. Fig. 4 gives a typical example of this inverse relationship between homing success and distance. Suppose that there is a steady relationship between the critical distance d (length of standard spontaneous excursions) and the maximal length that radial forays around the release site can have in homing experiments before unsuccessful animals stop their attempts to home. As a result of this relationship, this maximal length of radial forays can be expressed as kd. If all the individuals studied had the same critical distance, we would expect homing success to be nearly 100% for any experimental displacement distance below kd, and nearly 0% for any displacement distance over kd (Fig. 5a). The shape of the resulting homing success vs. distance curve obviously does not match the curve obtained empirically. Suppose now that the critical distance varies among individuals, due to some having performed longer spontaneous exploratory trips than others. Whether the distribution of the various individual critical distances is "normal" (as in Fig. 5b) or "skewed" (as in Fig. 5c), the curves of homing success vs. distance become as congruent with the empirical curves as does any competing theoretical curve based for instance on random radial scattering of animals around the release site (e.g., Wilson and Findley, 1971; Furrer, 1973), on probability of encountering an ecological barrier (Rodda, 1985) or on likelihood of being predated upon before reaching the home site (Bovet, 1978, 1982).

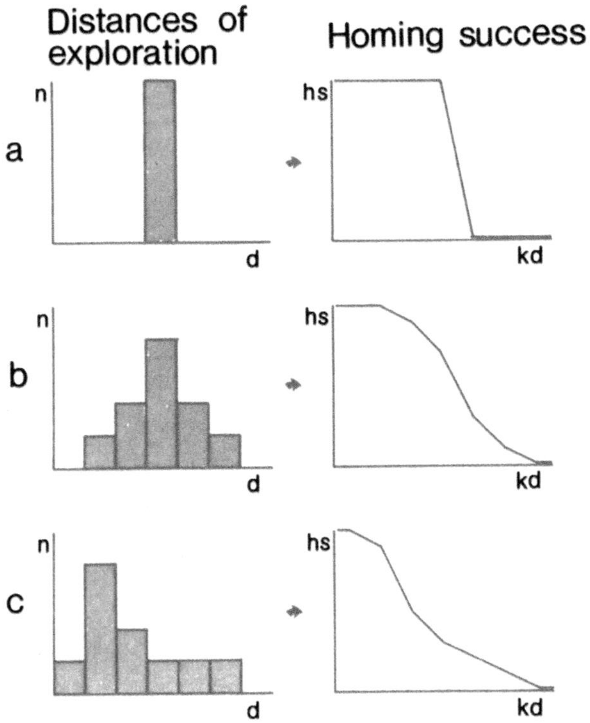

Fig. 5. Patterns of frequency distribution (n) of hypothetical individual maximal length (d) of exploratory forays (left) and resulting patterns of relationship between homing success (hs) and displacement distance (kd), as predicted by the critical distance model (right). See text for details. a) No interindividual variation of d. b) With interindividual variation of d, following a "normal" distribution. c) With interindividual variation of d, following a skewed distribution.

The case of homing pigeons

Lastly, I would like to examine whether the critical distance model might also apply to the homing performance of pigeons. At first sight, the answer should be no: "Pigeons home; they do so almost irrespective of what we do to them" (Kiepenheuer, 1982, p. 120). At second sight, however, it appears that Kiepenheuer's statement should be restricted to experienced pigeons, i.e., pigeons that have been repeatedly used in homing experiments, in other words pigeons for which being displaced and released in an unfamiliar area has become a normal event. Wallraff (1970) reported on experiments performed with inexperienced pigeons (used for the first time of their life in an experimental displacement) which had been given ample opportunities to fly freely, i.e., spontaneously, from and back to their loft. The relationship between homing success and displacement distance in these inexperienced pigeons is shown in Fig. 6. Obviously, many of these inexperienced pigeons *did not home*, and the curve in Fig.6 is analogous to the empirical curve of cotton rats (Fig. 4) as well as to the theoretical curves based on the critical distance model (Fig. 5, b and c). It is therefore not totally ludicrous to discuss the model in the context of pigeon homing.

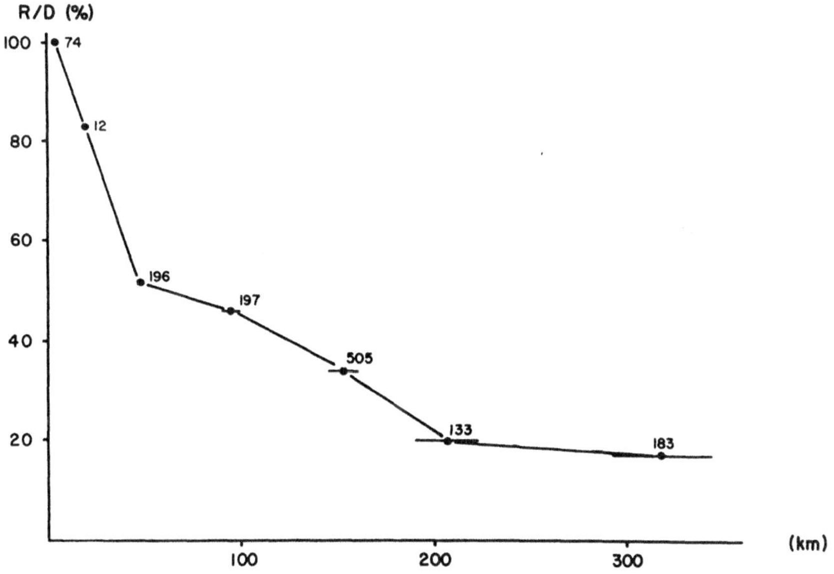

Fig. 6. Relationship between homing success (ordinate R/D, where D = number of animals displaced, R = number of animals which homed) and displacement distance (abscissa) in inexperienced free flight homing pigeons. Figures along the curve correspond to values of D for any one distance. (Reworked from Wallraff, 1970; some data obtained from different distances have been pooled, and the ranges of distances involved are indicated by the small horizontal lines).

Several sets of data in the literature on homing in pigeons point to the fact that a distance of about 20 km from the loft has something critical with respect to homing performance. As can be seen from Fig. 6, it is about the distance beyond which homing success of inexperienced pigeons drops to values well below 100%. This has dramatic effects during the standard training procedures usually undergone by pigeons before they are used in regular homing experiments. Typically, the training flights are performed over distances that increase stepwise from less than 10 km to 100 km or more (e.g., Grüter et al., 1982) and lead of course to the progressive elimination of subjects which fail to home from any one of those flights. The data produced by Sonnberg and Schmidt-Koenig (1970) clearly indicate that the first major losses are experienced as soon as the training procedure involves releases at about 20 km from the loft, irrespective of number of previous flights from lesser distances. Interestingly enough, the same distance is shown to be critical in rock doves (Visalberghi et al., 1978) and in feral pigeons (Edrich and Keeton, 1977), when subjected to similar training sequences. The same distance again has also been shown to be critical in the performances of homing pigeons trained to fly at night (Lipp and Frei, 1982). It should be stressed that the authors of these studies seem to agree on the fact that the losses were not due to increased difficulties in direction finding, but rather to an intrinsic effect of the distance *per se*.

Unfortunately, the homing pigeon literature remains vague, not to say silent, on the range of movements of the birds when they fly freely around the loft. According to incidental figures found in Matthews (1963) and in Papi (1982), this range could be of the order of 25-30 km.

According to Sonnberg and Schmidt-Koenig (1970) only 10-20% of the birds that undergo training "survive" through the whole procedure and are then used in experiments where they seem to home without problem over hundreds of kilometers. In the line of my critical distance model, the mastery of long distance homing flights by these happy few could be explained in two possible ways. (1) Assuming much interindividual variability in the range of free flights around the loft (as suggested by Wiltschko and Wiltschko, 1982), the effect of the training procedure would be to select for the few birds with a very long range of spontaneous movements (right-hand tails of Fig. 5b and 5c). (2) Alternatively, it could be that only a fraction of animals in a population use the critical distance strategy, the others relying on directional criteria only. It might be reminded here that this is perhaps the case in red squirrels: at least two of my squirrels made very long initial straight courses without U-turn (627 and 740 m), and the evidence available suggests that they eventually maintained their initial orientation (Bovet, 1984). In the case of pigeons, the effects of the training procedure could be to eliminate the animals adept in a critical distance strategy, and those with poor directional abilities.

In summary, this paper presents a "critical distance strategy" model which might explain why, in standard homing experiments with inexperienced animals, homing success can be very low, in spite of evidence for homeward oriented behavior. The model calls for some mental representation of space (cognitive map) on the part of the animal, based on past experience in spontaneous exploratory forays. The size of this cognitive map is the yardstick by which an animal, in a homing situation, evaluates whether it has made an orientation mistake. While totally inadaptive in the unnatural situation of an experimentally induced homing trip from a very distant area, the strategy is considered to be adaptive in the context of homing consecutive to a spontaneous exploratory foray: it ensures that, in case of an orientation mistake, the animal remains within "normal" distance from its goal. The model is based on observations made on red squirrels in the field, and has still to be tested. A survey of pertinent literature suggests that it might well apply to homing behavior in other species, including homing pigeons.

REFERENCES

Baker, R.R., 1984. "Bird Navigation". Hodder & Stoughton, London Sydney Auckland Toronto.

Bovet, J., 1978. Homing in wild myomorph rodents: current problems. in "Animal Migration, Navigation and Homing". K. Schmidt-Koenig and W.T. Keeton (Eds.), Springer-Verlag, Berlin Heidelberg New York.

Bovet, J., 1982. Homing behavior of mice: test of a "randomness"-model. Zeitschrift für Tierpsychologie, 58:301-310.

Bovet, J., 1984. Strategies of homing behavior in the red squirrel, *Tamiasciurus hudsonicus*. Behavioral Ecology and Sociobiology, 16:81-88.

DeBusk, J. and Kennerly, T.E.Jr., 1975. Homing in the cotton rat, *Sigmodon hispidus* Say and Ord. American Midland Naturalist, 93:149-157.

Edrich, W. and Keeton, W.T., 1977. A comparison of homing behavior in feral and homing pigeons. Zeitschrift für Tierpsychologie, 44:389-401.

Furrer, R.K., 1973. Homing of *Peromyscus maniculatus* in the channelled scablands of east-central Washington. Journal of Mammalogy, 54:466-482.

Grüter, M., Wiltschko, R. and Wiltschko, W., 1982. Distribution of release-site biases around Frankfurt a.M., Germany. in "Avian Navigation". F. Papi and H.G. Wallraff (Eds.), Springer-Verlag, Berlin Heidelberg New York.

Kiepenheuer, J., 1982. The effect of magnetic anomalies on the homing behaviour of pigeons: an attempt to analyse the possible factors involved. in "Avian Navigation". F. Papi and H.G. Wallraff (Eds.), Springer-Verlag, Berlin Heidelberg New York.

Lair, H., 1984. "Adaptations de l'Ecureuil roux (*Tamiasciurus hudsonicus*) à la Forêt mixte Conifères-Feuillus: Impact sur l'Ecologie et le Comportement des Femelles reproductrices". Doctoral Dissertation, Université Laval.

Lair, H., 1985. Mating seasons and fertility of red squirrels in southern Québec. Canadian Journal of Zoology, 63:2323-2327.

Lipp, H.P. and Frei, U., 1982. Variations of nocturnal homing performance in pigeons. in "Avian Navigation". F. Papi and H.G. Wallraff (Eds.), Springer-Verlag, Berlin Heidelberg New York.

Matthews, G.V.T., 1963. The orientation of pigeons as affected by the learning of landmarks and by the distance of displacement. Animal Behaviour, 11:310-317.

Mironov, A.D. and Kozhevnikov, V.S., 1982. {Character of migrations of *Clethrionomys glareolus* within the home range and outside it}. Zoologicheskii Zhurnal, 61:1413-1418. (in Russian, with English summary).

Mittelstaedt, H. and Mittelstaedt, M.L., 1982. Homing by path integration. in "Avian Navigation". F. Papi and H.G. Wallraff (Eds.), Springer-Verlag, Berlin Heidelberg New York.

Mittelstaedt, M.L. and Mittelstaedt, H., 1980. Homing by path integration in a mammal. Naturwissenschaften, 67:566.

Papi, F., 1982. Olfaction and homing in pigeons: ten years of experiments. in "Avian Navigation". F. Papi and H.G. Wallraff (Eds.), Springer-Verlag, Berlin Heidelberg New York.

Rodda, G.H., 1985. Navigation in juvenile alligators. Zeitschrift für Tierpsychologie, 68:65-77.

Saint-Paul, U. von, 1982. Do geese use path integration for walking home? in "Avian Navigation". F. Papi and H.G. Wallraff (Eds.), Springer-Verlag, Berlin Heidelberg New York.

Schmidt-Koenig, K., 1975. "Migration and Homing in Animals". Springer-Verlag, Berlin Heidelberg New York.

Sonnberg, A. and Schmidt-Koenig, K., 1970. Zur Auslese qualifizierter Brieftauben durch Übungsflüge. Zeitschrift für Tierpsychologie, 27:622-625.

Veen, J.F. de, 1970. On the orientation of the plaice (*Pleuronectes platessa* L.). I. Evidence for orientating factors derived from the ICES transplantation experiments in the years 1904-1909. Journal du Conseil international pour l'Exploration de la Mer, 33:192-227.

Visalberghi, E., Foà, A., Baldaccini, N.E. and Alleva, E., 1978. New experiments on the homing ability of the rock pigeon. Monitore zoologico Italiano (NS), 12:199-209.

Wallraff, H.G., 1970. Über die Flugrichtungen verfrachteter Brieftauben in Abhängigkeit vom Heimatort und vom Ort der Freilassung. Zeitschrift für Tierpsychologie, 27:303-351.

Wilson, D.E. and Findley, J.S., 1972. Randomness in bat homing. American Naturalist, 106:418-424.

Wiltschko, W. and Wiltschko, R., 1982. The role of outward journey information in the orientation of homing pigeons. in "Avian Navigation". F. Papi and H.G. Wallraff (Eds.), Springer-Verlag, Berlin Heidelberg New York.

THE HEURISTIC VALUE OF VISUAL SPATIAL ORIENTATION IN INSECTS

Guy BEUGNON

Centre de Recherches de Biologie du Comportement, U.A. - C.N.R.S. n°664
118, route de Narbonne - 31062 TOULOUSE Cédex - France

In the course of evolution, vertebrates and invertebrates emerged. The specification of the sensorimotor equipment in each branch led respectively to mammals and to insects. As a matter of fact, insects species appeared on earth long ago before dinosaures and even flowers and actually they represent more than 75 % of all the animal kingdom.

Flying, walking or swimming insects perform spatial orientation leading them from and to definite locations as nest, mating stations or feeding sites. Among several means of orientation, visual cues are mainly used.

USE AND LEARNING OF TERRESTRIAL CUES

At the present time, there is no evidence for true navigation in insects : when they are passively displaced and released at least from two different points located outside they home range area, insects do not return directly to a given place. Nevertheless, this home range can extend over 20 km in some South-American Orchid bees. Some of them, individually marked, returned within about 80 minutes to the capture site after a passive transport as large as 23 km (Janzen, 1971).

These returns are based on piloting. After some random or systematic searching, insects use known landmarks indicating the position of the goal or places leading to it. The terrestrial cues can be distant (mountains, hills) or closer landmarks (trees and bushes) (Campan, 1978), that the insect learns to use as a familiar terrestrial map (the "sentiment topographique" J.H. Fabre, 1879).

This map is not a three-dimentional representation of the external world with constant spatial relationships. Without accomodation and convergence during their visual processing, insects can only estimate relative distances between objects using motion parallax (Goulet et al., 1981) but are not capable of absolute depth perception beyond 20 cm, and true binocular triangulation (stereopsis) is limited to a few centimeters (Rossel, 1983). So, without constancy of size and absolute depth, insects could confound size and distance. Accordingly, the terrestrial map should be only a two-dimentional one. Nevertheless, individual insects can use this map with accuracy. Bees returning to their hive, wasps or ants to their nest and crickets to their burrow and whenever looking for food supply at a given location, these insects realize pin-point orientation.

Since the pioneer work of J.H. Fabre (1879) on solitary hymenopterans, it is well known (Tinbergen, 1932 ; Chmurzynski, 1963 ; Van Iersel and Van Assem, 1965) that insects learn the general spatial relation between natural features of their habitat. Landmark learning occurs during first orientation flights when the insect, prior leaving a given spot for the first time, realizes visual scanning with its eyes facing the panorama it will see, later on, when returning. In order to find again this spot, the flying insect need to update its visual memory if the relative position of the landmarks has been experimentally modified before any other outward trip. This is done during so-called re-orientation flights. Recent works provided by Cartwright and Collett (1982, 1983) and Wehner (1981) show that the insect stores a picture of the surroundings and then tries to match the current retinal image with the stored one to determine its current spatial location. For instance, ants and bees were trained to search for food near cylindrical guide-posts. If the size of the cues is modified, the insects will approach smaller guide-posts but move aside from bigger ones until the current angular size of these cues on the retina matches the memorized one during training.

In the same way, Hölldobler (1980) has shown that the pattern of the forest canopy (branches and foliage viewed against the sky above the nest) is learned by some African ants to find their way home. When reared afterwards in the laboratory, ants can still rely on a slide of the canopy projected onto the ceiling. Angular rotations of the slide are inducing equal angular deviations in the ant's tracks.

Thus insects may use memorized images of their habitat, to indicate a given place. Wood-crickets, Nemobius sylvestris, do not cross spontaneously bare forest trails. Yet, they are capable of return towards the side of the trail of capture, when passively released on the center of the forest path. Accuracy of returns, along the same straight trail, is better for wood-crickets released within the limits of their home range than outside it. In other words, crickets escape more easily towards familiar tree trunks than towards unknown ones (Morvan and Campan, 1976).

Development of landmark learning in crickets

Mastering of the significant visual cues is acquired during ontogeny by the individual young cricket (Beugnon et al., 1983). On the basis of elementary reactions to basic light stimulations namely scototaxis, the individual larvae is fleeing first towards the darker areas of its surroundings. This orientation is not merely a negative phototaxis but scototactic insects, attracted by low reflecting patterns, can display in the meantime photo-positive reactions (Weyrauch, 1936 ; Morvan et al., 1976). Then after a few weeks of maturation and experience, occuring during spontaneous daily migrations between forest and trail or wood and field (Lacoste et al., 1976; Beugnon, 1981) older larvae can orientate towards the closest dark outline of the forest and even learn more specific patterns of their living area such as familiar tree trunks (Campan and Gautier, 1975). This learning behavior from scototaxis to individual shape recognition appears through gradually induction processes (Campan and Beugnon, 1985).

USE AND LEARNING OF CELESTIAL CUES

On the other hand, when the insect moves spontaneously during active displacement, it can perform vector orientation independent of landmark orientation. From any point of their outward trip, even along very windy trails, some beetles (Frantsevitch et al., 1977), field crickets (Campan and Beugnon, in prep.) or ants (Wehner, 1981) are able to home in a quite straight line. The use of celestial cues allows return directly to the last bank they are coming from, in the swimming crickets of the Pteronemobius species (Beugnon, 1985 a,b). These compass orientations, sometimes with time compensation, are based on celestial cues as the sun and the polarized skylight of any patch of blue sky. The insect strategy was demonstrated in bees (K. von Frisch, 1967) but how UV polarization of light is perceived is still a question that has given rise to much controversy (Van der Glas, 1980 ; Rossel and Wehner, 1982 ; Gould, 1984). Bees and certainly other insects species do not localize the sun position before sunrise, after sunset or if hidden by cloud using some geometrical abstraction. In fact they seem to match any vector of polarization in the sky with an internal map based on the mean position of the vectors of maximum polarization (Wehner, 1983). This map could be derived from the peculiar anatomical geometry of the retina cells of the dorsal part of the bees, ants, beetles and crickets compound eyes. However, is this polarized skylight map resulting from unlearned (Wehner, 1983) or learned (Gould, 1984) components is still unsolved.

Development of celestial orientation learning in crickets

In some cases, the use of celestial cues appears to be learned by individuals more or less rapidly according to the life history of the insects. Some populations of wood crickets, Nemobius sylvestris, perform celestial orientation leading them towards their forests. After hatching, young larva are photonegatives, then it can be shown that they progressively associate the woodward return, towards the closest black forest, with its astronomical direction (Beugnon, 1984). Several weeks of behavioral development are necessary for the wood crickets to learn terrestrial and celestial cues (Beugnon et al., 1983). On the other hand, when selective pressure is more severe, duration of learning of a visual guiding system can be reduced. Thus, predation by lizards and birds sometimes force the shore-dwelling crickets of Pteronemobius species to jump onto water surface from where they swim back to the land (Beugnon, 1985 a). The use of celestial cues, guiding shoreward returns, appears as soon as the crickets perform evasive swimming. They can even learn a given landward astronomical direction only in one trial, whatever their past experience with water (Beugnon, 1985 b).

NEUROETHOLOGY OF SPATIAL ORIENTATION

A model of behavioural development of spatial orientation in crickets was recently proposed (Campan and Beugnon, 1985). Based on the induction processes starting from phototaxis and scototaxis, sometimes in relation with gravity (Beugnon et al., 1983), the developmental process leads gradually to complex learning of terrestrial and celestial cues.

With the knowledge of the biologically relevant information used by the crickets in natural conditions, and according to the relatively simple neural networks in insects compared to vertebrates, it is of high interest to complete such a developmental approach by a neuroethological investigation of the underlying neural mechanisms. The study of neural plasticity appearing during ontogeny of such small animals must be conducted in a comparative purpose. Similarities between mammals and honeybees appeared recently at behavioural level as well as at the neural organization of memory (Menzel, 1983). Furthermore the identifiable interneurons in the insect nervous system allow a deeper analysis of the interaction between sensory and motor systems as it is now developped for the study of the visuomotor behaviour of crickets.

Investigation of the electrophysiological properties, by extra-cellular recordings, and at the same time anatomical localization of some of the recorded units, by cobalt nitrate dyeing, indicate that the visual interneurons of the cricket are multimodal (Richard et al., 1982). This could allow plasticity of the behavioural responses for example in associative learning (Erber, 1983).

In the neck connective of the cricket Gryllus bimaculatus 80 % (of 192) of the Contralateral Descending Visual Interneurons have their receptive field lying in the fixation zone (Jeanrot et al., 1981) of visual space in which visual forms release orientation reactions with the most efficacy (Richard et al., 1985). Moreover, their axonal arborizations in the thoracic ganglia might overlap in two different medial regions of the dorsal neuropile with motoneurons innervating the proximal muscles of the foreleg which are involved in locomotion (Laurent and Richard, 1985).

Further research using double intracellular recording will allow a better understanding of the structure and function of the nervous network involved in the transfert of biologically relevant visual information from the receptive fields to the motoric output.

DISCUSSION, CONCLUSION

Invertebrate species offer a suitable model system for the study of spatial orientation (see Wehner's, 1981, landmark topic). This is particularly true with insects which are at the highest level of evolutionnary development.

However comparative studies, regarding strategies by insects species and higher vertebrate species to solve spatial orientation problems, are scarce.

The main reason can be due to the stereotypy of behavior that is generally attributed to insects. Such a theoretical viewpoint is nothing but a myth (Médioni, 1985). Capacities of learning and memory were both selected in vertebrate and invertebrate species, to provide the individual with the ability to behave adequately in different ecological constraints. The study of the ontogenetic development of spatial orientation in crickets, related in this paper and summarized Figure 1, demonstrates the very high behavioural plasticity as revealed in the natural habitat of the individual insects independently of their relative short lifetime (up to 2 years). Behavior is the production, here and now, of an individual process emerging from the dynamic interactions between the animal and its world. Influences of past experiences and of the species typical background are combined to contribute

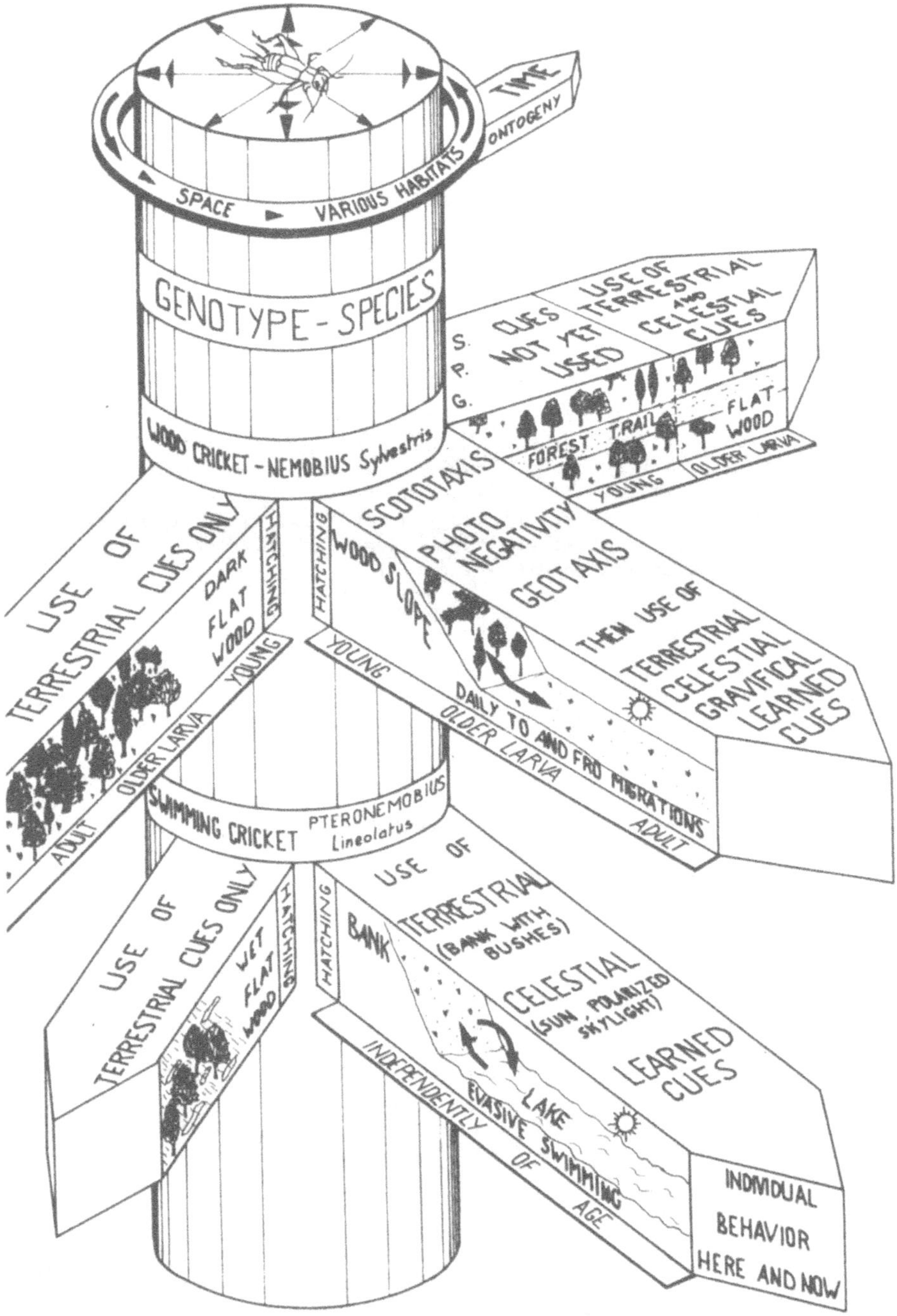

Synthesis of the ontogenetic development of spatial orientation in two cricket species : Individual behavior is expressed according to the insect's age (lenght of the arrow) and to its habitat (angular position of the arrow).

to the current individual spatial orientation which is, in fact, the hyphenate of the animal-world process. Furthermore, insects species are very well suited for neurobiological investigations and especially for the study of neural plasticity during behavioral development.

This short survey claims for more comparative studies regardless of species or, at least, to show the heuristic value of studies of spatial orientation in insects.

REFERENCES

Beugnon, G., 1980. Daily migrations of the wood-cricket Nemobius sylvestris (Bosc). Environmental Entomology, 9:801-805.

Beugnon, G., 1983. Terrestrial and celestial cues in visual orientation of the wood-cricket Nemobius sylvestris (Bosc). Biology of Behaviour, 8:159-169.

Beugnon, G., 1984. De la photonégativité à l'astroorientation chez le grillon des bois Nemobius sylvestris (Bosc). Monitore Zoologico Italiano (N.S), 18:185-197.

Beugnon, G., 1985a. Orientation of evasive swimming in Pteronemobius heydeni (Orthoptera : gryllidae, Nemobiinae). Acta Oecologica Oecologica generalis, 6:235-242.

Beugnon, G., 1985b. Learned orientation in the swimming cricket Pteronemobius lineolatus. Behavioural Processes.

Beugnon, G., Mieulet, F. and Campan, R. 1983., Ontogenèse de certains aspects de l'orientation du grillon des bois, Nemobius sylvestris, dans son milieu naturel, Behavioural Processes, 8:73-86.

Campan, R., 1978. L'utilisation des repères visuels terrestres dans l'orientation des insectes. Année Biologique, 8:61-90.

Campan, R. and Beugnon, G., 1985. The ontogeny of the visual orientation in crickets : A self-organisatory process (in press).

Campan, R. and Gautier, J. Y. 1975. Orientation of the cricket Nemobius sylvestris (Bosc) towards forest-trees. Daily variations and ontogenetic development. Animal Behaviour, 23:640-649.

Cartwright, B. A. and Collett, T. S., 1982. How honey bees use landmarks to guide their return to a food source. Nature, 295:560-564.

Cartwright, B. A. and Collett, T. S., 1983. Landmark learning in Bees. Journal of Comparative Physiology, 151:521-543.

Chmurzynski, J. A., 1963. The stages in the spatial orientation of female digger wasp Bembex rostrata (L.) (Hymenoptera : Sphegidae). Animal Behaviour, 11:607-608.

Erber, J., 1983. "The search of neural correlates of learning in the honeybee." In Neuroethology and Behavioural Physiology. F. Huber and H. Markl (Eds), Springer-Verlag : 216-227.

Fabre, J. H., 1879. "Souvenirs entomologiques". 1ère série. C. Delagrave Ed.

Frantsevich, L. I., Govardovski, V., Gribakin, F., Nikolajev, G., Pichka, V., Polanovsky, A., Shevchenko, V. and Zolotov, V., 1977. Astroorientation in Lethrus (Coleoptera, Scarabaeidae). Journal of Comparative Physiology, 121:253-271.

Frisch, K. von., 1967. "The dance language and orientation of bees." Oxford University Press, London.

Glas, H. W. van der., 1980. Orientation of bees Apis mellifera, to unpolarized Colour Patterns, simulating the polarized Zenith Skylight Pattern. Journal of Comparative Physiology, 139:225-241.

Gould, J. L., 1984. Natural history of Honey bee learning. In the Biology of Learning. P. Marler and H.S. Terrace (Eds), Dahlem Konferenzen Berlin, Heidelberg, New York, Tokyo, Springer-Verlag : 149-180.

Goulet, M., Campan, R. and Lambin, M., 1981. The visual perception of relative distances between objects in the cricket Nemobius sylvestris (Bosc). Physiological Entomology, 6:357-367.

Hölldobler, B. 1980. Canopy orientation : a new kind of orientation in ants. Science, 210:86-88.

Janzen, D. H., 1971. Euglossine bees as long-distance pollinators of tropical plants. Science, 171:203-205.

Jeanrot, N., Campan, R. and Lambin, M., 1981. Functional exploration of the visual field of the wood-cricket Nemobius sylvestris, (Bosc). Physiological Entomology, 6:27-34.

Lacoste, G., Campan, R. and Morvan, R., 1976. L'enchaînement journalier des activités chez le grillon Nemobius sylvestris (Bosc). Acrida, 5:27-44.

Laurent, G. and Richard, D. Organization of the proximal locomotor system in the cricket : I. Anatomy and innervation (submitted).

Médioni, J., 1985. La plasticité du comportement chez les insectes. Psychologie Française, 30:123-126.

Menzel, R., 1983. Neurobiology of learning and memory. The honeybee as a model system. Naturwissenschaften, 70:504-511.

Morvan, R. and Campan, R., 1976. Les déplacements du grillon des bois : conditions d'acquisition et de maintien d'une orientation dominante. Terre et Vie, 30:276-294.

Morvan, R., Goulet, M. and Campan, R., 1976. Etude préliminaire de l'orientation scototactique chez un thysanoure, lepismachilis targionii (grassi). Biology of Behaviour, 1:367-380.

Richard, D., Beugnon, G. and Williams, L., 1982. Données électrophysiologiques et anatomiques sur les interneurones impliqués dans le comportement à point de départ visuel chez le grillon Gryllus bimaculatus. Journal of Physiology, Paris, 78:473-476.

Richard, D., Preteur, V., Campan, R., Beugnon, G. and Williams, L., 1985. Visual interneurons of the neck connectives in Gryllus bimaculatus. Journal of Insect Physiology, 31:407-417.

Rossel, S., 1983. Binocular stereopsis in an insect. Nature, 302:821-822.

Rossel, S. and Wehner, R., 1982. The bee's map of the e-vector pattern in the sky. Proceedings of the National Academy of Sciences, USA, 79:4451-4455.

Tinbergen, N., 1932. Uber die orientierung des Bienenwolfes (Philanthus triangulum Fabr.). Zeitschrift für Vergleichende Physiologie, 16:305-325.

Van Iersel, J. J. A. and Van Assem, D. 1965. Aspects of orientation in the digger-wasp Bembex rostrata. Animal Behaviour (Suppl. I),145-162.

Wehner, R., 1981. Spatial vision in arthropods. In Handbook of sensory physiology. H. Autrum (Ed), vol. VII/6C, Springer-Verlag : 287-617.

Wehner, R., 1983. Celestial and Terrestrial navigation : Human strategies - insect strategies. In Neuroethology and Behavioral Physiology. F. Huber and H. Markl (Eds), Springer-Verlag : 366-381.

Weyrauch, W., 1936. Orientierung nach dunkeln Flächen. Zoologische Anzeige, 113:115-125.

PIGEON HOMING AND THE AVIAN HIPPOCAMPAL COMPLEX: A COMPLEMENTARY SPATIAL BEHAVIOR PARADIGM

Verner P. Bingman,[1,2,4] Paolo Ioale,[2,3] Giovanni Casini[1,2] and Paola Bagnoli[1,2]

1. Istituto di Fisiologia, Universita di Pisa and
2. Istituto di Neurofisiologia, C.N.R., Via San Zeno 31
3. Istituto di Biologia Generale, Via Volta 6
   I-56100 Pisa, Italy
4. Present Address: Department of Psychology, University of Maryland, College Park, Maryland 20742.

Despite the implied comparative nature of this volume, a brief glance at the papers presented shows that the vast majority of animal studies involved mammals to the exclusion of other vertebrate groups. In particular, those studies interested in examining the behavioral function of the hippocampal complex have focussed on rodent performance under laboratory conditions. This concentrated effort persists despite the questionable direct applicability of rodent results for other systems such as memory performance in humans following hippocampal related damage (Weiskrantz, 1978; Horel, 1978).

A large number of rodent lesion and electrophysiological studies have indicated an important role of the hippocampus in the organization of spatial behavior. But despite general agreement on the descriptive aspects of spatial behavior deficits following hippocampal lesion, a satisfactory characterization of the disrupted processes associated with the deficits has remained elusive (see the papers by Gerbrandt, Olton and Schenk).

Mammals are not the only vertebrates with a hippocampal complex nor are they the only vertebrates which engage in complex spatial behavior. It has been our contention that a study of the avian hippocampal complex in the homing behavior of pigeons would provide data relevant to a general discussion of hippocampal function in vertebrates. In addition, we feel the natural context of such a study would provide a broader description of the behavioral deficits one sees following hippocampal damage that could be used in the development of other laboratory studies. Following is a current summary of our work examining the role of avian hippocampal complex in the homing behavior of pigeons.

ANATOMY

For someone familiar with the anatomy of the rodent brain, a superficial look at the avian dorsomedial forebrain (referred to as HP Fig. 1) would leave him doubting the validity of the early anatomical description of this area as the homologue of the mammalian hippocampal complex (Craigie, 1930; Ariens-Kappers et al. 1936). However, even within mammals, the hippocampus has shown itself to be topologically maleable as a consequence

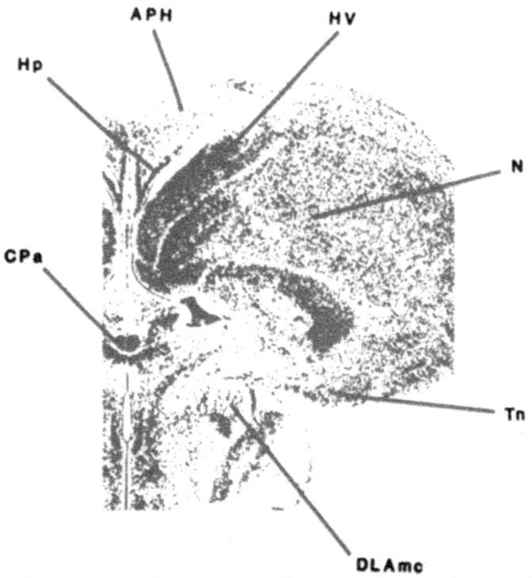

Fig. 1. Nissl-stained, coronal section through the pigeon forebrain. Abbr: Hp, Hippocampus; APH, Parahippocampus HV, Hyperstriatum ventrale; N, Neostriatum; Tn, Nucleus taeniae; DLAmc, Nucleus dorsolateralis anterior thalami pars magnocellularis; CPa, Commissura palli (Hippocampal commissure). Terminology of Karten and Hodos (1967).

of differing neo-cortical development (Elliot-Smith, 1910), and given the approximate 250 million years since the time the mammalian and avian lineages diverged from their common reptilian ancestor, such superficial differences should not come as a surprise. A more meaningful measure of homology would be similarities in pathway connections, and indeed, we have demonstrated remarkable similarity in the extrinsic connections of the avian HP when compared to the mammalian hippocampus (Casini et al., in press).

Using retrogradely transported wheat-germ bound horseradish peroxidase (WGA-HRP), we found HP afferents originating from, among other areas, the septum, hypothalamus and lower brain stem. With the use of anterogradely transported WGA-HRP and tritiated proline, HP efferents were found to project to, again among other areas, the septum and hypothalamus. There was also a reciprocal connection with a region of the basal telencephalon, the archistriatum, which is thought to correspond in part to the mammalian amygdala (Zeier and Karten, 1971). A complete summary of the connections are found in Table 1. It is interesting to note that the two components of HP identified by early comparative anatomists, the hippocampus and parahippocampus, were found to have somewhat different projection patterns that correspond in part with those found in the dentate gyrus - Ammon's horn and subiculum - entorhinal cortex, respectively. Based on the pathway data, HP appears to be the homologue of the mammalian hippocampal complex, and thus a study of its functional role in homing pigeons would seem to bear on a general discussion of hippocampal function.

Table 1. Afferent and Efferent Connections of the Dorsomedial Forebrain[A]

	Telencephalon	Diencephalon	Lower Brain Stem
		AFFERENT SOURCES	
Hippocampus and Parahippocampus	Hippocampus[B,C] Parahippocampus Hyperstriatum accessorium Nucleus of Diagonal Band Nucleus taeniae Area corticoidea dorsolateralis	Nucleus mammillaris lateralis Stratum cellulare internum Nucleus lateralis hypothalami Nucleus paramedianus internus thalami	Area ventralis of Tsai Nucleus reticularis pontis oralis Nucleus raphes Bilateral linearis caudalis
Hippocampus	Contralateral Hippocampus		
Parahippocampus	Hyperstriatum dorsale Archistriatum pars ventralis	Bilateral dorsolateralis anterior thalami Bilateral nucleus superficialis parvocellularis Nucleus subrotundus	Nucleus subceruleus dorsalis Nucleus centralis superior of Bechterew Locus ceruleus
		EFFERENT TARGETS	
Hippocampus and Parahippocampus	Septum Area fasciculus diagonalis Brocae Nucleus taeniae Area corticoidea dorsolateralis	Near Nucleus mammillaris lateralis	
Hippocampus	Contralateral Hippocampus Contralateral Septum		
Parahippocampus	Hyperstriatum dorsale Archistriatum		

A. Adapted from Casini et al., in press.
B. All areas are ipsilateral to injection site unless otherwise stated.
C. Terminology is that of Karten and Hodos (1967).

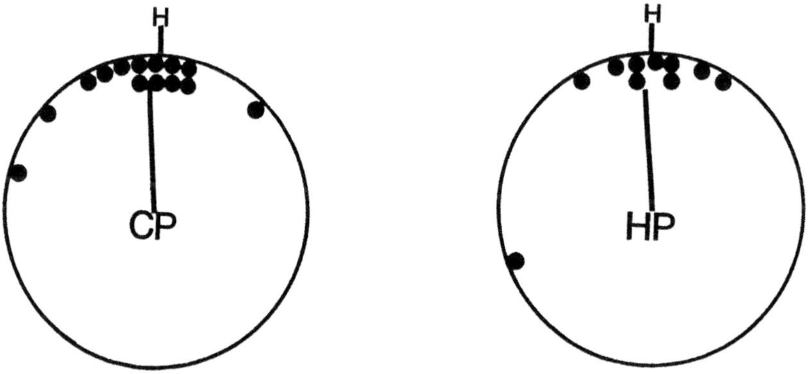

Fig. 2. Initial orientation of control (CP) and HP ablated pigeons (HP) from a site never before visited. Each dot on the periphery of the circle indicates the vanishing bearing of one bird. The inner lines represent mean vectors whose length is set in proportion to the radius of the circle. Homeward direction is set at the top of the circle. Adapted from Bingman et al. (1984).

HP AND DETERMINING THE HOME DIRECTION

The Navigational Map.

The pigeon navigational map, an excellent discussion of which can be found in the paper by Wiltschko and Wiltschko, is a general term used to describe the environmental information used by pigeons to determine their position in space relative to home. For pigeons at our loft in Pisa, atmospheric odors provide the most important source of map information (Papi, 1982), and pigeons are easily rendered unable to locate the home direction from sites never before visited by a variety of anosmic procedures. Pigeons whose HP was ablated via an aspiration procedure showed no impairment in determining the home direction from a site they had never been before and thus a site where their navigational map would have been the only source of homeward information (Fig. 2, Bingman et al., 1984). HP plays no necessary role in the functioning of the pigeon navigational map.

Familiar-Site Information.

In addition to their navigational map, pigeons can learn to use so-called familiar site information; i.e. local landmarks found in the area of a repeated training site, to determine the home direction when their navigational map is made dysfunctional (Hartwick et al. 1977). Homeward orientation based on familiar site information can be viewed as a simple task in stimulus-response association; local landmarks experienced during training releases serve as stimuli that permit the execution of a single, site specific orientation response. If HP corresponds to the mammalian hippocampus, HP ablated pigeons would be expected to show little or no deficit in the acquisition of a familiar-site information based homing response as hippocampal lesioned mammals show little or no deficit in solving simple

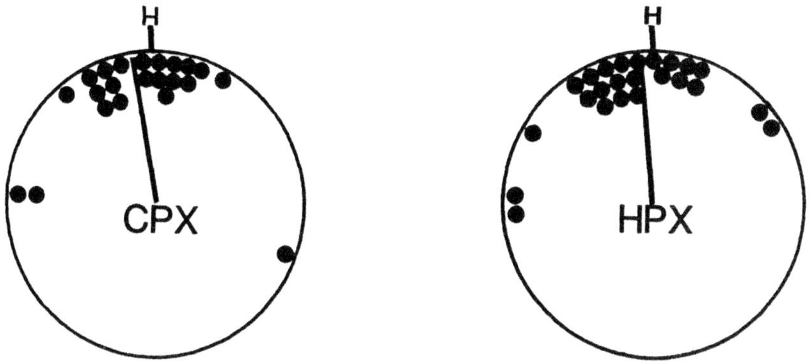

Fig. 3. Initial orientation of control (CPX) and HP ablated (HPX) pigeons rendered temporarily anosmic and released from two previous training sites. The data from the two releases are pooled and plotted with respect to the home direction. See Fig. 2 for an explanation of the diagram. Adapted from Bingman et al. (submitted ms.).

discrimination or spatial tasks (Olton and Papas, 1979; Morris et al., 1982; Berger and Orr; 1983). Indeed, after five training releases each from two directionally opposed sites located about 45 km from home, HP ablated pigeons had successfully learned a homeward orientation response based on familiar-site information. In the experimental releases from the two training sites, HP ablated pigeons rendered temporarily anosmic by the local anasthetic xylocaine (Ioale, 1983) were as successful as similarly treated controls in determining the home direction (Fig. 3, Bingman et al., submitted ms.). In a subsequent release from an unfamiliar site, both the HP ablated and control pigeons subjected to the anosmic treatment failed to orient homewards indicating that the anasthetic effectively rendered the birds' navigational map dysfunctional.

HOMING PERFORMANCE

From the above account, it is clear that HP plays no necessary role in the initial processes by which a pigeon determines the home direction from a release site. However, initial orientation is only the first step of a complex spatial task which may involve different behavioral-neural mechanisms. Behaviorally, homing has been hypothesized to consist of two spatial stages (Schmidt-Koenig and Walcott, 1978). The first stage includes release-site orientation plus the major part of the journey home which is mediated by a bird's navigational map, and although not specifically discussed by these authors, familiar-site information can be included here. The second stage is the actual location of the loft mediated by the use of local landmarks once a pigeon is within 5-10 km of home (the pigeon's navigational map, however, is probably functional at distances where birds may begin to use landmarks; Graue, 1963). Spatial deficits outside the release-site area would be suggestive of a deficit in the execution of the hypothesized second stage of the homing process and would be reflected in an increased amount of time needed to return home; i.e. poorer homing

	CPX	HPX	HPC
Massa	0h 52m (11)	1h 30m (13)*	0h 58m (12)
Bolgheri	1h 03m (11)	1h 11m (13)	1h 05m (11)
Villamagna	1h 40m (10)	4h 10m (12)	1h 05m (11)

() - number of test pigeons * - from Bolgheri, 10 of 13 HPX pigeons arrived home with a later departing CX or HPC pigeon.
Adapted from Bingman et al. (submitted ms).

Table 2. Median Homing Speed of Control (CPX) and HP ablated (HPX) pigeons rendered temporarily anosmic and a second group of HP ablated pigeons (HPC) not rendered anosmic when released from two previous training sites (Massa and Bolgheri) and a site never before visited (Villamagna).

performance. Indeed, in the experimental releases following training, the anosmically treated HP ablated pigeons, which successfully used familiar-site information to determine the home direction, showed a significant deficit in the time needed to return home, and did so from both one training site (from the second training site they tended to follow later departing controls and no comparison could be made) and the unfamiliar site where, once the anasthetic had worn off, both the HP ablated and control pigeons would have used their equally functional navigational maps to locate the home direction (Table 2, Bingman et al., submitted ms). The performance of a second group of HP ablated pigeons not subject to the anosmic procedure (Table 2), as well as that of the anosmic HP ablated pigeons from the training site where they tended to follow controls, indicates that HP ablated pigeons are capable of returning home at speeds similar to controls and that they are motivated to do so. In another experiment (Bingman et al., in press; unpubl. data), we studied pigeons with ablations in another forebrain area, the wulst, who were limited to using familiar-site information to guide their initial orientation (partial olfactory nerve section rendered their navigational map dysfunctional at release). Whether released from a training or unfamiliar site, wulst ablated pigeons returned home at speeds which did not differ from controls with partial olfactory nerve sections only. The homing performance deficit reported above appears specific to HP ablations.

The performance deficit following HP ablation appears spatial in origin and not related to initial orientation guided by either of the two mechanisms described. It seems conceivable then, that the impairment is related to a spatial mechanism employed to locate the home loft once in its vicinity. Like initial orientation, a pigeon can in principal find the home loft once in its vicinity by different ways. Fig. 4 displays three hypothetical approaches, excluding home finding via their navigational map. In A, random searching is employed, and it can be considered the crudest of the approaches described, effectively ignoring local landmarks as cues. In B, fixed orientation responses from the location of specific landmarks is employed. The responses can direct a pigeon toward home or toward another landmark where a second response would be affected. This approach appears to be that implied in the mosaic map of Wiltschko and Wiltschko, complicated by the possibility of response chaining. The mechanism would appear to differ little from the use of familiar-site information to guide initial orientation and thus HP ablation should not affect the acquisition

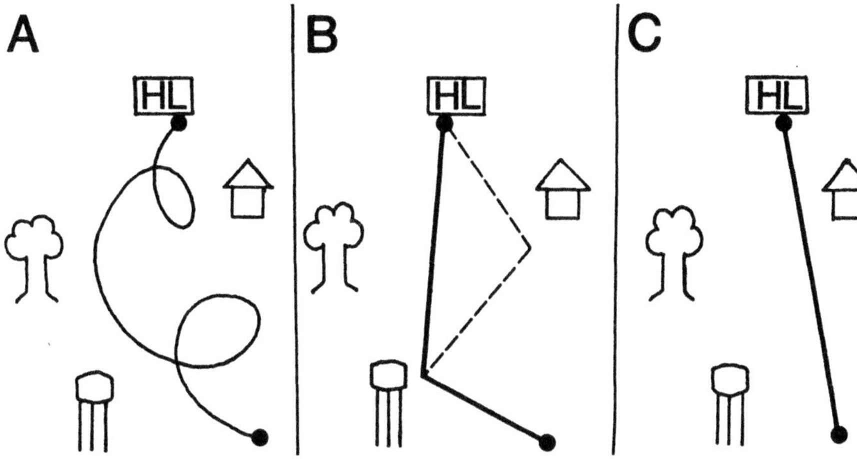

Fig. 4. Schematic representation of three hypothetical strategies that could be employed to locate the home loft once in its vicinity. HL identifies the home loft.

of such a strategy.

The strategy in C would be the most efficient in minimizing the distance covered to reach home and it would be the conceptually most difficult. Here, a pigeon employs a strategy of position finding using distant landmarks or sets of landmarks followed by a position specific orientation response. Thus, if the landmarks themselves are used to guide the orientation response, the response would vary depending on the pigeons position relative to them and home. We wish to suggest that the deficit reported here reflects an impairment in this type of spatial strategy. Should our interpretation prove correct, there is no direct evidence to support it at this time, our results would be consistent with those from mammals indicating deficits in position finding via a posited "cognitive map" (O"Keefe and Nadel, 1978; Morris et al., 1982) and, if broadly defined, possibly working memory (Olton et al., 1980) and conditional responding (Hirsh, 1980) following hippocampal lesion.

In any event, the data demonstrate a homing performance deficit in HP ablated pigeons which appears spatial in origin. As suggested for mammals with hippocampal lesion (Schenk and Morris, 1985), the deficit is limited to a part of the spatial task. We hypothesize that the deficit may be in part explained by an impairment in the birds' ability to locate their position in space using distal landmarks. The data are consistent with the idea that at the level of the hippocampus, birds and mammals share some characteristics in the neural organization of spatial behavior.

References

Ariens-Kappers, C., Huber, G. and Crosby, E., 1936. The Comparative Anatomy of the Vertebrates, including Man. Macmillan, New York.

Berger, T. and Orr W., 1983. Hippocampectomy selectively disrupts discrimination reversal conditioning of the rabbit nictitating membrane response. Behavioral Brain Research, 8: 49-68.

Bingman, V., Bagnoli, P., Ioale, P. and Casini, G., 1984. Homing behavior of pigeons after telencephalic ablations. Brain Behavior Evolution, 24: 94-108.

Bingman, V., Casini, G., Ioale, P. and Bagnoli, P., in press. The avian dorsomedial forebrain: evidence for structural homology and similar function with the mammalian hippocampus. Bolletino della Societa italiana della Biologia sperimentale.

Casini, G., Bingman, V. and Bagnoli, P., in press. Connections of the pigeon dorsomedial forebrain studied with WGA-HRP and 3-H proline. Journal of Comparative Neurology.

Craigie, E., 1930. Studies on the brain of the Kiwi (Apteryx australis). Journal of Comparative Neurology, 56:223-357.

Elliot-Smith, G., 1910. Some problems relating the evolution of the brain. The Aris and Gale Lectures. Lancet, 11:1-6, 147-153, and 221-226.

Graue, L., 1963. The effect of phase shifts in the day-night cycle on pigeon homing at distances of less than one mile. The Ohio Journal of Science, 63: 214-217.

Hartwick, R., Foa, A. and Papi, F., 1977. The effect of olfactory deprivation by nasal tubes upon homing behavior in pigeons. Behavioral Ecology and Sociobiology, 2: 81-89.

Hirsch, R., 1980. The hippocampus, conditional operations, and cognition. Physiological Psychology, 8: 175-182.

Horel, J., 1978. The neuroanatomy of amnesia: A critique of the hippocampal memory hypothesis. Brain, 101:403-445.

Ioale, P., 1983. Effect of anasthesia of the nasal mucosae on the homing behaviour of pigeons. Zeitschrift fuer Tierpsychologie, 61: 102-110.

Karten, H. and Hodos, B., 1967. A Stereotaxic Atlas of the Brain of the Pigeon (Columba livia). Johns Hopkins University, Baltimore.

Morris, R., Garrud, P., Rawlins, J. and O"Keefe, J., 1982. Place navigation impaired in rats with hippocampal lesions. Nature, London, 297: 681-683.

O''Keefe, J. and Nadel, L., 1979. The Hippocampus as a Cognitive Map. Clarendon Press, Oxford.

Olton, D., Becker, J. and Handelmann, G., 1980. Hippocamapal function: working memory or cognitive mapping? Physiological Psychology, 8: 239-246.

Olton, D. and Papas, B., 1979. Spatial memory and hippocampal function. Neuropsychologia, 17: 669-682.

Papi, F., 1982. Olfaction and homing in pigeons: Ten years of experiments. in "Avian Navigation", F. Papi and H. Wallraff, (Eds.), Springer, Berlin.

Schenk, F. and Morris, R., 1985. Dissociation between components of spatial memory in rats after recovery from the effects of retrohippocampal lesions. Experimental Brain Research, 58: 11-28.

Schmidt-Koenig, K. and Walcott, C., 1978. Tracks of pigeons homing with frosted lenses. Animal Behaviour, 26: 480-486.

Weiskrantz, L., 1978. A comparison of hippocampal pathology in man and other animals. in "Functions of the Septal-Hippocampal Complex". Ciba Foundation symposium Vol. 58. Elsevier, Amsterdam.

Zeier, H. and Karten, H., 1971. The archistriatum of the pigeon. Organization of afferent and efferent connections. Brain Research, 31: 313-326.

EFFECTIVENESS AND LIMITATION OF RANDOM SEARCH IN HOMING BEHAVIOUR.

Marc JAMON

Departement de Psychologie animale
Institut de Neurophysiologie et Psychologie
C.N.R.S.- INP 9
31 chemin Joseph-Aiguier
13402 Marseille cedex 9
France.

A common feature of the animal kingdom, and particularly among vertebrates, is the fact that animals confine their movements to a relatively stable area. This common pattern doubtless reflects a fundamental constraint encountered in the course of evolution and it is fairly likely that processes leading animals to find their way back have evolved concurrently with their life history milieu. Although, the remarkable homing capacity of some flying animals like bees or pigeons suggests a navigational mechanism which still eludes our understanding, no completely efficient homing strategies have been demonstrated in the majority of species. On the other hand, the idea that animals use simple search procedures, chiefly random searches, is generally considered worthless, although the theory in many cases shows a good fit with experimental data. This paper reviews some applications of the random search hypothesis to homing studies in order to show that random search is a somewhat more complicated problem than one might think at first glance, and deserves to be given more consideration in studies on animal orientation.

AN OPTIMAL SYSTEMATIC SEARCH : THE SPIRAL.

The most efficient way for an animal to home with absolute certainty without having any information at all about its location would be to move in a spiral until it encounters a familiar area. In a classical study, Griffin (1952) suspected that such a spiral pattern might be used in the search flights of transplanted Gannets that he tracked by plane. Theoretically, any animal using such a systematic search is bound to home successfully, but even a slight inaccuracy in the angular measurement will result its being hopelessly lost, because a spiral will continue to expand indefinitely. Moreover, the distance travelled in a spiral is very considerable in comparison with the distance of displacement. Therefore this search procedure will be time-consuming in comparison with the distance covered, and animals clearly need a way of homing in a shorter time even if it entails some risk of missing their goal.

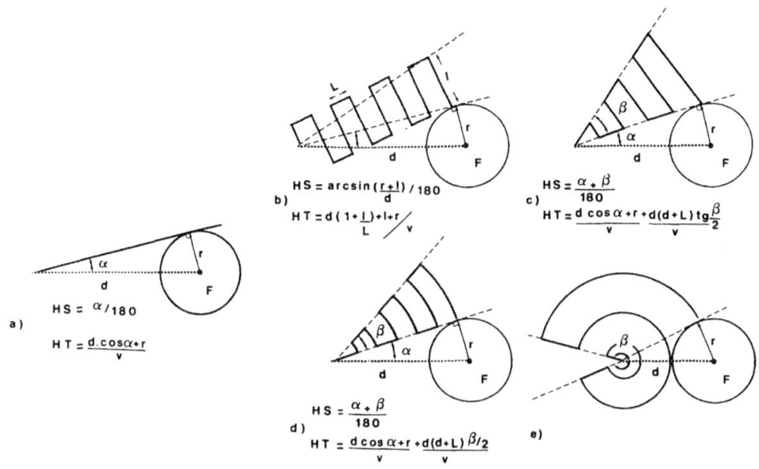

fig. 1: Homing performance of uniform radial scattering model with various movement patterns : 1a) Straight line path ; 1b) rectangular scanning ; 1c,d) Expanding search pattern ; 1e) Scanning with concentric curves (1d) tending towards the spiral performance when β tends towards 180°.
F= familiar area ; r = radius of the familiar area ; d= displacement distance ; v = movement speed ; α = Arcsin(r/d) ; β = scanning angle ; HS = homing probability ; HT = maximum homing time. (b,c are drawn after Griffin 1952).

THE SIMPLEST RANDOM MODEL: THE RANDOM RADIAL SCATTERING

Linear radial scattering

Animals can use the opposite strategy : choosing their direction at random and moving in a straight line during the whole journey (fig. 1a). This is the worst type of random search but it enables homers to reach their goal within a very short time. In this model, the homing success from a given distance depends only on the assumed size of the familiar area (fig. 1a). For instance, Furrer (1973) in a study on homing of rodents considers that random search is consistent with homing success, assuming a value for r of about 500 m. This corresponds to an area of 78 hectares, and greatly exceeds the usual size of the home range of small rodents, which varies roughly between 0.3 and 2 hectares. But the assumption that rodents may have a familiar area map could possibly account for such a large area (Baker, 1981).

In spite of its low efficiency, the radial scattering model was found to fit the homing rate of several species, such as Salmon (Harden Jones, 1968), Bats (Wilson and Findley, 1972) and Rodents (Furrer, 1973). But, it is not consistent with other homing characteristics such as home speed and path straightness. For instance, Salmon homed faster than predicted by linear radial scattering (Harden Jones, 1968) ;the homing flights of Bats released outside their familiar area were very tortuous instead of being straight (Williams and Williams, 1970) ; and

Bovet (1982) showed that the pattern of wood mouse recaptures at different distances around the release point, differed greatly from the radial scattering model predictions.

Systematic exploratory search pattern.

The efficiency of radial scattering can be increased if walking along a straight line is replaced by a zigzagging search pattern which enables the animal to scan a wider band of territory. Griffin (1952) has given two examples of this exploratory search pattern (figure 1b,c). With this assumption, the probability of the animal's reaching home increases with the width of the scanning band, but the distance travelled and the homing time increase as well. One might also imagine an expanding search pattern with concentric curves rather than parallel lines (fig. 1d). As a borderline case, when the searching angle tends towards 180° (fig. 1e), this exploratory pattern tends towards the spiral search pattern. Such systematic search patterns have never been applied to actual data, since their stereotypy is irrelevant to the animal behaviour. But they interestingly show that the use of systematic procedures is a valid means of maximizing a random search, although it is costly in terms of search time and keeping one's course.

THE RANDOM WALK MODEL

Search efficiency can be improved in another way by turning randomly instead of scanning systematically. Such a walk was first formulated by physicists who gave it the name of random walk. In this model, the walk is a succession of steps with a constant length and rotations uniformly distributed in all directions. The probability density function is approximated by the normal distribution. Therefore, the homing probability from a given distance (d) is dependent on the step length, the speed of movement, and the duration of search, and tends towards 1 when the searching time (t) becomes infinite. Wilkinson (1952) first applied this model to animal homing in a historic publication and showed that random walk fitted the actual homing performance of seabirds with regard to both frequency of returns and homing speed. Some years later, Saila (1963) successfully applied the random walk model to the winter coastal migration of the Flounder. The efficiency of the random walk is low however and its uniform distribution in space is at variance with the commonly observed non-uniform distribution of step orientation. A more directional trend may be applied to the random walk by incorporating a small amount of orientational bias consistent with a random search. This problem was first studied by physicists in connection with diffusion with drift, when particles are exposed to an externally applied force. Wilkinson (1952), reviewed some mathematical aspects of such biased random walks in connection with the transoceanic migrations of birds and, later, Saila and Shappy (1962, 1963) analysed biased random search efficiency in the transoceanic homing of Salmon by simulating walk by means of the Monte-Carlo process.

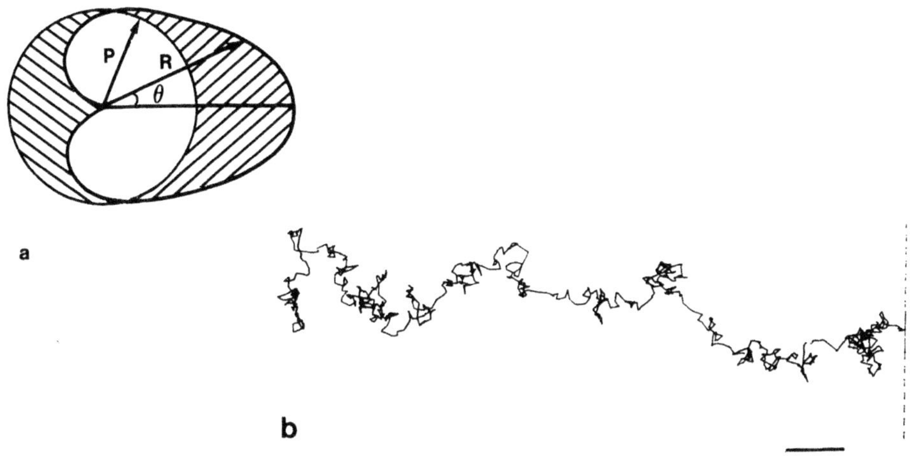

fig. 2 : The biased random walk.
2a) Mode of calculation of the step length. The biased step length R= P(1+A.Cos θ) ; where A is the bias intensity and P is the step length parameter. Hatched area represents the biased component of the step length.
2b)Example of computed biased random walk in Salmon migration, with a bias intensity of 0.25 (Computed from Saila and Shappy 1963).

The biased random walk model.

In the case of a biased random walk, orientations are always uniformly distributed, but the step length increases in relation to the step's orientation towards the preferred direction (fig. 2). The biased random hypothesis is consistent with the dispersion of coastal recaptures of Salmon and with both the general home orientation of fishes in the open sea and the high variability of movement orientation revealed by ultra-sonic and sonar tracking (Harden Jones, 1968 ; Legett, 1977).

The search loop model.

In a completely different situation, when no external orienting cues are available, random walk efficiency can be increased by incorporating a systematic component. Such models have been developed for the homing of Ants by Wehner and Srinivasan (1981) and that of Woodlice by Hoffman (1983). In both cases, an Arthropod homed by a dead-reckoning system, but it had not actually arrived at the nest when the vector integrator was reset to zero, because of a small inaccuracy in the vector integration. Thus, during the final stage of its homing travel, it had to use another searching strategy, which mainly consisted of the increasing random search loops typical of the "search loop" behaviour of invertebrates. Here, the problem was to optimize the search for a target assuming that

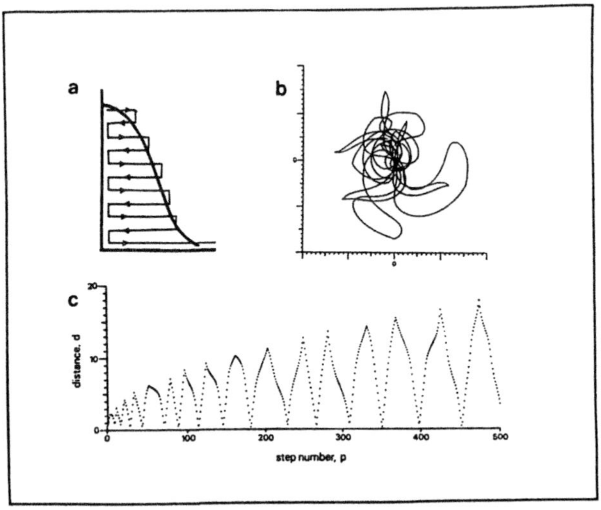

fig. 3 : Characteristics of the optimized search model of Wehner and Srinivasan (From Wehner and Srinivasan 1981)
3a) Sketch of the search pattern in relation to the spatial profile of the a priori probability density function.
3b) Simulated path
3c) Variation of the radial distance from the starting point in relation to the distance travelled, expressed in step numbers.

it is probably located at a small distance from the search starting point. It was optimally solved by using a randomized systematic search that Hoffman called "area-restricted uniform random search". In order to simulate such a search pattern, Wehner and Srinivasan (1981) have incorporated an oscillating component into the radial diffusion of the random search. In their model, the tangential path component was positive until a given radial distance was reached, then negative until the starting point was encountered again. At that point, the radial path component was set to positive and the radial distance incremented (fig. 3a). This procedure produced a randomized systematic search (fig. 3b) having the same oscillating characteristics as the search loop pattern (fig. 3c), although the systematic aspect of the simulated path seemed very considerable in comparison with the actual walk (fig. 3b).

THE "RANDONNEE" MODEL

The previous applications of the random walk to various homing situations show that this model is a convenient tool for the measurement of random search performance. It is less well adapted, however, to the description of animal motion than to Brownian motion in the physical world, because it cannot account for the general tendency to go ahead which is shared by living organisms due to their body polarity and self-powered motion.

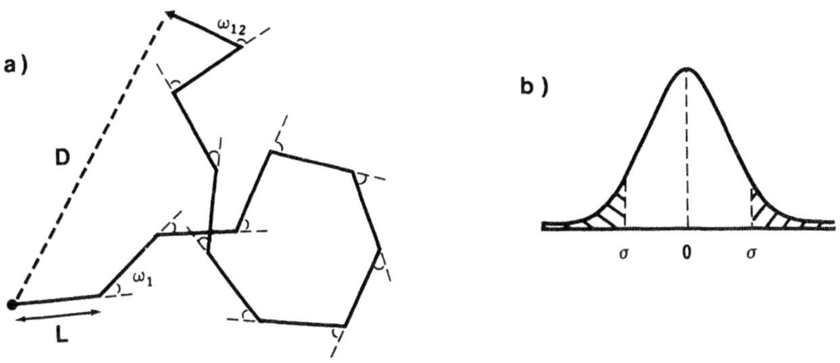

fig. 4 : Determination of the σ of rotations used in the "Randonnée" model.
4a) Example of a path with 13 steps having length L and rotation ω_i.
4b) Probability distribution function of rotations ω_i

Bovet (1985) has studied the properties of a correlated walk model, which is much better adapted to animal motion. He has called this model "Randonnée". Basically, in this model, the walk is a succession of steps with length L and rotation ω_i. But, contrary to the random walk model, rotations are not uniformly distributed, but has a Gaussian distribution with mean zero and standard deviation σ (fig. 4). Bovet (1985) have shown that any travel, from random walk to straight line, may be described by a single parameter, the sinuosity, which is the ratio of σ to the square root of the path length.

Thus, the "Randonnée" model is a more universal descriptor of paths than the random walk model and, hence, agrees better with the structure of animal paths. It has been used to study the travel structure of several species such as Rotifera, Ants and Rodents (Bovet and Benhamou, 1984).

Drift along an axial gradient

The "Randonnée" model may be biased, like the random walk model. But the bias acts on the sinuosity of the path instead of modifying the step length. In order to quantify the effect of an axial gradient on the random dispersion of homing Woodmice (Apodemus sylvaticus) in an heterogeneous environment, I simulated a biased "Randonnée" by modifying σ of rotations in relation to the location of the trajectory projected on the gradient axis. In spite of the small variation of σ, the travels drifted in the gradient direction (fig. 5) and were consistent with the observed orientations of the Woodmice's travels at various increasing distances from the release point, indicating that the general orientational trend resulted from the accumulation of local responses to the space heterogeneity.

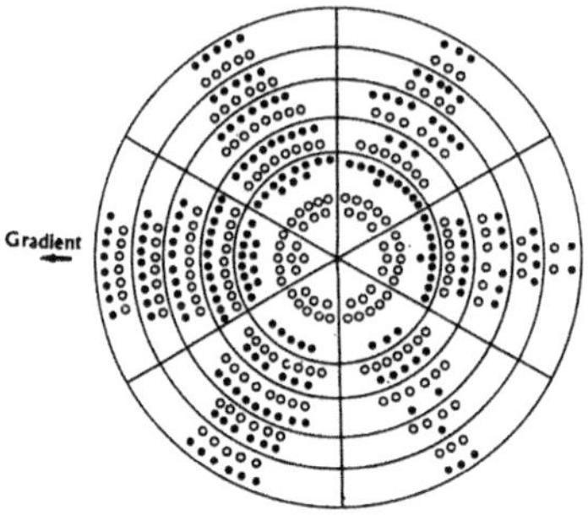

DISTANCE	10m	20m	30m	40m	50m
a	0.52	0.2	0.26	0.3 *	0.38 **
b	0.49	2.47	2.38	3.3	4.48
d.o.f	2	2	2	2	2

significance level * : .10 ; ** : .05

fig. 5 : Progressive bearings at increasing distance from the release point. Black dots : observed bearings ; open circles : theoretical bearings. Concentric circles represent 10m successive radial distances. On right hand side, χ^2 values in the case of orientation along a gradient (a) and under non-graded conditions (b).

Movements in a centered gradient.

In a theoretical study, Bovet and Benhamou (1984) simulated a biased "Randonnée" on a radial gradient. The aim of their study was to demonstrate that a centrally biased random procedure is a possible explanation for home attachment, as previously suggested by Holgate (1969) and Siniff and Jessen (1969). In this case, a gradient effect was centered on the starting point of the simulation, in such a way that σ increased with the gradient variation encountered at each step. Simulation showed that the travel was a succession of loops centered on the starting point and the probability of presence distribution decreased monotonically around the starting point (fig. 6).

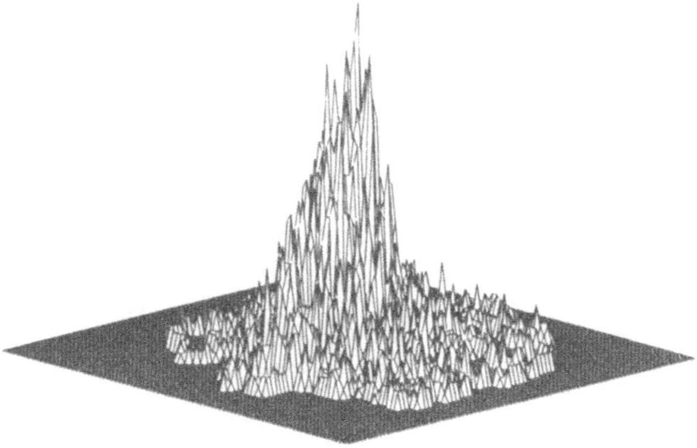

fig. 6 : Space use pattern obtained with a simulated centrally biased "Randonnée" of 1000 steps (From Bovet et al 1984). Height of the figure represents the presence frequency on each set of x,y coordinates on the plane.

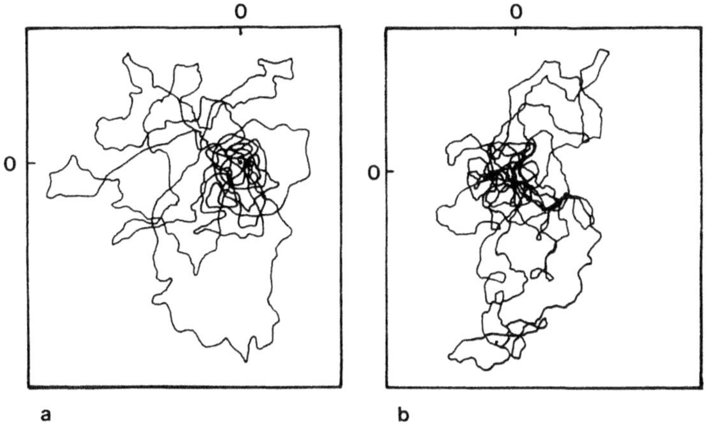

fig. 7 : Examples of actual and simulated search paths
7a) 21mn record of a "search loop" Ant's path (From Wehner and Srinivasan 1981).
7b) Simulated path computed from the centrally biased "Randonnée" model . Step number = 1000 ; σ of rotations =0.6(1+ 0.3•ΔD).

Clearly, such random search is consistent with the search loop behaviour of insects, as shown by a comparison with Wehner and Srinivasan's data (fig. 7) and could be extended to that situation.

DISCUSSION

It emerges from the random search models reviewed here, that the random hypothesis is efficient enough to account for animals' orientation in many situations, but the choice of a random search process cannot be made lightly, since it implies several prerequisites which have to be fulfilled by the animal's behaviour if the model is to be valid. Hence, many random search models must be rejected on the grounds of their lack of biological validity, although they are consistent with actual homing performances. For instance, homing by chance in the course of a radial scattering after an initially haphazard orientation implies strong assumptions about the range of the familiar area or the use of unrealistic systematic searches. Random walk is a more efficient model, especially when it is biased by an external or internal information source, but it provides only a rough estimate of actual animals' movements. The "Randonnée" model is better suited to animal motion, because it describes the path as well as predicting the performance. It therefore provides a biologically appropriate measurement of randomness.

In many species the efficiency of random searching is at variance with the evidences of available orienting cues. Animals behave as if they were not motivated, or not able, to use these orienting cues. For instance, although several authors argue in favour of a route-based navigation ability in rodents (probably based on geomagnetic cues - Baker 1980), these animals' homing performance can also be accounted for by random processes such as the one reported here. This paradox could be explained as follows : in many experiments, animals are transported from the trapping area to the laboratory and kept there for some hours before they are carried to the release point. This might disrupt information collected during the outward journey. The weakness of the route based navigation system, as shown by A. Etienne (this issue), is consistent with this hypothesis.

Animals' orientation mechanisms are constructed in the course of spontaneous exploratory movements, so that they are not necessarily suited to accidental or experimental displacements. In the case of such events, the ability to optimize random processes will ensure successfull homing and it is possible that natural selection may have maintained the adaptative function of stochastic processes during evolution : in any case there is no reason to reject them a priori. Compass bearings are known to be used by many species, but maps are more hypothetical. Investigation of the association of optimized random searches with compass bearings might be a fruitful approach to homing ability in many species. Even in species using navigation, randomness is present, however, not as background noise, but as a component of the system. A good illustration of this point is provided by the homing of Arthropods which, in addition to the use of a vector navigation system, have to compensate for small navigation inaccuracies by conducting an optimized random search at the end of their journey. Here, randomness is what gives the orientation system its plasticity, and homing animals are more likely to use probabilistic than deterministic strategies. Indeed, the stochastic nature of homing behaviour is reflected in homing paths, even in species reputed for their homing abilities, as shown by Pigeons' initial flight tracks recorded by Elsner (1978).

Therefore, the analysis of actual homing paths is a good means of studying animals' orienting behaviour. But these are extremely stochastic structures and "Randonnée" is a useful tool for investigations of this kind.

REFERENCES

BAKER, R. R., 1981. "Human navigation and the sixth sens". Hodder and Stoughton, London.

BOVET, J., 1982. Homing behaviour of mice : test of a randomness model. Zeitschrift fur Tierpsychologie, 58:301-311.

BOVET, P., 1985. "Les déplacements au hasard chez les êtres vivants". Thèse de Doctorat d'Etat, Université d'Aix-Marseille I.

BOVET, P. and BENHAMOU, S., 1984. La clinocinèse : un mécanisme élémentaire de direction. In "La lecture sensori-motrice et cognitive de l'expérience spatiale" (J. Paillard, Ed.), Comportements, 1:171-178.

FURRER, R. K., 1973. Homing of Peromyscus maniculatus in the channelled scablands of east-central Washington. Journal of Mammalogy, 54:466-482.

GRIFFIN, D. R., 1952. "Bird navigation". Biological Review, 27:359-389.

HARDENJONES, F. R., 1968. "Fish migration". St-Martin's Press,New York.

HOFFMAN, G., 1983. The random element in the systematic search behaviour of the desert Isopod Hemilepistus reaumuri. Behavioral Ecology and Sociobiology, 13:81-92.

HOLGATE, P., 1969. Random walk models for animal behavior. in " Sampling and modelling biological populations and population dynamics". G.P. Patil, E.C. Piélou and W.E. Waters (Eds.), Statistical ecology, 2. Pennsylvania University Press:1-12.

LEGETT, W. C., 1977. The ecology of fish migrations. Annual Review of Ecology and Systematic, 8:285-308.

SAILA, S. B., 1961. A study of winter Flounder movements. Limnology and Oceanography 6:292-298.

SAILA, S. B. and SHAPPY, R. A., 1962. Migration by computer. Discovery, 23:23-26.

SAILA, S.B. and SHAPPY, R.A., 1963. Random movement and orientation in salmon migration. Journal du conseil permanent international pour l'exploration de la mer, 28:153-166.

SINIFF, D. B. and JESSEN, C. R., 1969. A simulation model of animal movement patterns. Advances in Ecological Research, 6:185-219.

WEHNER, R. and SRINIVASAN, M.V., 1981. Searching behaviour of desert ants, genus Cataglyphis (Formicidae, Hymenoptera). Journal of Comparative Physiology A, 142:315-338.

WILLIAMS, J. C. and WILLIAMS, J. M., 1970. Radiotracking of homing and feeding flights of a neotropical bat, Phyllostomus hastatus. Animal Behavior, 18:302-309.

WILKINSON, D.H., 1952. The random element in bird "navigation". Journal of experimental Biology, 29:532-560.

WILSON, D. E. and FINDLEY, J. S., 1972. Randomness in bats' homing. American Naturalist, 106:418-424.

FORAGING AND SOCIAL BEHAVIOUR OF UNGULATES: PROPOSALS FOR A MATHEMATICAL MODEL

Stefano Focardi and Silvano Toso

Istituto Nazionale di Biologia della Selvaggina. Via Stradelli Guelfi, 23/A; I - 40064 Ozzano dell' Emilia. Italy.

INTRODUCTION

One of the topics more deeply discussed by the modern theories on the social evolution of ungulates is the role of the ecological constraints (Eisenberg, Lockart, 1972; Estes, 1974; Jarman, 1974; Jarman, Jarman, 1979). Moreover, Geist (1974) has included in the theory the effects that social behaviour and ecology of ungulates have on the population dynamics and on the inclusive fitness of individuals. These theoretical papers select the ecological parameters which more deeply influence the social evolution of ungulates. They are the degree of selectivity of the food resources and the spatial distribution pattern of the food items.

Studying african antelopes, Jarman (1974) identifies five distinct feeding styles, ranging from the browser which feeds on high quality food items, such as fruits, seeds and so on (type 1 of his classification) to those species, such as the eland (Taurotragus derbianus), which feed unselectively on low quality food sources, both grass or browse (type 5). It is clear that different feeding styles are related to different patterns of food dispersion, the latter being often controlled by external drives, such as the changing seasons. Browsing species exploit rich food sources which are gathered in small patches far from each other, while grazing species use very large patches of low quality food.

These evolutionary trends in the selection of the food sources are generally reflected by the pattern of aggregation found among ungulates; the coevoluted behavioural mechanisms permit the maintenance of herds of suitable dimension with respect to the food distribution and furthermore the possibility to adapt the herd size to the variation of the available food biomass and to its distribution. This kind of behavioural plasticity (Jarman, Jarman, 1979) is an important component of the ecological adaptations of many ungulates and it is greatly connected with the development of a complex social organization.

Typically, many of the field studies on the feeding habits of ungulates deal with the long-term variations of the spatial and social population structure. These studies show how aggregation patterns are correlated to the environmental variables, suggesting the existence of causal relationships. These ecological parameters influence global biological features of ungulate populations acting on the spatial behaviour of the individuals. In particular they modify the probability of encounter among animals and the response of the individuals to the presence of conspecifics. This thoretical framework points out the importance of studying the short-term variations of the social and foraging behaviour. (Underwood, 1983; Risenhoover, Bailey, 1985).

At short temporal scale both the vegetation structure of the ecosystem and the foraging habilities and requirements of the animals may be considered as constants. Their interaction determines the distribution of the food accepted by each species. So every species, living in a given environement, detects a specific pattern of food dispersion and density. For instance, selective browsers detect a scattered food distribution, while generalists detect a much more homogeneus environment, where food is distributed in large patches (Jarman, 1974). The distribution of the accepted food deeply influences the observed short-term foraging behaviour and so the actual path followed by each individual is the result of both this factor and of the search strategy of the species.

The foraging path of an animal may be modified by the number of conspecifics which are inside its communication range and so the movement pattern is also a function of the social contest. The level of communication is an important component of the social behaviour of ungulates. The ecological and evolutive role of communication mechanisms is increased by the importance that structure and size of the herds have in the antipredatory strategy of these animals. In fact scanty information is available about the communication system which permits the existence of large and long-lasting herds. It has been shown that the communication between conspecifics is influenced both by the environmental structure and by the peculiar features of the species. For instance, consistent with the fact that the range of visual communication is lower in wooded than in open environments, several authors — e.g. Estes (1969) — have shown a prevalence of chemical communication in wood dwelling species and a prevalence of vocalization and, overall, of visual signals, in species living in open environments. Actually, it has been observed that group size decreases going from prairies to woods (e.g. Hirth, 1977). Both Geist (1974) and Jarman (1974) have pointed out that wood dwelling species are incospicuous, while those living in open spaces have phaneric mantles. This fact allows the former species to reduce the probability to be detected by predators while the latter increase the communication rate between conspecifics and, consequently, the stability of the herd. The interactions between individual behaviour and external constraints — such as the phenology of the vegetation (Hanley, 1982) -- produce the seasonal or inter-habitat variations in the aggregation level which have been described for many species.

In a theoretical study of animal grouping, Suzuki and Sakai (see: Okubo, 1980) have simulated the development and the movements of groups of animals using three deterministic forces. The forward thrust is the tendency of each animal to continue to move forward at its own speed. The mutual interaction is the tendency of an animal to be attracted by conspecifics (this force decreases with the distance and is negative for very short distances). The third one is the arrayal force by which close animals tend to assume the same direction and speed. Jarman and Jarman (1979) state that this last kind of interaction mechanism between the members of a group is the main factor allowing the herd to be a stable and structured organization. The model of Suzuki and Sakai is a study on animal grouping in general, but it is especially conceived for schooling in fishes. This limit is reflected by the nature of the medium where the animals move, which is considered homogeneus. As previously discussed, this is not the case of ungulates in which a model must simulate the interaction between the social behaviour and the environmental features.

The aim of the present paper is to formulate a mathematical model of the social behaviour of ungulates which would permit a quantitative test of validation of theory versus field results.

THE MODEL

The assumptions of this model regard both the distribution of food and the movement pattern of the animals. Let S be a square surface and P the number of animals living on that area. The model is based on the assumption that the population density P/S is time-independent and that the surface is topologically a torus, (i.e. left and right borders and upper and lower borders are geometrically coincident). A certain part of S is covered with food and is subdivided in a certain number of patches. The animal' speed depends on its position on the surface. We introduce two reference speeds:

$$w(x,y) = V_1 \qquad \text{outside food patches} \qquad (1a)$$
$$w(x,y) = V_2 \qquad \text{inside food patches} \qquad (1b)$$

with $V_1 \geqslant V_2$. It is reasonable to assume that V_2 is a decreasing function of the food density within the patches.

Let x(i,t) and y(i,t) be the coordinates of the i-th animal (i=1....P) at time t. The following system of equations is assumed to govern the time evolution of the animal' s position:

$$x(i,t+1) = x(i,t)+v(i,t+1)\cos(\beta(i,t+1)) \qquad (2a)$$
$$y(i,t+1) = y(i,t)+v(i,t+1)\sin(\beta(i,t+1)) \qquad (2b)$$

Here v(i,t+1) denotes the speed of the i-th animal during the interval (t,t+1), whereas $\beta(i,t+1)$ denotes tha angle between the direction of the animal during (t,t+1) and the X direction. Note that ever when the speed of animal i is zero we assume it has a specific direction β determined by the orientation of its body axis.

Let us consider two distinct situations concerning the motion of animal i during (t,t+1):

1. THE ANIMAL IS NOT AFFECTED BY THE PRESENCE OF OTHER CONSPECIFICS.

In this case we conjecture that:

$$\beta(i,t+1) = \beta(i,t) \tag{3a}$$

and that v(i,t+1) is equal to:

$$\tilde{v}(i,t+1) = v(i,t) + a(w(x,y)-v(i,t)) \tag{3b}$$

where $0 \leq a \leq 1$ and where x and y denote the coordinates of animal i at time t.

2. THE MOTION OF THE ANIMAL IS AFFECTED BY THE PRESENCE OF CONSPECIFICS.

We assume the the motion of animal i during (t,t+1) is affected only by those conspecifics which at time t were located at a distance less than a given threshold distance D (the communication range of the species in a given environment) and within a frontal sector of 180° (Fig. 1).

Let $N(i,t) \geq 1$ be the number of such animals and let Σ denotes a summation extended only to them. We define:

$$d(i,t) = \sqrt{(\bar{x}(i,t)-x(i,t))^2+(\bar{y}(i,t)-y(i,t))^2} \tag{4a}$$

$$\bar{x}(i,t) = \sum x(j,t)/N(i,t); \; \bar{y}(i,t) = \sum y(j,t)/N(i,t) \tag{4b}$$

$$\underline{r}(i,t) = \sum \underline{v}(j,t)/N(i,t) \tag{4c}$$

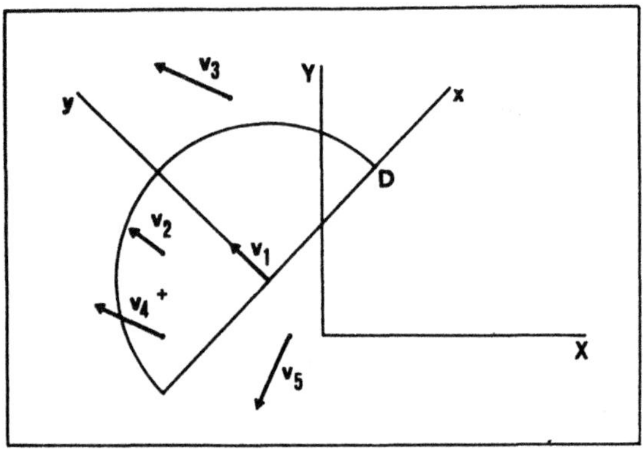

FIGURE 1. Points and arrows indicate the position of animals at time t and their velocity vector during (t-1,t) respectively. (For brevity, v_k stands for v(k,t)). Animal 1 "sees" only animals 2 and 4. The cross indicates the point of coordinates $\bar{x}$ and $\bar{y}$. x and y axis are the reference system of animal 1 and X and Y axis are the absolute reference system on the surface, D is the communication range.

When the magnitude of $\underline{r}(i,t)$ is not zero, we let $\gamma(i,t)$ denote the angle between the direction of $\underline{r}$ and the X direction. We consider three different subcases:

a. When $\underline{r}(i,t)$ has zero magnitude $v(i,t+1)=0$ and $\beta(i,t+1)=\beta(i,t)$.

b. When $\underline{r}(i,t)$ has not zero magnitude and has an alignement which is opposite to that of animal i, this animal stops and takes a direction which is rotated counterclockwise of an amount of $b\pi$ with respect to its original direction:

$$\beta(i,t+1) = \beta(i,t)+ b\pi \tag{5}$$

where:

$$0 < b = (1-d/D)(1-e^{-sN}) < 1 \tag{6}$$

where $N=N(i,t)$ and $d=d(i,t)$ and the parameter s describes the intensity of social behaviour of the species. Moreover, we assume that:

$$v(i,t+1) = 0 \tag{7}$$

c. When $\underline{r}(i,t)$ has non zero magnitude and has not opposite allignment with respect to animal i, we assume that animal i at time t+1 takes a direction whose angle with respect to X direction is:

$$\beta(i,t+1) = \beta(i,t)+b(\gamma(i,t)-\beta(i,t)) \tag{8}$$

Where b is defined by eq. (6). Since β and γ are defined module 2π we select them in such a way that $|\beta-\gamma| < \pi$.
Finally, we assume (for $\gamma-\beta \neq 0$) that:

$$v(i,t+1) = \tilde{v}r\sin(\gamma-\beta)/(r\sin(\gamma-\beta)+(\tilde{v}-r\cos(\gamma-\beta))\mathrm{tg}\vartheta)\cos\vartheta \tag{9}$$

Where $\vartheta=\beta(i,t+1)-\beta(i,t)$, $\tilde{v}=\tilde{v}(i,t+1)$, $r=r(i,t)$, $\gamma=\gamma(i,t)$ and $\beta=\beta(i,t)$ as defined by eqs. 3b and 8.
The interpretation of eq. 9 is given in Fig.2. When $\gamma-\beta$ is zero we take the limiting value:

$$v(i,t+1) = \tilde{v}r/(r+(\tilde{v}-r)b) \tag{10}$$

Finally we impose the additional condition that the new position of the animal must not overlap that of some other individuals. Each animal is considered to have a circular shape of diameter c. When two animals overlap their positions the i-th animal (i.e. the animal which moves) comes back of a distance equal to c along the direction $\beta(i,t+1)+\pi$.

In order to give an idea of how the model simulates the animals' behaviour, the path of two animals, twenty five time units long, are plotted for three different s values (Fig. 3).

When s=0 the movement is not affected by the presence of the other animals, when s=0.1 there is a reciprocal deflection of the path, but the two animals do not form a stable herd. For a s value ten times greater

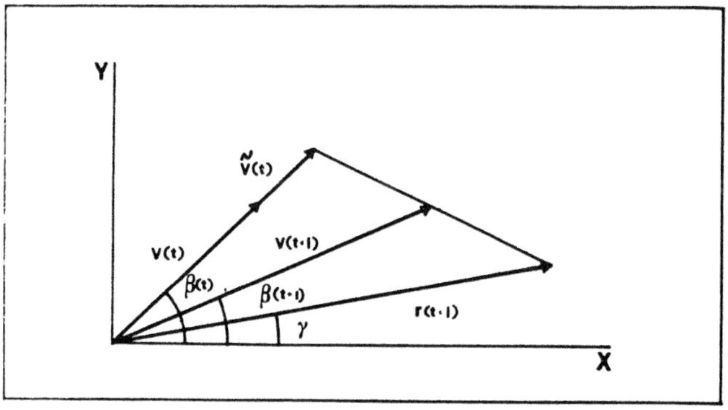

FIGURE 2. The notation relative to the animal i is omitted for brevity. X and Y are the reference axis on the surface.

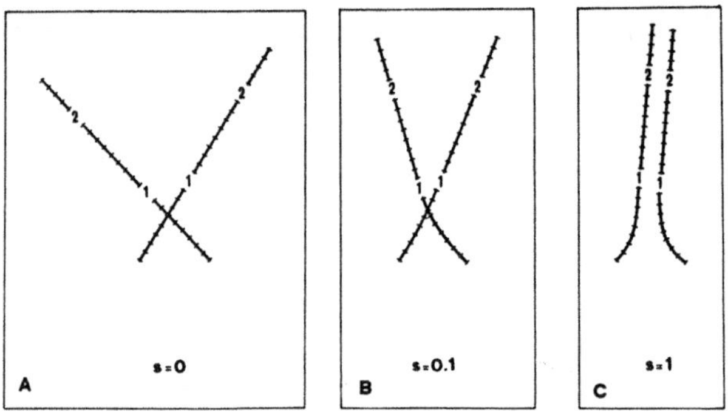

FIGURE 3. The simulated paths of 2 animals in the case of absence of social behaviour (s=0), low (s=0.1) and high sociality (s=1). In these simulations the animals move on a homogeneus surface. v(x,y)=0.1, D=0.5, a=0.5.

(s=1) the two paths became parallel.

To show a possible use of the model, the positions of 20 animals moving on a surface covered by a given food distribution are shown in Fig. 4. The simulations start with the same initial distribution of animals (Fig. 4a) and run for 500 steps. The reciprocal positions of non-social ungulates (s=0) are shown in Fig. 4b while a social species (s=1) is shown in Fig. 4c. The effect on the spatial distribution is evident: the second population is much more aggregated than the first one. The comparison of the aggregation levels of Fig. 4b and Fig. 4c, using the distribution of animals in classes of herd size, gives significative difference (X^2 test, $P < .01$). The comparison between Fig. 4a and 4b does not reach a significative level ($P > .2$).

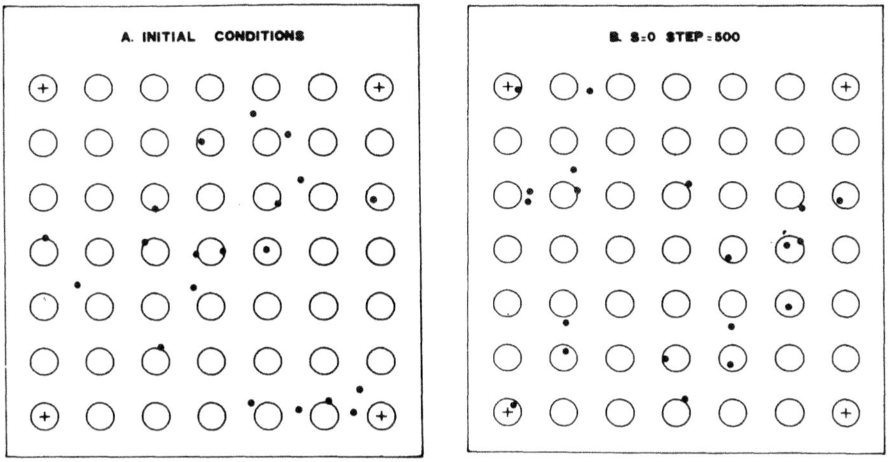

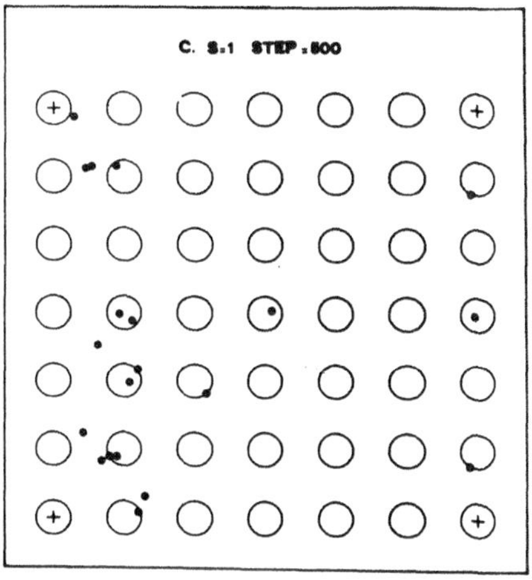

FIGURE 4. Comparison of the aggregation pattern of a population of non social and one of social ungulates. A. The initial position and direction is given at random and it is the same for the two populations. B. The position of non social ungulates after 500 time units. C. The same for social ungulates. D=0.5, V_1=0.1, V_2=0.01, the radium of a patch is 0.26, the distance between patches is 1.04, a=0.5.

Some small herds can be also detected in the case of solitary animals, but these herds are a simple by-product of the food distribution.

DISCUSSION

The model presented in this paper allows the analysis of some of the main biological parameters and variables involved in the social behaviour of ungulates. In particular the model permits a discussion of the interactions between social behaviour and structure of the environment.

The environment is described by the spatial distribution and size of the food patches, by the density of food inside the patches (represented by V_2) and by the visibility degree inside the ecosystem (D). The response to the presence of conspecifics takes into account the social drive itself, the distance among animals and the population density. With respect to the assumptions of Suzuki and Sakai (see: Okubo, 1980) this model uses only the arrayal force. In fact, the arrayal force considered in the present paper is different from that proposed by Suzuki and Sakai because it depends on the distance and on the group size. The definition of a true centripetal force has been neglected because of the scarcity of biological information about it.

The definition of the arrayal force given in the model respects some logical constraints acting in the field. First, the force is null outside the communication range of the animals, otherwise it is always positive. Second, it depends on the herd size; the antipredatory effect of the herd is size-dependent (a review of the existing literature on this topic can be found in Jarman and Jarman, 1979) and it is unreliable that a small herd has the same attraction than a big one.

Although the model simulates realistically the foraging behaviour of ungulates, it introduces some simplifications. The more important of them are:

- the age and sex classes of the animals are not taken into account.
- only two behavioural patterns are considered: moving between food patches and foraging, while other behaviours, such as rumination or escaping, are neglected.
- the movement pattern of the animals is simplified; the only variations of the direction are produced by deterministic interactions between conspecifics. The stochastic fluctuations of the direction produced by the presence of natural obstacles and by explorative attempts have been omitted.

One of the main features of the model is that it is possible to estimate its parameters and variables in the field. In fact the analysis of the foraging paths (Underwood, 1983) allows the estimation of the distribution and density of food and of the speed of the ungulates. Even the visibility degree within each environment is quantitatively estimable, for instance, using physical models of the animals (Risenhoover, Bailey, 1985). So the only unestimated parameter is the social drive (s); its evaluation is the aim of the comparison between model's outputs and field data. The model may help in the study of the complex relationships between ecological constraints, social behaviour and population density, revealing the relative importance of these factors in determining the aggregation pattern of the ungulates. Moreover the model is able to simulate the dynamic response in the space use of a population of ungulates to the seasonal and long-term changes of the environment.

ACKNOWLEDGEMENTS

The authors wish to thank Prof. S. L. Paveri - Fontana for helpful discussions.

REFERENCES

Eisenberg, J. F. and Lockart, M., 1972. An ecological recognissance of Wilpattu National Park, Ceylon. Smithsonian Contributions to Zoology, 101:1-118.

Estes, R. D., 1969. Territorial behaviour of the wildbeest (Connochaetes taurinus Burchell, 1823). Z. Tierpsychol., 26:284-370.

Estes, R. D., 1974. Social organization of african Bovidae. in "The behaviour of ungulates and its relation to menagement". Geist, V. and F. Walther (Eds.), IUCN Publications new series 24, Morges.

Geist, V., 1974. On the relationship of social evolution and ecology in ungulates. Am. Zool., 14:205-220.

Hanley, T. A., 1982. Cervid activity patterns in relation to foraging constraints: western Washington. Northwest Sciences, 56:208-217.

Hirth, D. H., 1977. Social behaviour of white-tailed deer in relation to habitat. Wildlife Monographs, 53.

Jarman, P. J., 1974. The social organization of antelope in relation to their ecology. Behaviour, 48:215-266.

Jarman, P. J. and Jarman, M. V., 1979. The dynamic of ungulate social organizatión. in "Serengeti. Dynamics of an ecosystem". Sinclair, A. R. E. and M. Northon-Griffiths, (Eds.), The University of Chicago Press, Chicago.

Okubo, A., 1980. "Diffusion and ecological problems: mathematical models". Biomathematics Volume 10, Springer-Verlag, Berlin.

Risenhoover, K. L. and Bailey, J. A., 1985. Foraging ecology of mountain sheep: implications for habitat management. J. Wildl. Manage., 49:797-804.

Underwood, R., 1983. The feeding behaviour of grazing african ungulates. Behaviour, 84:195-243.

SPATIAL MEMORY IN FOOD-STORING BIRDS

David F. Sherry

Department of Psychology, University of Toronto

A variety of animals store food. Eastern chipmunks (Tamias striatus), and acorn woodpeckers (Melanerpes formicivorus), gather a large amount of food together in a single place, often called a larder hoard. Chipmunks, like many other rodents, store food in their underground burrow system (Elliot, 1978), while acorn woodpeckers store acorns in "granaries" - trees, fence posts, and wooden buildings drilled with thousands of small holes (MacRoberts & MacRoberts, 1976).

Other animals, such as red foxes (Vulpes vulpes; Macdonald, 1976) and the subjects of this paper, marsh tits and black-capped chickadees (Parus palustris and P. atricapillus), store food in a scattered distribution within their home range or territory. Marsh tits store single food items, such as insects, seeds and pieces of beech mast in mossy logs, hollow nettle stems, and crevices in the bark of trees. They never re-use cache sites, and may establish up to several hundred dispersed storage sites in a single day (Cowie, Krebs & Sherry, 1981; Sherry, Avery & Stevens, 1982). Marsh tit storage sites are spaced widely apart. In Wytham Woods, Oxford, the mean distance between neighbouring storage sites is about 7 m, with little variation around this mean either within or between individual birds (Cowie, Krebs & Sherry, 1981).

Why these birds scatter stored food in this way is not fully understood. It may be that their small size, around 10 g, makes them unable to defend a concentrated larder of food. Chipmunk and acorn woodpecker larder hoards are attacked and sometimes completely emptied by other animals (Shaffer, 1980; MacRoberts & MacRoberts, 1976). To maintain their granaries, acorn woodpeckers must vigorously defend them against tree squirrels, fox squirrels, crows, jays and other acorn woodpeckers (MacRoberts, 1970). Although they cannot physically defend their caches, marsh tits and chickadees could space their caches in a way that reduces the probability that another animal finding one cache by chance will be able to find others by searching the surrounding area (Stapanian & Smith, 1978). Field studies have shown that the spacing used by the birds does minimise the loss of stored food to other animals (Sherry, Avery & Stevens, 1982).

Whatever the reason for scattering storage sites, doing so presents scatter-hoarders with a problem in spatial orientation that animals maintaining a larder hoard do not face: how to relocate and recover the supply of stored food. A number of hypotheses have been advanced as to how this problem is solved. Some have suggested that caches are not recovered at all (Nichols, 1958). Gibb (1960) felt it "inconceivable that tits can long remember exactly where they have hidden even a small fraction of their stored foods (p. 178)". and proposed, along with others (Källander. 1978; Haftorn. 1974; Macdonald. 1976), that food is stored in the kind of sites that would be re-encountered in the course of normal foraging. Löhrl (1950), in contrast, held that marsh tits are able to relocate caches by accurately remembering large numbers of dispersed spatial locations.

CACHING IN THE WILD

To obtain a better understanding of the spatial problem, and clues to how marsh tits solve it, basic data on food storage was collected in the field (Cowie, Krebs & Sherry, 1981). Pairs of marsh tits, which defend territories year round (Southern & Morley, 1950) were colour-banded for individual identification and provided with sunflower seeds to store. Seeds were numbered and radioactively labelled with technetium (Tc99m). When a bird under observation approached a special feeder placed in its territory, seeds were dispensed one at a time, in numbered order. After 50 seeds had been taken, the bird's territory was searched with a portable scintillation counter to locate stored seeds. Seeds were found stored in a variety of sites, including moss, hollow stems, bark, curled dry leaves, and in the ground - the latter a seemingly odd storage site, but one also reported by Löhrl (1950). Seeds were stored at varying distances from the feeder, with a mean equal to 43.3 m, though some were carried over 100 m from the feeder for storage. Individual storage sites were never found to be re-used. either in this or subsequent studies (Sherry, Avery & Stevens, 1982).

To discover how often marsh tits successfully recovered their stored food, and how soon after storage they did so, an experiment was performed using each of the storage sites located. Two control seeds were placed near each radioactive seed found stored, at distances of 10 cm (the near control) and 100 cm (the far control). These seeds were placed in sites identical to that used for storage by the marsh tit, for example in hollow nettle stems. Control sites were in randomly chosen directions from the storage site, and on some trials the radioactive seed was removed and randomly re-assigned to either the near control, far control or storage position. The sites were then inspected at 3 hour intervals during daylight for the following three days. The logic behind this experiment was that if marsh tits rarely succeeded in recovering their own caches, then seeds at all three sites would be expected to be taken by other animals at about the same rate. This would occur for example if rodents or birds other than the food-storer removed seeds. Similarly, if marsh tits merely re-encountered their caches by chance in the course of normal foraging, then all seeds would also be expected to disappear from their sites at the same rate. If, however, the food storing marsh tit returned accurately to its caches, the seed in the storage site should disappear at a higher rate than the seeds in the two control positions. This was in fact the outcome of the experiment.

Seeds hoarded by marsh tits survived in place a mean of 7.7 daylight hours, near controls 13.5 daylight hours, and far controls 20.4 daylight hours. This overall difference in survivorship is significant ($F_{2,13}$ = 7.49, $p < .01$), and _post hoc_ tests showed that the difference between storage sites and far controls is significant, while that between storage sites and near controls is not (Cowie, Krebs & Sherry, 1981). On 120 occasions we were able to determine on successive inspections whether the seed at the storage site or at the near control site was taken first, and 93 times it was the seed at the storage site.

These results reveal two interesting points about cache recovery. First, the seed at the storage site would only be expected to disappear more rapidly than seeds at control sites if the bird that had stored it returned to collect it. Pilfering by other animals, and chance re-encounter of caches by the food-storer, would not be expected to produce this outcome. Although bias could have been introduced by the fact that storage sites were chosen by marsh tits and control sites by us, this seems unlikely. Seeds stored by the birds were sometimes placed out of sight, but more often were partly or wholly visible, and there were no obvious features of storage sites that could not easily be duplicated. Random re-assignment of the stored seed to the three sites ought to have controlled for any biases due to the stored seed itself. Secondly, the marsh tits returned fairly precisely to the cache site, and seemed not to merely search the general area around a cache, because on most recoveries they overlooked a seed stored in an identical site 10 cm away. Whether they accomplished this by remembering the locations of caches sites or by some other means, however, is not shown by this experiment.

Finally, the interval between the caching and recovery of food was short. At the latitude of Oxford at the time of year these experiments were performed an interval of 7.7 daylight hours between caching and recovery meant that most seeds were taken the day following caching, or the next day. Why the birds should store food at all for such short periods seems a puzzle. Other scatter storing birds, such as Clark's nutcracker and the Eurasian nutcracker (Nucifraga columbiana and N. caryocatactes), store autumn crops of seeds and nuts and recover their stores months later, to counter winter food scarcity and to begin breeding in spring before other food sources are available (Swanberg, 1981: Vander Wall & Hutchins, 1983). But perhaps the advantages of long term food storage - prolonging the benefit of a temporary abundance, and avoiding scarcity - apply to short term storage too, especially if abundances and scarcities also occur on a short time scale (Sherry, 1985). The benefit of a short-lived food surplus, for example a rich patch of food that is rapidly depleted by other animals, can be prolonged by caching, and scarcities may well occur on a diurnal as well as an annual cycle. Many species, for example, cache preferred food types at times of day when foraging conditions are good, and recover them at the end of the day when prey are unavailable (Collopy, 1977; Powlesland, 1980; Rijnsdorp, Daan & Dijkstra, 1981).

MEMORY AND CACHE RECOVERY

The field study just described showed that marsh tits can relocate their caches, but not how they do it. To discover by what means caches are relocated, a number of experiments have been conducted in the laboratory (Sherry, Krebs & Cowie, 1981; Sherry, 1982, 1984a; Shettleworth & Krebs, 1982). North American black-capped chickadees are closely related to marsh tits and store food in the wild in an almost identical fashion. Both species adapt easily to captivity and were observed storing and recovering food in large indoor aviaries connected to the birds' home cages. In a typical experiment birds were released singly into the aviary to store food, then returned to their home cages. After an interval ranging from 3 hours to several days in different experiments, they were allowed to return to the aviary to search for cache sites.

Cache Recovery

To determine whether marsh tits and chickadees could indeed relocate cache sites in the lab, birds were offered sunflower seeds to store. For marsh tits three trays of moss, each divided into quadrants, were provided as storage sites while chickadees stored food in six tree branches each with twelve small holes for storage sites (Sherry, Krebs & Cowie, 1981; Sherry, 1982, 1984a).

Data were collected in the same way for both species. At the beginning of each trial, behaviour was recorded during a 15 min pre-storage control period, during which no food was available, as the bird moved through the aviary searching the floor, walls, and potential storage sites, now empty of food. The purpose of this pre-storage control data was to obtain an estimate of any initial biases or preferences to search particular sites or to spend time in particular regions of the aviary. The bird was then allowed 15 min in which to store sunflower seeds. Once a bird had arrived at a cache site, putting the seed in place required at most a few seconds, followed by a rapid visual examination of the site, and return to the feeder for the next seed. The bird was then returned to its home cage for an interval of either 3 or 24 h. During this time all seeds stored by the bird and all other remaining seeds were removed so that the stored food itself could give no visual, olfactory or other cue to the location of caches when the bird returned to the aviary for a 15 min search test.

The results of several experiments are shown in Figure 1, comparing search performance to behaviour during the pre-storage control period. Both marsh tits and chickadees perform the task of relocating storage sites quite accurately in the lab. Marsh tits perform as well with a 24 h delay between caching and recovery as with a 3 h delay, and the performance of chickadees after a 24 h delay is equally accurate.

This result indicates that simple re-encounter of cache sites while foraging is not sufficient to account for the accuracy of recovery, since behaviour during the pre-storage period gives an estimate of how frequently particular sites would be visited in this way. This pre-storage rate lies near the chance rate calculated by assuming that all potential storage sites are equally likely to be visited during the search period (Figure 1). It was also apparent from the results of these experiments that the route used while establishing caches was not re-traced during recovery, and so did not provide the basis for accurate cache recovery. As mentioned earlier, direct cues from stored seeds had been eliminated by removing all seeds before the search test.

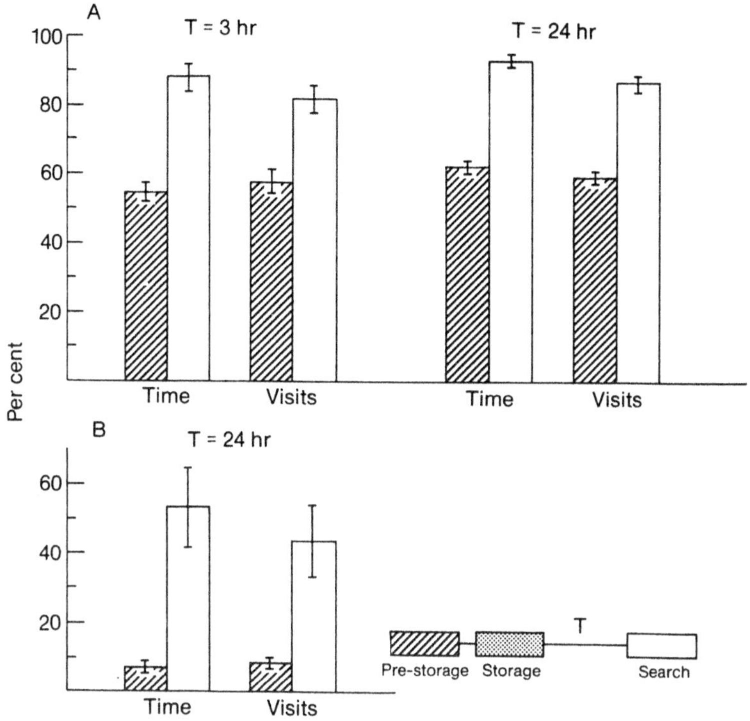

Figure 1. Time and visits at sites used for storage, as a percentage of time and visits at all potential storage sites. A: marsh tits, B: black-capped chickadees. Values differ between A and B because 12 potential storage sites were available for marsh tits and 72 for chickadees. Chance rates of re-encountering caches are : A 3h: 37%, 24h: 46%, B: 6%. Error bars equal $\pm$ 1 SEM and all differences between Pre-storage and Search differ significantly ($p < 0.05$, paired t-tests on arc-sin transformed data: Sherry, Krebs & Cowie, 1981; Sherry, 1984a).

Site Preferences

A further experiment with marsh tits was performed by Shettleworth & Krebs (1982) to investigate the importance of cache site preferences. Cowie, Krebs & Sherry (1981) had found that in the wild, marsh tits have individual preferences about the type of site used for storing. Birds might be able to relocate caches by selectively using only storage sites that conformed to some particular preference and then following this preference to relocate caches, rather than remembering the precise spatial location of each cache (Andersson & Krebs, 1978). Shettleworth & Krebs (1982) used hemp seeds as storable food and quantified recovery performance in a different way - by comparing the frequency of revisiting a cache, given that a seed had been stored there, to the frequency of revisiting given that no seed had been stored. The results showed that the birds were able to relocate caches more accurately than expected by chance, whatever the preference level for particular storage sites.

Interocular Transfer

In a further study, Sherry, Krebs & Cowie (1981) used the properties of interocular transfer in birds to gain additional information on how cache recovery occurs, and what role is played by memory. It has been shown that in pigeons there is little interocular transfer of some kinds of learned discriminations (Goodale & Graves, 1980; Graves & Goodale, 1977; Levine, 1945). That is, if a discrimination between two stimuli is learned to criterion by a pigeon with one of its eyes occluded by an opaque cover, the bird may show no evidence of having learned the task when tested with the "naive" eye. Learning by the naive eye proceeds at the same rate as the original learning of the task, and by alternating occluders on the two eyes it is possible to train two conflicting discriminations simultaneously (Goodale & Graves, 1980). On some tasks, however, good interocular transfer of learning does occur in pigeons (Catania, 1965). Retinal locus of the stimuli to be discriminated appears to be the critical factor in whether or not transfer occurs (Goodale & Graves, 1982). During monocular viewing, stimuli falling on the retinal area serving binocular vision transfer interocularly, while stimuli falling outside this region do not. The optic paths of the pigeon cross completely at the chiasm, and while information from the binocular areas crosses commissures to the other hemisphere, information from the monocular regions appears not to. The greatest part of the pigeon retina is without binocular overlap, and this probably explains the lack of transfer on such tasks. In mammals with laterally placed eyes, such as rabbits, failure of interocular transfer may also occur (Van Hof & Van der Mark, 1976). Interestingly, Wehner & Müller (1985) have recently reported interocular transfer for some, but not all, kinds of visual stimuli used in orientation by the desert ant *Cataglyphis fortis*, though the neural basis of the effect is surely quite different in this case.

If marsh tits, like pigeons and rabbits, show little interocular transfer for some kinds of visually-acquired information, then it would be possible to use this effect to determine whether information acquired during caching must be available during recovery for the accurate relocation of caches to occur. If by covering one of the bird's eyes during caching, and then comparing recovery performance using this naive eye to performance using the eye used for caching, it should be possible to determine whether viewing the caching site, and retaining this information in memory, is necessary for cache recovery.

Marsh tits stored sunflower seeds with one of their eyes covered by a small opaque plastic cap (Sherry, Krebs & Cowie, 1981). After caching, the bird left the aviary, as in other experiments, and all food that had been stored was removed. The eye cap was then taken off and replaced at random on either the naive eye, or the eye used during storing. Three hours later the birds were allowed to search for their storage sites.

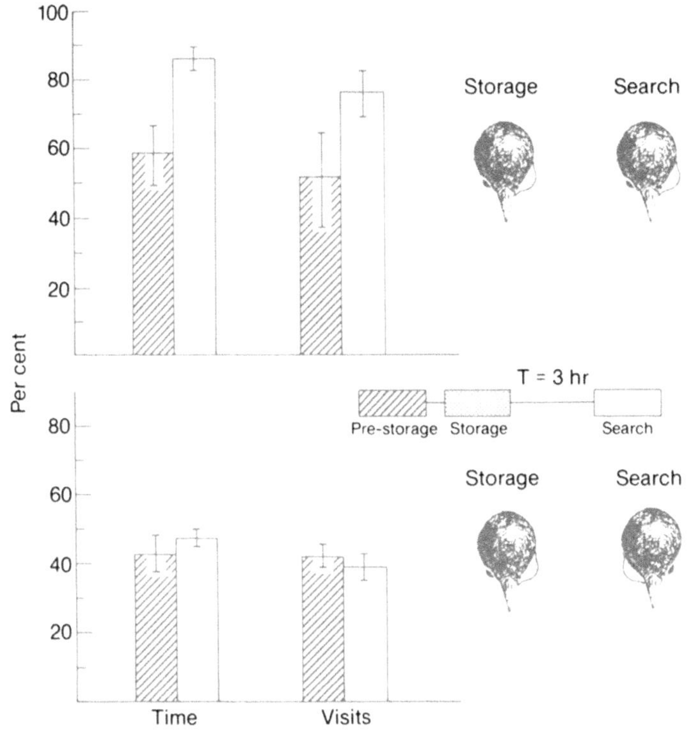

Figure 2. Interocular transfer and cache recovery in marsh tits. Time and visits at sites used for storage, as a percentage of time and visits at all 12 potential storage sites. Upper panel: results for marsh tits using the same eye for Pre-storage, Storage and Search. (Time, $t = 8.72$, $p < 0.05$. Visits, $t = 9.49$, $p < 0.05$). Lower panel: results for marsh tits using a different eye for Search than for Pre-storage and Storage. (Time, $t = 0.94$, NS. Visits, $t = 0.61$, NS. Paired t-tests on arc-sin transformed percent data. Error bars equal $\pm$ 1 SEM. Sherry, Krebs & Cowie, 1981).

Birds using the eye that was open during storing performed as accurately as birds using both eyes, while birds using the naive eye during recovery performed at the pre-storage control level (Figure 2). This result shows that information acquired during storing must be retained and be accessible during recovery for accurate relocation of caches. Behaviour of birds using the "naive" eye provides a good estimate of the frequency of encountering caches by chance, and this rate in fact lies close to that expected from pre-storage control behaviour (Figure 2). There was no indication that one eye, when used for both storage and search, was better than another, and no indication that use of one eye during storage biased the placement of caches or the choice of storage sites. Monocular vision alone is not responsible for the effect, since caches were accurately relocated monocularly if the same eye was used for both caching and recovery (Sherry, Krebs & Cowie, 1981).

A number of lines of evidence thus point to memory for the spatial location of caches as the major means of cache recovery by marsh tits and chickadees. In the wild, caches are recovered in preference to food placed in artificial caches in identical sites nearby. Caches are recovered in the lab at a higher rate than chance encounter could produce and in the absence of direct cues from the stored food itself. The route used when establishing caches is not retraced during recovery and preferences for particular cache sites are not sufficient to account for recovery performance. The results obtained in the interocular transfer study show that unless visual information acquired during caching is accessible during recovery, successful cache recovery does not occur.

MEMORY AND CACHE RECOVERY IN OTHER BIRDS

Studies of another genus of food-storing birds, the nutcrackers, have also shown the importance of spatial memory in cache recovery. Clark's nutcracker and the Eurasian nutcracker store seeds, primarily pine seeds, in caches they make in the ground. Each cache may contain a number of seeds and nutcrackers search for caches, usually months after making them, by probing in the soil with the bill (Swanberg, 1951; Tomback, 1977). In the field, Tomback (1980) analysed the spatial distributions of successful and unsuccessful probes and concluded that Clark's nutcrackers relocate their caches more accurately than could be accounted for by random search. Vander Wall (1982) has shown in a laboratory study that local landmarks like logs and rocks are used by nutcrackers to relocate their caches. Experimentally moving these landmarks produced errors in the placement of recovery probes by the birds that were equal to the distance the landmarks had been moved.

Kamil and Balda (1985) also found accurate memory for the location of caches in the lab, and investigated the importance of the choice of cache sites for successful recovery. The choice of sites might aid cache recovery if the birds used only sites that conformed to some rule, and by remembering this rule rather than the locations of all caches were able to search effectively for stored seeds. Kamil & Balda (1985) found however, that cache recovery was no more accurate when the nutcrackers had free choice of cache sites than when the choice of sites was randomly determined by the experimenters.

Memory for cache sites has also been suggested or demonstrated in studies of other food-storing corvids, such as Northwestern crows (Corvus caurinus: James & Verbeek, 1983) and European jays (Garrulus glandarius: Bossema, 1979; Bossema & Pot, 1974).

PROPERTIES OF MEMORY

If memory is indeed used by marsh tits, chickadees, and other birds to relocate scattered caches, then it might be expected to possess a number of additional properties, given the nature of food-caching by these birds. For example, since caches are recovered shortly after storage, the birds are faced on subsequent recovery attempts with a set of cache locations, all established at the same time, some of which still contain food and some of which do not because of the bird's own recovery behaviour. Similarly, a considerable amount of food stored by birds is lost to other animals, especially rodents (Sherry, Avery & Stevens, 1982: Tomback, 1980). Caches may be discovered to be empty at the time of the first recovery attempt, and are therefore not worth revisiting. Finally, a variety of food types are stored by chickadees and tits (Haftorn, 1956, 1974). These foods may differ in nutrient and caloric content, and in the efficiency with which they can be handled and consumed. Effective use of the stored food supply would be assisted by retaining the contents of caches in memory, along with their location and whether or not they have already been recovered or found to have been pilfered.

To test whether the birds can indeed add this information to the remembered spatial locations of caches, several variants of the basic laboratory experiment were conducted (Sherry, 1982, 1984a). Birds were allowed to enter the aviary as before for a pre-storage control period, followed by food storing. Their caches were left undisturbed and on their next return to the aviary they were allowed to recover approximately half of their stored food. For the final search test, 24 h later, all remaining food had been removed by the experimenter, and the birds' ability to avoid caches they had emptied themselves and to return to caches they had not, was assessed. Caches they had emptied were visited at the pre-storage control rate, while caches they had left undisturbed on their previous visit were located reliably (Sherry, 1982, 1984a). This ability of marsh tits and chickadees to avoid caches already emptied appears not to be shared by nutcrackers (Balda, 1980: Kamil & Balda, 1985). The reason for this lie be in differences in the cache recovery behaviour of these birds in the wild (Balda, 1980; Sherry, 1984b).

In a further experiment, to simulate the effects of pilfering of stored food by other animals, one-third of the bird's storage sites, chosen at random, were emptied by the experimenter before the birds' first return visit to the aviary. Some of those sites the bird inspected, and at some of the intact sites the bird recovered food. Thus on the final search test (for which all remaining food was removed), the bird was confronted with several categories of storage site: sites it had not visited since the first episode of caching, sites it had inspected and found to be empty, and sites where it had itself recovered food. It was as successful at not visiting sites it had discovered empty as it was at not visiting sites it had emptied itself. This result shows not only that the act of recovery is unnecessary in order for birds to avoid sites known to be empty, but also that to successfully avoid certain sites the bird need not choose them (as it does when recovering food), because sites emptied in the simulated pilfering experiment were chosen at random by the experimenter (Sherry, 1984a). This result is reminiscent of that reported earlier for nutcrackers (Kamil & Balda, 1985), that the spatial information acquired by the birds need not be selected by them in order to be effectively remembered.

To examine whether or not the kind of food stored at a site has any lasting effect on recovery behaviour, birds were offered two kinds of seeds to store, sunflower (*Helianthus*) and safflower (*Catharmus*) seeds. Sunflower seeds were offered with the shell removed, and were highly preferred to safflower seeds which were offered with the shell intact.

Both were stored, however, if small numbers of each kind, about six, were offered simultaneously. After storing, birds left the aviary and all stored food was removed as in previous experiments. After 24 h, birds searched for their cache sites. They returned at the usual high rate to sites where they had cached sunflower seeds but did not revisit safflower cache sites above the pre-storage control level. Thus the kind of food stored does influence subsequent recovery behaviour, and in the direction expected from known preferences for these foods. There was no indication that the two types of seed were stored in different parts of the aviary, in different kinds of site, or at different heights (Sherry, 1984a).

Thus the status of a cache - whether it was empty or intact on the bird's last visit - and some information about its contents, are retained in memory along with the spatial location of the cache.

Other aspects of memory for cache sites, including the retention interval and the occurrence of serial position effects in recovery are discussed in Sherry (1984b).

THE HIPPOCAMPUS AND CACHE RECOVERY

Recent studies with mammals have shown that the hippocampus plays an important role in memory for spatial (O'Keefe & Nadel, 1978; Olton, 1982) and other kinds of information (Olton, Becker & Handelmann, 1982; Winocur, 1982). The chapters by Olton, Bingman, Gerbrandt and Smith in volumes I and II of this book discuss some of this research with animals and humans. The analogous structure in the avian brain, the dorsomedial forebrain (Casini, Bingman & Bagnoli, in press; Krayniak & Siegel, 1978), may have an analagous function (Bingman, Ioalè, Casini & Bagnoli, 1985;Sahgal, 1984). If so, what role does it play in cache recovery by food-storing birds? A suggestive early study was that of Krushinskaya (1966) which showed that Eurasian nutcrackers with bilateral hippocampal lesions fed and stored food normally, but were unable to relocate their caches. They searched for caches but their probes were misplaced with respect to cache sites.

Anthony Vaccarino and I have recently carried out a study on the role of the hippocampus in cache-recovery by black-capped chickadees and obtained essentially the result described by Krushinskaya (1966). Caching data were collected for five trials as in previous experiments. The hippocampus was then bilaterally aspirated and two days after surgery five more caching trials were run. Operated birds cached normally but were unable to successfully relocate their caches (Figure 3). This effect was equally pronounced whether the search period followed storage by 3 h or by 15 min. The failure to relocate caches was not merely due to inactivity by birds with hippocampal aspiration, because searches occurred as frequently as in control birds and were of comparable duration, but were misdirected.

To determine what kind of disruption of memory occurred following hippocampal damage in chickadees, or indeed if memory was the aspect of cache recovery that was disrupted, we conducted a study of the effects of hippocampal aspiration on the performance of two non-caching problems, modelled on the cue and place tasks of Jarrard, Okaichi, Steward &

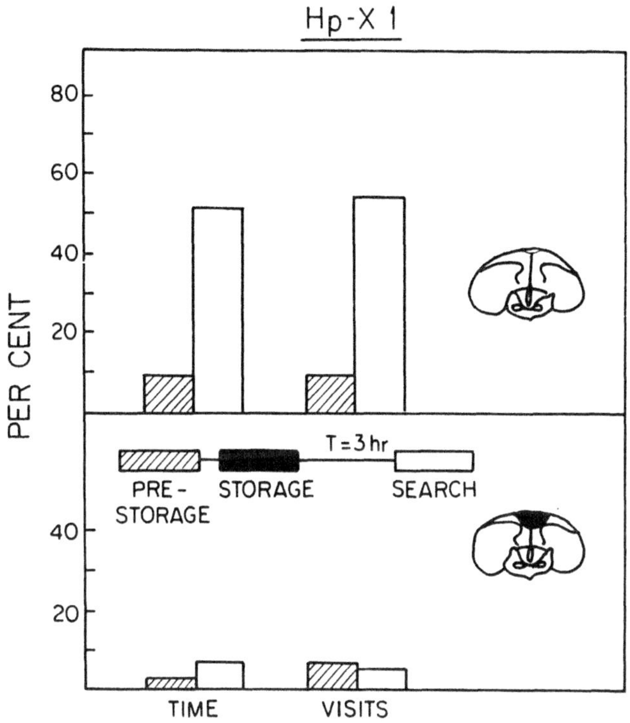

Figure 3. Hippocampal aspiration and cache recovery in black-capped chickadees. Time and visits at sites used for storage, as a percentage of time and visits at all 72 potential storage sites for a single black-capped chickadee. Upper panel: mean for 5 trials prior to surgery. Lower panel: mean for 5 trials after bilateral hippocampal aspiration. A mean of 7.4 seeds were stored per trial before surgery and a mean of 6.3 per trial after surgery.

Goldschmidt (1985). Birds in the place condition were required to visit the same six places, out of 72 available, to collect a small piece of sunflower seed at each site. One trial was run daily and the six places were always in the same spatial locations, that is, the same holes in a set of six artificial trees. Birds in the cue condition were required to visit six sites marked with small coloured cardboard squares to collect food. The spatial locations of the cues changed from one trial to the next however, so that the cue, not the place, indicated where food was available. The cue task was more difficult for birds to learn before any surgery was performed, but both groups eventually reached a criterion of 4 or more correct responses in the first 6 choices for 5 consecutive trials. Following bilateral hippocampal removal, birds in the place group showed considerable disruption of performance, compared to birds with control lesions placed rostral to the hippocampus, and unoperated birds. Birds in the cue condition showed relatively little disruption compared to controls. When errors were partitioned into two categories, however, a further pattern emerged. "Reference" memory errors, defined as searching incorrect sites, occurred frequently in the place task but not in the cue task. "Working" memory errors to correct sites, defined as revisiting a correct site previously visited on that trial, occurred with approximately equal frequency in both conditions, and more frequently than in control birds.

In addition to showing that damage to the avian hippocampus has effects analagous to those found in mammals, (see also Bingman, in this volume), this result sheds light on the kind of information used by chickadees to recover their caches. If the birds relocated caches by relying on some cue shared by all caches we would expect to find relatively little disruption of cache recovery from hippocampal damage, because performance was not disrupted in the cue condition. The fact that cache recovery was disrupted while the cue task was not, indicates that the capacity to perform the cue task is not sufficient to accurately relocate caches. The observation that hippocampal aspiration disrupted both cache recovery and the place task does not, of course, indicate that they are the same memory task, but it does show that they share a depedence on an intact hippocampus. The consequences for cache recovery of "working" memory errors of the kind found in both the cue and the place tasks remains to be determined.

CONCLUSIONS

Marsh tits, black-capped chickadees, nutcrackers, and probably many other food-storing animals, retrieve their caches by accurately remembering large numbers of scattered spatial locations. For tits and chickadees this requires learning and remembering up to a hundred novel locations every few days. Over the course of a year, thousands of spatial locations are involved. Caching a food item takes at most a few seconds at the cache site, so although the general area where caching occurs may be well-known, the time available to learn the features of each new site is very brief.

In addition to retaining in memory the spatial locations of caches, further information is incorporated. Caches which have been recovered or discovered empty are not re-visited, and the kind of food stored can be shown to influence recovery behaviour, such that sites containing preferred food are re-visited preferentially.

Preliminary studies have shown that hippocampal damage disrupts cache recovery in chickadees, as might be expected from previous studies of mammals and birds. But in addition, there is evidence that tasks requiring place learning are more affected by hippocampal damage than are tasks involving the use of previously learned cues. If this is correct, it confirms earlier work showing that caches are recovered by remembering their places, rather than any cue common to all caches.

Many questions about the spatial and cognitive abilities of food-storing birds remain unanswered. If the birds can remember whether caches have been pilfered and what kind of food they contained, can this information be combined to influence subsequent storage behaviour? Do the birds, for example, refrain from storing foods that tend to attract other animals, or avoid storage sites that are easily pilfered? Is the decision whether to eat or cache a particular item influenced by the numbers and kinds of food items already stored, and the probability of recovering them?

Perhaps the most interesting question of all is whether the spatial abilities of food-storing birds are the result of unique properties in the mechanisms of memory, or are instances of spatial and cognitive abilities that are widespread in animals, but made more conspicuous (and easier to investigate) by their role in solving the problem of relocating scattered food stores.

References

Andersson, M. and Krebs, J., 1978. On the evolution of hoarding behaviour. Animal Behaviour, 26: 707-711.

Balda, R.P., 1980. Recovery of cached seeds by a captive Nucifraga caryocactes. Zeitschrift für Tierpsychology, 52: 331-346.

Bingman, V.P., Ioalè, P., Casini, G. and Bagnoli, P., 1985. Dorsomedial forebrain ablations and home loft association behavior in homing pigeons. Brain Behavior & Evolution, 26: 1-9.

Bossema, I., 1979. Jays and oaks: an eco-ethological study of a symbiosis. Behaviour, 70: 1-117.

Bossema, I. and Pot, W., 1974. Het terugvinden van verstopt voedsel door de vlaamse gaai (Garrulus g. glandarius L.). De Levende Natuur, 77: 265-279.

Casini, G., Bingman, V.P. and Bagnoli, P., in press. Connections of the pigeon dorsomedial forebrain studied with WGA-HRP and 3H-proline.

Catania, A.C., 1965. Interocular transfer of discriminations in the pigeon. Journal of the Experimental Analysis of Behavior, 8: 147-155.

Collopy, M.W., 1977. Food caching by female American kestrels in winter. Condor, 79: 63-68.

Cowie. R.J.. Krebs, J.R. and Sherry, D.F., 1981. Food storing by marsh tits. Animal Behaviour, 29: 1252-1259.

Elliott, L., 1978. Social behavior and foraging ecology of the eastern chipmunk (Tamias striatus) in the Adirondack mountains. Smithsonian Contributions to Zoology, 265: 1-107.

Gibb, J. A., 1960. Populations of tits and goldcrests and their food supply in pine plantations. Ibis, 102: 163-208.

Goodale, M.A. and Graves, J.A., 1980. Failure of interocular transfer of learning in pigeons Columba livia trained on a jumping stand. Bird Behaviour, 2: 13-22.

Goodale, M.A. and Graves, J.A., 1982. Retinal locus as a factor in interocular transfer in the pigeon. in "Analysis of Visual Behavior". D. J. Ingle, M.A. Goodale and R.J.W. Mansfield (Eds.), MIT Press, Cambridge MA.

Graves, J.A. and Goodale, M.A., 1977. Failure of interocular transfer in the pigeon (Columba livia). Physiology and Behavior, 19: 425-428.

Haftorn, S. (1956). Contribution to the food biology of tits especially about storing of surplus food. Part IV. A comparative analysis of Parus atricapillus L., P. cristatus L. and P. ater L. Det Kgl. Norske Videnskabers Selskabs Skrifter, 4: 1-54.

Haftorn, S., 1974. Storage of surplus food by the boreal chickadee Parus hudsonicus in Alaska, with some records on the mountain chickadee Parus gambeli in Colorado. Ornis Scandinavica, 5: 145-161.

James, P.C. and Verbeek, N.A.M., 1983. The food storage behaviour of the northwestern crow. Behaviour, 85: 276-291.

Jarrard, L.E., Okaichi, H., Steward. O. and Goldschmidt, R.B., 1984. On the role of hippocampal connections in the performance of place and cue tasks: Comparisons with damage to the hippocampus. Behavioral Neuroscience, 98: 946-954.

Källander, H., 1978. Hoarding in the rook Corvus frugilegus. Anser Supplement, 3: 124-128.

Kamil, A.C. and Balda, R.P., 1985. Cache recovery and spatial memory in Clark's nutcrackers (Nucifraga columbiana). Journal of Experimental Psychology: Animal Behavior Processes, 11: 95-111.

Krayniak, P.F. and Siegel, A., 1978. Efferent connection of the hippocampus and adjacent regions in the pigeon. Brain Behavior & Evolution, 15: 372-358.

Krushinskaya, N.L., 1966. Some complex forms of feeding behaviour of nut-cracker Nucifraga caryocatactes, after removal of old cortex. Zhurnal Evoluzionni Biochimii y Fisiologgia, 11: 563-568.

Levine, J., 1945. Studies in the interrelations of central nervous structures in binocular vision: I. The lack of bilateral transfer of visual discriminative habits acquired monocularly by the pigeon. Journal of Genetic Psychology, 67: 105-129.

Löhrl, H., 1950. Beobachtungen zur Soziologie und Verhaltensweise von Sumpfmeisen (Parus palustris communis) im Winter. Zeitschrift für Tierpsychology, 7: 417-424.

Macdonald, D.W., 1976. Food caching by red foxes and some other carnivores. Zeitschrift für Tierpsychology, 42: 170-185.

MacRoberts, M.H., 1970. Notes on the food habits and food defense of the acorn woodpecker. Condor, 72: 196-204.

MacRoberts, M.H. and MacRoberts, B.R., 1976. Social organization and behavior of the acorn woodpecker in central coastal California. Ornithological Monographs, 21: 1-115.

Nichols, J.T., 1958. Food habits and behavior of the gray squirrel. Journal of Mammalogy, 39: 376-380.

O'Keefe, J. and Nadel, L., 1978. "The hippocampus as a cognitive map". Clarendon Press, Oxford.

Olton, D.S., 1982. Spatially organized behaviors of animals: behavioral and neurological studies. in "Spatial Abilities: Development and Physiological Foundations" M. Potegal (Ed.), Academic Press, New York.

Olton. D.S., Becker. J.T. and Handelmann. G.E., 1979. Hippocampus, space, and memory. Behavioral and Brain Sciences, 2: 313-365.

Powlesland. R.G., 1980. Food-storing behaviour of the south island robin. Mauri Ora, 8: 11-20.

Rijnsdorp, A., Daan, S. and Dijkstra, C., 1981. Hunting in the kestrel, Falco tinnunculus, and the adaptive significance of daily habits. Oecologia, 50: 391-406.

Sahgal, A., 1984. Hippocampal lesions disrupt recognition memory in pigeons. Behavioural Brain Research, 11: 47-58.

Shaffer, L., 1980. Use of scatterhoards by eastern chipmunks to replace stolen food. Journal of Mammalogy, 61: 733-734.

Sherry, D.F., 1982. Food storage, memory, and marsh tits. Animal Behaviour, 30: 631-633.

Sherry, D.F., 1984a. Food storage by black-capped chickadees: memory for the location and contents of caches. Animal Behaviour, 32: 451-464.

Sherry, D.F., 1984b. What food-storing birds remember. Canadian Journal of Psychology, 38: 304-321.

Sherry, D.F., 1985. Food storage by birds and mammals. Advances in the Study of Behavior, 15: 153-188.

Sherry, D., Avery, M. and Stevens, A., 1982. The spacing of stored food by marsh tits. Zietschrift für Tierpsychology, 58: 153-162.

Sherry, D.F., Krebs, J.R. and Cowie, R.J., 1981. Memory for the location of stored food in marsh tits. Animal Behaviour, 29: 1260-1266.

Shettleworth, S.J. and Krebs. J.R., 1982. How marsh tits find their hoards: The roles of site preference and spatial memory. Journal of Experimental Psychology: Animal Behavior Processes, 8: 354-375.

Southern, H.N. and Morley, A., 1959. Marsh tit territories over six years. British Birds, 43: 33-47.

Stapanian, M.A. and Smith. C.C., 1978. A model for seed scatterhoarding: Coevolution of fox squirrels and black walnuts. Ecology, 59: 884-896.

Swanberg, P.O., 1951. Food storage, territory and song in the thick-billed nutcracker. Proceedings of the 10th International Ornithological Congress, Uppsala: 545-554.

Swanberg, P.O., 1981. Kullstorleken hos nötkraka *Nucifraga caryocatactes* i Skandinavien, relaterad till förengaende ars hasselnöttillgang. Var Fagelvärld, 40: 399-408.

Tomback, D.F., 1977. Foraging strategies of Clark's nutcracker. The Living Bird, 16: 123-161.

Tomback, D.F., 1980. How nutcrackers find their seed stores. Condor, 82: 10-19.

Vander Wall, S.B., 1982. An experimental analysis of cache recovery in Clark's nutcracker. Animal Behaviour, 30: 84-94.

Vander Wall, S.B. and Hutchins, H.E., 1983. Dependence of Clark's nutcracker, *Nucifraga columbiana*, on conifer seeds during the postfledging period. Canadian Field-Naturalist, 9: 208-214.

Van Hof, M.W. and Van der Mark, F., 1976. Monocular pattern discrimination in normal and monocularly light-deprived rabbits. Physiology & Behavior, 16: 775-781.

Wehner, R. and Müller, M., 1985. Does interocular transfer occur in visual navigation by ants? Nature, 315: 228-229.

Winocur, G., 1982. Radial-arm-maze behavior by rats with dorsal hippocampal lesions: Effects of cuing. Journal of Comparative and Physiological Psychology, 96: 155-169.

SUBJECT INDEX

AUTHOR INDEX